Ur
S

h

Light and Plant Development

Light and Plant Development

Edited by

GARRY C. WHITELAM
Department of Botany
School of Biological Sciences
University of Leicester
Leicester
UK

and

KAREN J. HALLIDAY
School of Biological Sciences
The University of Edinburgh
Edinburgh
UK

Blackwell
Publishing

Editorial Offices:
Blackwell Publishing Ltd, 9600 Garsington Road, Oxford OX4 2DQ, UK
  Tel: +44 (0)1865 776868
Blackwell Publishing Professional, 2121 State Avenue, Ames, Iowa 50014-8300, USA
  Tel: +1 515 292 0140
Blackwell Publishing Asia Pty Ltd, 550 Swanston Street, Carlton, Victoria 3053, Australia
  Tel: +61 (0)3 8359 1011

First published 2007 by Blackwell Publishing Ltd

ISBN: 978-1-4051-4538-1

Library of Congress Cataloging-in-Publication Data

Light and plant development / edited by Garry C. Whitelam and Karen J. Halliday.
    p.   cm.
  Includes bibliographical references and index.
  ISBN: 978-1-4051-4538-1 (hardback : alk. paper)
  1. Phytochrome.   2. Plants—Photomorphogenesis.   3. Plants—Development.
I. Whitelam, Garry C.   II. Halliday, Karen J.

  QK 898. P67L54 2007
  571.8′2—dc22

                                                              2006024268

A catalogue record for this title is available from the British Library

Set in 10/12 pt Times
by TechBooks, New Delhi, India
Printed and bound in Singapore
by Fabulous Printers Pte Ltd

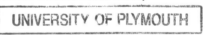
The publisher's policy is to use permanent paper from mills that operate a sustainable forestry policy, and which has been manufactured from pulp processed using acid-free and elementary chlorine-free practices. Furthermore, the publisher ensures that the text paper and cover board used have met acceptable environmental accreditation standards.

For further information on Blackwell Publishing, visit our website:
www.blackwellpublishing.com

Contents

ANDREAS HILTBRUNNER, FERENC NAGY AND
EBERHARD SCHÄFER

ALFRED BATSCHAUER, ROOPA BANERJEE AND
RICHARD POKORNY

The colour plate section follows page 30

Contributors

Dr Trudie Allen Department of Biology, University of Leicester, Leicester LE1 7RH, UK

Roopa Banerjee Philipps-University, Biology-Plant Physiology, Karl-von-Frisch-Strasse 8, 35032 Marburg, Germany

Professor Dr Alfred Batschauer Philipps-University, Biology-Plant Physiology, Karl-von-Frisch-Strasse 8, 35032 Marburg, Germany

Bobby A. Brown Plant Science Group, Division of Biochemistry and Molecular Biology, Institute of Biomedical and Life Sciences, Bower Building, University of Glasgow, Glasgow G12 8QQ, UK

Dr John M. Christie Plant Science Group, Division of Biochemistry and Molecular Biology, Institute of Biomedical and Life Sciences, University of Glasgow, Glasgow G12 8QQ, UK

Dr Xing Wang Deng Department of Molecular, Cellular and Developmental Biology, Yale University, New Haven, CT 06520-8104, USA

Dr Paul Devlin School of Biological Sciences, Royal Holloway, University of London, Egham, Surrey TW20 0EX, UK

Suhua Feng Department of Molecular, Cellular and Developmental Biology, Yale University, New Haven, CT 06520-8104, USA

Dr Keara A. Franklin Department of Biology, University of Leicester, Leicester LE1 7RH, UK

Dr Karen J. Halliday School of Biological Sciences, The University of Edinburgh, Daniel Rutherford Building, The King's Buildings, Mayfield Road, Edinburgh EH9 3JR, UK

Dr Andreas Hiltbrunner Albert-Ludwigs-Universität Freiburg, Institute of Biology II/ Botany, Schänzlestrasse 1, 79104 Freiburg, Germany

Dr Matthew Hudson Department of Crop Sciences, University of Illinois, Urbana, IL 61801, USA

Professor Gareth I. Jenkins Plant Science Group, Division of Biochemistry and Molecular Biology, Institute of Biomedical and Life Sciences, Bower Building, University of Glasgow, Glasgow G12 8QQ, UK

Dr Eve-Marie Josse School of Biological Sciences, The University of Edinburgh, Daniel Rutherford Building, The King's Buildings, Mayfield Road, Edinburgh EH9 3JR, UK

Professor Catherine Lillo Department of Mathematics and Natural Sciences, University of Stavanger, 4036 Stavanger, Norway

Dr Simon Geir Møller Department of Mathematics and Natural Sciences, University of Stavanger, 4036 Stavanger, Norway; Department of Biology, University of Leicester, Leicester LE1 7RH, UK; and Laboratory of Plant Molecular Biology, Rockefeller University, New York, NY 10021-3699, USA

Dr Ferenc Nagy   Biological Research Centre, Institute of Plant Biology, Hungarian Academy of Sciences, P.O. Box 521, 6701 Szeged, Hungary

Dr Richard Pokorny   Philipps-University, Biology-Plant Physiology, Karl-von-Frisch-Strasse 8, 35032 Marburg, Germany

Dr Peter H. Quail   UC Berkeley, Plant Gene Expression Center, United States Department of Agriculture (USDA), 800 Buchanan Street, Albany, CA 94710, USA

Dr Nihal C. Rajapakse   Department of Horticulture, Clemson University, 168 Poole Agricultural Center, Box 340319, Clemson, SC 29634-0319, USA

Dr Eberhard Schäfer   Albert-Ludwigs-Universität Freiburg, Institute of Biology II/Botany, Schänzlestrasse 1, 79104 Freiburg, Germany

Dr Yosepha Shahak   Department of Fruit Tree Sciences, Agricultural Research Organization, The Volcani Center, P.O. Box 6, Bet Dagan 50250, Israel

Professor Garry C. Whitelam   Department of Biology, University of Leicester, Leicester LE1 7RH, UK

Preface

Living organisms are subject to fluctuating environmental conditions. While most animals are able to move away from unfavourable conditions, plants are sessile and so must cope with whatever comes their way. As part of their coping strategy, plants have evolved an exquisite array of mechanisms to sense environmental signals coupled with an extraordinary degree of developmental plasticity that enables them to modulate their growth and development in response to external cues.

Of all the environmental cues that challenge the developing plant, light can probably be considered to be the most important. In addition to its key role in plant metabolism, and hence almost all life on Earth, where it drives the process of photosynthesis, light energy also acts to regulate plant growth and development. Light quantity, quality, direction and diurnal and seasonal duration regulate processes from germination, through seedling establishment to the architecture of the mature plant and the transition to reproductive development. These developmental responses of plants to light constitute photomorphogenesis.

Regulatory light signals are detected by an array of specialised, information-transducing photoreceptors, including the red/far-red light-absorbing phytochromes, the blue/ultraviolet-A light-absorbing cryptochromes and phototropins and one or more, as yet unidentified, ultraviolet-B-absorbing photoreceptor molecules. Light-mediated signal transduction in plants starts with the perception of light by these specialised photoreceptors leading to altered expression of up to several thousand genes, thus enabling the plant to respond at the physiological level. In recent years, the application of genetic, biochemical and molecular studies, particularly in the model *Arabidopsis thaliana*, has led not only to the identification and characterisation of the photoreceptors and their genes, but also many of the components that act downstream of photoreceptor activation. It is evident that the photoreceptors operate through interactions with one another and with other signalling systems thus forming complex response networks.

This volume is designed to provide the reader with state-of-the-art accounts of our current knowledge of the major classes of higher plant regulatory photoreceptors and the signal transduction networks that comprise plant developmental photobiology. Consideration is also given to the ways in which knowledge of plant photoreceptors and their signalling networks can be exploited, for instance to improve the quality and productivity of commercially grown plants. The book is aimed at advanced students and new researchers requiring up-to-date accounts of the major themes in higher plant photomorphogenesis research.

Garry C. Whitelam
Karen J. Halliday

Annual Plant Reviews

A series for researchers and postgraduates in the plant sciences. Each volume in this series focuses on a theme of topical importance and emphasis is placed on rapid publication.

Editorial Board:

Prof. Jeremy A. Roberts (Editor-in-Chief), Plant Science Division, School of Biosciences, University of Nottingham, Sutton Bonington Campus, Loughborough, Leicestershire, LE12 5RD, UK; **Dr David Evans**, School of Biological and Molecular Sciences, Oxford Brookes University, Headington, Oxford, OX3 0BP, UK; **Prof. Hidemasa Imaseki**, Obata-Minami 2419, Moriyama-ku, Nagoya 463, Japan; **Dr Michael T. McManus**, Institute of Molecular BioSciences, Massey University, Palmerston North, New Zealand; **Dr Jocelyn K.C. Rose**, Department of Plant Biology, Cornell University, Ithaca, New York 14853, USA.

Titles in the series:

Part I Photoreceptors

1 Phytochromes

Andreas Hiltbrunner, Ferenc Nagy and Eberhard Schäfer

1.1 Introduction

In addition to some minerals, plants need only water, air and light to grow and develop. To take up minerals they evolved a root system and became sessile. As a consequence they cannot move away from unfavourable conditions. To overcome this problem they evolved a fascinating ability to adapt, especially to changes in the variable light conditions. Plants now possess a series of photoreceptors that monitor light quality, quantity and temporal/spatial patterns of light. Prominent amongst these are the red/far-red reversible photoreceptors the phytochromes, which are the focus for this chapter.

1.2 Historical aspects

In their seminal work Garner and Allard (1920) discovered the phenomenon of photoperiodism. Based on action spectroscopy of wild type and albino seedlings together with analysis of many different photomorphogenic responses, this group at Beltsville concluded that a unique pigment controlled photoperiodism and photo-morphogenesis (Parker *et al.*, 1945; Borthwick *et al.*, 1948, 1951). The red/far-red reversibility of seed germination (Toole *et al.*, 1953) and many other photomor-phogenic responses indicated that either a photoreversible pigment or two antago-nistic pigments generated the signal. Quantitative action spectroscopy, still a very powerful tool in plant physiology, could not distinguish between these two possi-bilities. However, the partial purification of a pigment named phytochrome solved the problem (Butler *et al.*, 1959).

In the late 1950s and early 1960s the HIR (high irradiance response) was dis-covered for many photomorphogenic responses under continuous irradiation. These responses, which did not exhibit red/far-red reversibility, had action spectra peaks both in blue and far-red light (Hartmann, 1967). Although Hartmann (1966) demon-strated that the HIR peak in far-red was mediated by phytochrome, the question of how phytochrome function could produce this type of action spectra still remains a controversial subject despite many proposed models (Hartmann, 1966; Schaefer, 1975; van der Woude, 1987; Hennig *et al.*, 2000). It became evident that in addition to the photochemical reaction, phytochrome also exhibits complex dark reaction kinetics, which may contribute to the HIR phenomenon. Further progress was made with the discovery by Sharrock and Quail (1989) and later Clack *et al.* (1994) that, at least in *Arabidopsis*, there are five phytochrome genes which encode a small gene family designated *PHYA-E*. Subsequent isolation of phyA mutant alleles clearly showed that the HIR is mediated by phyA and that most of the previous spectroscopical measurements were, in fact, assaying phyA reactions.

With the help of mutants of different *PHY* genes, it became possible to identify which phytochrome is responsible for different modes of action observed previously. The three standard modes of action are the classical red/far-red reversible induction responses, named low fluence responses (LFR), the far-red HIR and the very low fluence responses (VLFR). Although the LFR is mediated primarily by phyB and to a lesser extent the other light-stable phytochromes the far-red HIR and the VLFR are mediated solely by phyA (Nagy and Schaefer, 2002).

1.3   Properties of phyA *in vitro*

Phytochrome which was first isolated from oat seedlings in a partially purified form had a molecular weight of ca 60 kDa. This turned out to be a proteolytically degraded N-terminal fraction which contained the chromophore (Butler *et al.*, 1959; Siegelman and Firer, 1964). The chromophore, phytochromobilin, is an open-chain tetrapyrrole, which is covalently linked to the apoprotein by a thio-ether-bond. Later it was shown that the native phytochrome has a molecular mass of ca 120 kDa (slightly varying between different phytochrome species) and that it is present in two forms, Pfr, which is considered to be the active form, and Pr. It was shown that *in vitro* photoconversion between Pr and Pfr follows first-order kinetics in both directions and that it is triggered by a configuration change between 15Z and 15E isomers of phytochromobilin, upon FR or R absorption, respectively. It was also established that the Pfr form undergoes (light-independent) dark relaxation to Pr, known as dark reversion (see below). However, this appears to be a biophysical property of a phytochrome molecule *in vitro* as it was not observed *in vivo*.

Despite many attempts, no crystal structures of higher plant phytochromes have been solved. The very recent first crystal structure from two domains around the chromophore binding site of a bacterial phytochrome allowed the first view of the chromophore configuration and its interaction between the protein backbone. The structure of the Pfr form still remains illusive. Therefore, we still do not know how the physiological inactive Pr and its active Pfr form differ. The primary mode of Pfr action has not been solved despite numerous *in vitro* studies. One of the most promising observations has been that purified and recombinant reassembled phytochromes have kinase activity (Yeh and Lagarias, 1998). This topic will be described in a separate chapter. The production of *in vitro* data for other phys (B-E) has been relatively rare due to the much lower abundance of these molecules.

1.4   Properties in yeast cells

Pioneering work by Lagarias and Lagarias (1989) made it possible to express phytochrome in yeast cells. They demonstrated phyA holoprotein assembly in yeast cells expressing phyA to which chromophore had been added. This illustrated that the phyA apoprotein had an intrinsic bilin ligase activity that facilitated the autocatalytic attachment of its chromophore. Interestingly, the chromophore assembly has been demonstrated for phyA and phyB N-terminal fragments. Thus the bilin ligase activity appears to be intrinsic to the N-terminal domain. Indeed, comparative analysis of *Arabidopsis* phyA, phyB, phyC and phyE expressed in yeast cells (phyD could not be expressed in sufficient amounts) showed that these phytochromes had bilin ligase activity. In each case photoreversibility was detected. Although the spectral properties of these phytochromes were shown to be similar, some significant spectral differences were also observed between phytochrome species (Eichenberg *et al.*, 2000). It remains to be tested whether these differences have significant physiological functions, though it is possible that they lead to altered shade avoidance function (see Chapter 9). In yeast cells, each of the reconstituted phytochromes exhibited partial dark reversion after red light irradiation and transfer to darkness. Previous data from different *Arabidopsis* accessions indicated that native phyA did not dark revert *in vivo*. This suggests that the capacity for dark reversion is a property of the phytochrome molecule. The Pfr form, even in its ground state, has a higher energy state than the Pr form. Thus a thermal relaxation from the Pfr form to the Pr form is energetically possible. The yeast data and the *in vivo* measurements show that this reaction is regulated *in planta* (see below). Thus, for a molecular understanding of phytochrome function these photoreceptors must be studied *in vivo*.

1.5   *In vivo* properties of phytochromes

1.5.1   In vivo *spectroscopy*

The electronics engineer Karl Norris together with Warren Butler built a 'Ratiospect' which allowed *in vivo* measurement of phytochrome (Butler *et al.*, 1959). With this instrument, and later on with more modern, custom-built instruments, it became possible to measure phytochrome properties *in vivo*.

Despite some debates in the late 1960s and early 1970s (see Schmidt *et.al.*, 1973) photoconversion of phyA shows a first-order kinetics in both directions Pr → Pfr and Pfr → Pr (Schmidt *et al.*, 1973). PhyA is synthesized with zero-order kinetics (Schaefer *et al.*, 1972) it is degraded in its Pfr form and shows partial dark reversion, though this characteristic varies between species (Kendrick and Hillmann, 1971). Since neither the degradation mechanism nor the dark reversion mechanism is solved, it is unclear why some species exhibit strong dark reversion, whilst others do not dark revert. One speculation assumes that the dark reversion has different kinetics for phytochrome PfrPfr and PrPfr dimers (Brockmann *et al.*, 1987). This speculation was supported by experiments with recombinant PrPfr dimers expressed in yeast. In this case the dark reversion was complete, i.e 100% and much faster than the control (Hennig and Schaefer, 2001).

Astonishingly in the RLD ecotype of *Arabidopsis* a partial dark reversion of phyA is detected, whereas in the Col ecotype, no dark reversion can be found even though their phyA sequences are identical. This demonstrates that dark reversion is not an intrinsic property of the phyA molecule but a reaction regulated *in vivo* (Eichenberg *et al.,* 2000).

The destruction of phyA Pfr in darkness follows first-order kinetics *in vivo* (Marmé *et al.,* 1971; Hennig *et al.,* 2000). Detailed studies showed that under continuous irradiation the destruction of phyA has a complex wavelength and fluence rate dependence. Surprisingly, the wavelength and fluence rate dependency is quite different in the various dicotyledonous and monocotyledonous plant systems tested (Schaefer, 1975; Schaefer *et al.,* 1976; Hennig *et al.,* 2000). An additional complication arises from the finding that Pr is not always as stable as it was thought to be, based on turnover measurements in dark-grown seedlings (Quail *et al.,* 1973b, c). Pr undergoes a rapid degradation (with a similar rate to Pfr) when first cycled through Pfr. This Pfr induced Pr degradation was first described by Stone and Pratt (1979) and then confirmed by others (Hennig *et al.,* 2000). It should be mentioned that the Pfr-induced degradation of Pr is at maximum only 20 to 30% due to a competing relaxation reaction which brings the Pr to a more stable form (Hennig *et al.,* 2000).

In vivo spectroscopical measurements showed that the synthesis of Pr follows a zero-order kinetics, i.e. the rate is independent on the Pr level and on pre-irradiation (Schaefer *et al.,* 1971; Gottmann and Schaefer, 1982). This holds for most of the dicotyledonous seedlings. In oat seedlings, and later in pea, Pr synthesis was shown to be light regulated. This is a phytochrome-mediated response that occurs at the transcriptional level (Lissemore and Quail, 1988). The extent of this regulation, i.e. the sensitivity to light, is species specific being very strong in monocots and less strong in dicots (Otto *et al.,* 1983, 1985; Sharrock and Quail, 1989).

In vivo measurements of the other phytochromes is technically challenging owing to their much lower abundance. Indeed, measurements of phyB have only been accurately performed in *Arabidopsis* phyB overexpressor lines which lack phyA (Sweere *et al.,* 2001). In these experiments phyB exhibits a fast and strong, but incomplete Pfr to Pr dark reversion. Thus, in this respect phyA and phyB have different molecular properties *in vivo*. In addition, *in vitro* and *in vivo* assays demonstrated that this dark reversion is regulated by the response regulator ARR4 (Sweere *et al.,* 2001). Unpublished data show that this regulation requires ARR4 in its phosphorylated state (V. Mira Rodado, K. Harter and E. Schäfer, unpublished). Thus, it appears that the stability of phyB Pfr is regulated *in vivo*. The levels of ARR4 and its phosphorylation state are regulated by several hormones. Therefore, hormones, especially cytokinins, appear to influence light signalling, at least partly, by regulating the levels of active phyB Pfr. This type of control plays an important role in phyB inactivation after light/dark transition or in low light conditions.

It should be mentioned that the interpretation of the kinetic properties of phytochromes is complicated by the fact that phytochrome molecules form dimers, both *in vitro* and *in vivo* (Sharrock and Clack, 2004). *In vitro* studies show that the dark reversion of the heterodimer PrPfr to PrPr is much faster than that from the homodimers PfrPfr to PrPr (see above; Hennig and Schaefer, 2001). This may explain why dark reversion *in vivo* is faster than destruction but not complete, and that optimal levels of dark reversion are obtained at a photo equilibrium of 50% Pfr (Brockmann *et al.*, 1987).

1.6 Intracellular localisation of phytochromes

1.6.1 Classical methods

To characterise the intracellular localisation of phytochromes, classical studies have employed spectroscopic, immunocytochemical and cell biological/biochemical techniques. For more details of these studies and methods see Chapter 4 in *Photomorphogenesis in Plants* (Kendrick and Kronenberg, 1994).

1.6.2 Spectroscopic methods

Prior to the onset of the molecular era, micro-beam irradiation was a major tool for obtaining information about the intracellular localisation of phytochromes. In their pioneering experiments, Etzold (1965) and Haupt (1970) observed an action dichroism for phototropism and polarotropism in the chloronemata of ferns and for chloroplast orientation in the green alga *Mougeotia*. These findings suggested that the absorption dipole moment was parallel to the cell surface for Pr but perpendicular for Pfr. Moreover, the responses induced by a micro-beam pulse appeared to be local, since they could only be reversed by a subsequent far-red pulse given to the same spot. Thus, it was concluded that the intracellular mobility of phytochrome was very limited in these cases.

The group led by M. Wada further refined these experiments and clearly demonstrated that the micro-beam must hit a region including the cell wall, the plasma membrane and part of the cytosol to initiate the response (Kraml, 1994). This suggested that phytochromes mediating these responses are not associated with plastids, mitochondria or nuclei, but localise close to the plasma membrane. Although these experiments clearly indicate an ordered localisation of phytochromes, their physical association with the membrane could not be proven by this method.

Attempts to use similar techniques in higher plants failed primarily for two reasons: First, light scatters within the tissue making it impossible to irradiate a clearly defined area. Second, no strictly localised responses mediated by phytochromes were known in higher plants. In contrast, results obtained by Marmé and Schaefer, who used polarised light to induce photoconversion of phytochrome *in vivo*, indicated partial action dichroism, i.e. an ordered localisation of the photoreceptor (Marmé and Schaefer, 1972). In these experiments oat coleoptiles were lined up on a microscope cover slip and irradiated with vertically or horizontally polarised light. The rate of photoconversion was measured by *in vivo* spectroscopy measuring the Pfr formation in response to these two polarised beams. Although the differences between irradiation sources were statistically significant, contribution from light attenuation and scattering could not be ruled out. This complicated the interpretation of these experiments.

1.6.3   Cell biological methods

In addition to spectroscopic studies, cell fractionation has also been considered an efficient tool to determine whether phytochromes are bound to membranes (Kraml, 1994). In summary, these studies indicated that phytochromes can be associated with various organelles, as well as with the plasma membrane. The biological significance of these findings, however, has not yet been demonstrated and there is considerable doubt whether these observations indeed reflect phytochrome localisation *in vivo*. Yet, using cell fractionation, Quail *et al.* (1973a) observed red/far-red reversible pelletability of phytochrome (with subcellular constituents) which was confirmed later on by immunocytochemical methods (MacKenzie *et al.*, 1975; Speth *et al.*, 1986).

1.6.4   Immunocytochemical methods

Because of the technical problems inherent to the cell fractionation method, the next approach, pioneered by the Pratt laboratory, was immunocytochemistry. McCurdy and Pratt (1986) showed that the immunodetectable phytochrome (phyA) in dark-grown oat coleoptiles is homogenously distributed throughout the cytoplasm and does not associate with organelles or membranes. Irradiation very rapidly – within a few seconds – induced formation of sequestered areas of phytochrome (SAPs). In darkness these SAPs disappeared with a half-life of about 30 minutes (Speth *et al.*, 1986). Co-localisation of SAPs and ubiquitin indicated that the SAPs might be the site of phyA degradation (Speth *et al.*, 1987) and that the 26S proteasome may be involved in this process. However, no further evidence has yet been provided to support this hypothesis. Work by Moesinger and Schaefer in 1984 and 1985, demonstrated that red light irradiation could induce transcription of light-regulated genes in isolated nuclei. This suggested that at least a fraction of phytochrome was localised in the nucleus during signal transduction. As these findings were not compatible with the considered opinion on phytochrome localisation at that time, they were ignored and forgotten for the following 10 years.

1.6.5   Novel methods

When the genes encoding phytochromes were cloned the derived amino acid sequences were searched for localisation motifs (Clack *et al.*, 1994; Sharrock and Quail, 1989). These analyses suggested that phytochromes were not integral membrane proteins and that they probably did not localise to the nucleus as they did not possess canonical nuclear localisation signals (NLS). Thus, it became generally accepted that phytochromes were soluble cytosolic proteins that may associate with membranes by binding to membrane localised helper proteins. Pioneering work performed by Sakamoto and Nagatani (1996) seriously challenged this view. These authors reported for the first time the enrichment of phyB in nuclear extracts isolated from light-grown *Arabidopsis* seedlings. Moreover, the same authors demonstrated that a fusion protein consisting of the C-terminal part of *Arabidopsis* phyB fused to the GUS reporter constitutively localised to the nucleus in transgenic plants. These data obviously contradicted the membrane model, providing an alternative site of action for phytochromes. Although the light signalling community was initially little sceptical of these data the situation changed dramatically two years later, when Ni *et al.* (1998) reported interaction of phyA and phyB with phytochrome interacting factor 3 (PIF3), a basic helix-loop-helix (bHLH) transcription factor. This finding implied that phyA and phyB have to localise to the nucleus at least temporarily in order to interact with PIF3 and mediate light-induced signal transduction. In 1999, Nagatani's group and we, ourselves, demonstrated beyond reasonable doubt that light can induce nuclear import of a phyB-GFP (green fluorescent protein) fusion protein in transgenic *Arabidopsis* plants (Yamaguchi *et al.*, 1999; Kircher *et al.*, 1999). Expression of phyB-GFP results in a characteristic phyB overexpression phenotype in wild type plants (Kircher *et al.*, 1999) and complements phyB deficient mutants of *Arabidopsis* (Yamaguchi *et al.*, 1999) and *Nicotiania plumbaginifolia* (Gil *et al.*, 2000). This suggests that GFP does not impair phytochrome function and that the fusion protein represents a photobiologically active photoreceptor.

Kircher *et al.* (1999) also studied nucleo-/cytoplasmic partitioning of a phyB mutant form that cannot incorporate the chromophore due to a Cys → Ala mutation at the chromophore attachment site. This chromophoreless phyB version fused to GFP localised constitutively to the cytosol. Based on the hypothesis that chromophoreless phytochromes have a conformation similar to the Pr form, it was concluded that the Pr conformer of the photoreceptor is not compatible with nuclear import. An N-terminal 651 aa fragment of phyB fused to GFP localises to the cytosol as well (Matsushita *et al.*, 2003), whereas a fusion protein of the C-terminal half of phyB and GFP is constitutively in the nucleus suggesting that phyB contains a functional NLS in the C-terminal half (Nagy and Schaefer, 2000). With the various transgenic lines in-hand expressing easily detectable, biologically functional phytochrome-GFP photoreceptors, it is possible to analyse the molecular mechanism regulating intracellular localisation of phytochrome.

1.7 Intracellular localisation of phyB in dark and light

In six-day-old dark-grown seedlings the phyB-GFP fusion protein localises predominantly to the cytosol. However, strong overexpression of the transgene occasionally results in weak diffuse nuclear fluorescence in etiolated seedlings (Kircher *et al.*, 1999; Matsushita *et al.*, 2003; Yamaguchi *et al.*, 1999; Kircher *et al.*, 2002). Results obtained by Kircher *et al.* (2002) suggest that light treatment of imbibed seeds to promote homogenous germination can induce nuclear import of phyB. Thus, it is conceivable that the weak nuclear staining detected in six-day-old etiolated seedlings is due to phyB-GFP molecules that have been imported into the nucleus during this early phase of development. However, independent of the occasional diffuse staining in the nucleus of dark-grown seedlings, irradiation with either red or white light induces nuclear import of phyB-GFP. Nuclear localised phyB is not distributed homogenously in the nucleoplasm but rather accumulates in characteristic structures termed speckles or nuclear bodies (Kircher *et al.*, 1999, 2002, Yamaguchi *et al.*, 1999; Chen *et al.*, 2003).

Detailed studies have shown that nuclear import of phyB-GFP and the formation of phyB-GFP containing speckles are relatively slow processes and that they are fluence rate dependent (Gil *et al.*, 2000). The wavelength dependency of these processes, tested under six-hour continuous irradiation, paralleled that of phyB mediated seed germination (Shinomura *et al.*, 1996). The almost complete lack of responsiveness to wavelengths longer than 695 nm establishing a Pfr/Ptot ratio of about 40% was, however, quite surprising. Tests with light pulses have shown that a single light pulse was almost ineffective whereas three consecutive five-minute pulses given at hourly intervals induced import and formation of speckles containing phyB-GFP. The inductive signal was reversible by a subsequent far-red light pulse, indicating that nuclear import of phyB has the characteristics of a typical low fluence response (LFR) (Kircher *et al.*, 1999; Nagy and Schaefer, 2000). Physiological experiments have shown that the responsiveness to an inductive light pulse is often poor in etiolated seedlings but can be strongly enhanced by pre-irradiation to activate phyB (red light), phyA (far-red light) or cry1/cry2 (blue light). Gil *et al.* (2000) reported that pre-irradiation with red and blue, but not with far-red light enhanced nuclear import of phyB and the formation of phyB containing speckle. The so-called signal amplification by pre-irradiation disappears slowly, with a half-life of about 3 to 6 h for most physiologically tested responses. Also the effectiveness of a second light treatment to induce nuclear speckle formation disappeared slowly after the red light pre-treatment and the inductive effect of a five-second red light pulse was completely lost after a 24 h dark period.

After light-induced nuclear accumulation the phyB-GFP fusion protein disappears slowly (half-life of about 6 h) in seedlings transferred back to darkness. The first step in this process is the dissolution of speckles and the appearance of diffuse nuclear GFP fluorescence and a more homogenously distributed phyB. This is followed by a complete loss of nuclear staining which takes about 10 h (Gil *et al.*, 2000). Whether the slow disappearance of nuclear phyB is due to export or turnover of the photoreceptor is currently unknown. We observed, however, that the disappearance of nuclear phyB-GFP can be accelerated by irradiating the seedlings with a 2 h far-red pulse before transfer to darkness. This observation may underlie the so called 'end-of-day responses', which are triggered by such light treatments and include enhanced hypocotyl elongation.

In summary, recent work has shown that light-induced nuclear import of phyB exhibits the characteristics of a typical phyB-mediated physiological response. Namely, it displays low responsiveness to single light pulses, red/far-red reversibility of multiple pulses (LFR), sharp decline of responsiveness to wavelengths longer than 695 nm, fluence rate dependence and responsiveness amplification. Light-induced nuclear import of phyB is followed by the rapid formation of large sub-nuclear complexes, termed speckles or nuclear bodies, which have been shown to comprise the bulk of the photoreceptor detectable in nuclei.

1.8   Intracellular localisation of phyA in dark and light

Immunocytological experiments performed in the 1970s and 1980s characterised the localisation of phyA primarily in monocotyledonous plants. These experiments have shown that light treatment results in a rapid rearrangement of cytosolic phyA and leads to the formation of phyA-containing cytosolic complexes (SAPs). As for phyB localisation studies, transgenic tobacco and *Arabidopsis* lines expressing a phyA-GFP fusion protein have been employed to reinvestigate the intracellular localisation of phyA (Kim *et al.*, 2000; Kircher *et al.*, 1999, 2002). The functionality of the phyA-GFP fusion protein was verified by complementation of a *phyA* null mutant. Both rice and *Arabidopsis* phyA fused to GFP localise exclusively to the cytosol in dark-grown transgenic tobacco and *Arabidopsis* seedlings. In contrast to phyB-GFP, however, the intracellular distribution of phyA-GFP fusion proteins changes within minutes after irradiation. A single far-red light pulse is sufficient to induce rapid (within seconds) formation of cytosolic speckles reminiscent of the SAPs previously described in monocotyledonous plants (MacKenzie *et al.*, 1975, McCurdy and Pratt, 1986, Speth *et al.*, 1986) and translocation of the phyA-GFP fusion protein to the nucleus. Similar to phyB-GFP, accumulation of phyA-GFP in the nucleus is followed by the formation of nuclear speckles. However, the phyA speckles appear very rapidly and both their size and number are much reduced as compared to those of phyB (Kim *et al.*, 2000; Kim, 2002). These data demonstrate that light-induced nuclear import of phyA is a typical very low fluence response (VLFR), which is mediated by phyA. The far-red high irradiance response (HIR) is another phyA-mediated response. In both, transgenic tobacco and *Arabidopsis* seedlings, continuous far-red light induces nuclear import of phyA-GFP. The import process is fluence rate and wavelength dependent (Kim *et al.*, 2000) and, therefore, reflects a typical far-red HIR. Another characteristic of the far-red HIR is that it is diminished after a pre-treatment with red light (Beggs *et al.*, 1981; Holmes and Schaefer, 1981). Accordingly, nuclear import of phyA-GFP is almost completely inhibited by a 24-h red light pre-treatment (Kim *et al.*, 2000). These experiments demonstrate that the nuclear localisation properties of phyA are consistent with its characterised physiological roles in the VLFR and HIR. Similar results were obtained by Hisada *et al.* (2000), who used cytochemical methods to analyse the intracellular localisation of phyA in pea seedlings after exposure to continuous far-red light or pulse irradiation.

It can be concluded that (i) phyA-GFP localises exclusively to the cytosol in dark-grown seedlings, (ii) irradiation initiates rapid formation of cytosolic SAPs and (iii) import into the nucleus is followed by formation of nuclear speckles containing the phyA-GFP fusion protein. These processes display complex dynamics and are mediated by VLFR and HIR.

1.9   Intracellular localisation of phyC, phyD and phyE
in dark and light

To complete the characterisation of the nucleo-/cytoplasmic partitioning of all members of the *Arabidopsis* phytochrome gene family, Kircher *et al.* (2002) produced transgenic *Arabidopsis* lines expressing phyC, phyD and phyE fused to GFP. Similar to phyA and phyB the GFP fusion proteins of phyC, phyD and phyE localised primarily to the cytosol in dark-grown seedlings and accumulated in the nucleus after irradiation. Translocation to the nucleus was followed by formation of nuclear speckles as previously observed for phyA and phyB. Light-induced import into the nucleus and speckle formation are therefore common features of all phytochromes analysed so far. The kinetics and light dependency of these processes, however, appear to be specific for each type of phytochrome. Nuclear transport of phyC, phyD and phyE is red and white light inducible. Interestingly, although phyB and phyD are the most closely related phytochromes in *Arabidopsis* (Mathews and Sharrock, 1997), they showed the largest difference regarding speckle formation. phyD-GFP displayed a very slow nuclear import and even after an eight-hour white light irradiation only one or two large speckles were detectable per nucleus (Kircher *et al.*, 2002). The speckle formation of all phytochromes, except that of phyD, is subject to robust diurnal regulation under light/dark cycles. In addition, speckle formation is under circadian control as it starts even before the light-on signal. This phenomenon is most clearly seen for phyB-GFP (Kircher *et al.*, 2002).

1.10   Phytochrome/PIF3 co-localisation and nuclear speckles

Although PIF3 localises constitutively to the nucleus, it exhibits a highly dynamic behaviour. In the dark PIF3 is homogenously distributed in the nucleus. A short red or far-red light pulse rapidly induces the formation of PIF3 nuclear speckles followed by phyA/B/D dependent degradation of PIF3 (Bauer *et al.*, 2004). After a few hours in the dark PIF3 starts to re-accumulate in the nucleus and finally reaches similar levels as before the light treatment (Bauer *et al.* 2004). Both phyA and phyB co-localise with PIF3 in these rapidly formed speckles (Bauer *et al.*, 2004). However, the early PIF3 containing phyA and phyB speckles were only detectable with phytochrome yellow fluorescent protein (YFP) but not GFP fusion proteins which is, most probably, due to the improved fluorescence properties of YFP compared to GFP. For these reasons, the rapid, early, speckles were overlooked in the first experiments. The observation that nuclear import of phyB is a quite slow process implies that the light-dependent formation of these early phyB speckles is due either to small amounts of phyB already present in the nucleus of dark-grown seedlings or to rapid nuclear transport of a small fraction of phyB. This is at the moment an unanswered question.

Co-localisation studies clearly demonstrate that these very fast phyA and phyB speckles contain PIF3 and that at least the phyB speckles do not form in *pif3* mutant background (Bauer *et al.*, 2004). After light absorption phyA, phyB and phyD induce a rapid degradation of PIF3 thus explaining why these complexes are transient and disappear again after a few minutes (Bauer *et al.*, 2004). Under prolonged irradiation (a few hours) new phyB containing nuclear speckles can be detected (Kircher *et al.*, 2002; described above), which however do not contain PIF3 and also appear in *pif3* mutant background (Bauer *et al.*, 2004). These data indicate that there is not just one type of phytochrome containing speckle, and that different speckle types may have specific physiological functions (Kircher *et al.*, 2002; Bauer *et al.*, 2004; Chen *et al.*, 2003). Moreover, these findings also demonstrate that the speckles are not static but rather highly dynamic structures.

Two main questions regarding the different speckles are still open: What is the protein composition of these nuclear complexes and what is their precise physiological function? It has been known for years, for instance, that phyA is polyubiquitinated and rapidly degraded in dark-grown plants exposed to light (Sharrock and Clack, 2002; Clough and Vierstra, 1997). COP1 has recently been shown to co-localise with phyA in nuclear speckles when transiently expressed in onion epidermal cells (Seo *et al.*, 2004). Moreover, ubiquitination assays suggest that COP1 has E3 ubiquitin ligase activity towards phyA and *cop1* mutant plants exhibit a reduced phyA degradation rate after exposure to light (Seo *et al.*, 2004). Thus, the COP1/phyA speckles may be involved in phyA desensitisation, an essential step in terminating phyA signal transduction.

In their phyB localisation studies, Kircher *et al.* (2002) also analysed the intracellular localisation of two phyB signalling mutants carrying mutations in the Quail-box. Nuclear accumulation of the mutant phyB molecules fused to GFP was indistinguishable from GFP-tagged wild-type phyB. Both mutant versions of phyB, however, were unable to form light induced nuclear speckles indicating that the late phyB speckles are involved in phyB signalling. How exactly these speckles are involved in phyB signalling and what components they contain besides phyB is currently under investigation.

Over the past few years a whole suite of factors involved in phytochrome signalling has been reported to form nuclear speckles, including, LAF1, HFR1, COP1, HY5, PIF3, SPA1, EID1, PAPP5, and FHY1 (Ballesteros *et al.*, 2001; Jang *et al.*, 2005; Seo *et al.*, 2003; Ang *et al.*, 1998; Bauer *et al.*, 2004; Marrocco *et al.*, 2006; Ryu *et al.*, 2005; Hiltbrunner *et al.*, 2005; Yamaguchi *et al.*, 1999; Chen *et al.*, 2003). To understand phytochrome signalling at the molecular level, it will be crucial to define which of these components co-localise in the same speckles and to link these speckles to specific steps in signalling.

1.11   Regulation of intracellular localisation of phytochromes

The intracellular localisation of phytochromes has been shown to depend on light. The inactive Pr form localises to the cytosol, whereas the Pfr form, which is considered the biologically active form, is transported to the nucleus. Although this finding suggests that light-induced nuclear accumulation is an essential step in phytochrome signalling, it does not strictly prove this hypothesis. To test whether nuclear localisation of the photoreceptor molecules is a prerequisite for phytochrome signalling, Huq *et al.* (2003) employed a glucocorticoid receptor-based fusion protein system, which allowed them to control the intracellular localisation of phyB independent of its Pr/ Pfr state. Irrespective of the light treatment, phyB did not complement a *phyB* null mutant when trapped in the cytosol. The same line, however, was indistinguishable from the wild-type control when grown in red light on medium supplemented with Dex, which allows the Pfr form of phyB to enter the nucleus. Both activation by light and nuclear localisation are therefore essential for phyB signalling (and most probably for phytochrome signalling in general).

If nuclear transport of phytochromes is indeed an essential step in phytochrome signalling, mutants affected in this step may be expected to exhibit a phenotype similar to loss of the photoreceptor. Hiltbrunner *et al.* (2005) therefore analysed the localisation of phyA-GFP in the *fhy1* mutant, one of the most severe hyposensitive phyA signalling mutants. In fact, light-induced nuclear accumulation of phyA-GFP is strongly reduced in *fhy1* mutant background whereas it is only slightly affected in another strong phyA signalling mutant. *In vitro* pull-down and yeast-two-hybrid analysis further demonstrate that FHY1 and phyA interact with each other in a light-dependent manner. Moreover, co-expression of YFP-FHY1 and phyA-CFP in transiently transformed mustard seedlings confirmed that FHY1 and phyA co-localise in light-induced nuclear bodies. These findings therefore indicate that light-induced nuclear accumulation of phyA depends on FHY1 and that it is not an intrinsic property of the phyA molecule itself. In contrast, nuclear accumulation of phyB-GFP is not affected in *fhy1* mutant background suggesting that phyB relies on an FHY1-independent mechanism for nuclear accumulation.

Chory and co-workers (Chen *et al.*, 2005) have shown that the N- and C-terminal halves of phyB physically interact with each other. As this interaction is stronger in dark than in light, it was suggested that in the inactive Pr form of phyB the N-terminal half may mask a putative NLS in the C-terminal half. Upon activation by light, the switch from the Pr to the active Pfr form may unmask the NLS and allow nuclear import of phyB. This attractive hypothesis implies that under saturating light conditions nearly all phyB should localise to the nucleus. Moreover, it predicts that nuclear transport of phyB is not saturable and that any mutant specifically affected in phyB nuclear transport must be due to amino acid changes in the phyB molecule itself. This model also suggests that light-induced nuclear transport of phyB would work in any eukaryotic organism able to synthesise or take up the chromophore. It is worth noting that the regulation of phyA and phyB nuclear translocation could be quite different, if this hypothesis holds true.

Acknowledgements

This work was supported by grants from the Deutsche Forschungsgesellschaft to E.S. (SFB388 and SFB 592) and the Human Frontier Science Program (HFSP) to A.H. (LT00631/2003-C).

References

Ang, L.H., Chattopadhyay, S., Wei, N., Oyama, T., Okada, K., Batschauer, A. and Deng, X.W. (1998) *Mol. Cell* **1**, 213–222.
Ballesteros, M.L., Bolle, C., Lois, L.M., Moore, J.M., Vielle-Calzada, J.P., Grossniklaus, U. and Chua, N.H. (2001) *Genes Dev.* **15**, 2613–2625.
Bauer, D., Viczian, A., Kircher, S., Nobis, T., Nitschke, R., Kunkel, T., Panigrahi, K.C., Adam, E., Fejes, E., Schaefer, E. and Nagy, F. (2004) *Plant Cell* **16**, 1433–1445.

Beggs, C.J., Geile, W., Holmes, M., Jabben, M., Jose, A.M. and Schaefer, E. (1981) *Planta.* **151**, 135–140.

Borthwick, H.A., Hendricks, S.B. and Parker, M.W. (1948) *Bot. Gaz.* **110**, 103–118.

Borthwick, H.A., Hendricks, S.B. and Parker, M.W. (1951) *Bot. Gaz.* **113**, 95–105.

Brockmann, J., Rieble, S., Kazarinova-Fukshansky, N., Seyfried, M. and Schaefer, E. (1987) *Plant Cell Environ.* **10**, 105–111.

Butler, W.L., Norris, K.H., Siegelman, H.W. and Hendricks, S.B. (1959) *Proc. Natl. Acad. Sci. USA* **45**, 1703–1708.

Chen, M., Schwab, R. and Chory, J. (2003) *Proc. Natl. Acad. Sci. USA* **100**, 14493–14498.

Chen, M., Tao, Y., Lim, J., Shaw, A. and Chory, J. (2005) *Curr. Biol.* **15**, 637–642.

Clack, T., Mathews, S. and Sharrock, R.A. (1994) *Plant Mol. Biol.* **25**, 413–427.

Clough, R.C. and Vierstra, R.D. (1997) *Plant Cell Environ.* **20**, 713–721.

Eichenberg, K., Hennig, L., Martin, A. and Schaefer, E. (2000) *Plant Cell Environ.* **23**, 311–319.

Etzold, H. (1965) *Planta.* **64**, 254–280.

Garner, W.W. and Allard, H.A. (1920) *J. Agric. Res.* **18**, 553–606.

Gil, P., Kircher, S., Adam, E., Bury, E., Kozma-Bognar, L., Schafer, E. and Nagy, F. (2000) *Plant J.* **22**, 135–145.

Gottmann, K. and Schaefer, E. (1982) *Photochem. Photobiol.* **35**, 521–525.

Hartmann, K.M. (1966) *Photochem. Photobiol.* **5**, 349–366.

Hartmann, K.M. (1967) *Z. Naturforsch.* **22b**, 1172–1175.

Haupt, W. (1970) *Z. Pflanzenphysiol.* **62**, 287–298.

Hennig, L., Bueche, C. and Schaefer, E. (2000) *Plant Cell Environ.* **23**, 727–734.

Hennig, L. and Schaefer, E. (2001) *J. Biol. Chem.* **276**, 7913–7918.

Hiltbrunner, A., Viczian, A., Bury, E., Tscheuschler, A., Kircher, S., Toth, R., Honsberger, A., Nagy, F., Fankhauser, C. and Schaefer, E. (2005) *Curr. Biol.* **15**, 2125–2130.

Hisada, A., Hanzawa, H., Weller, J.L., Nagatani, A., Reid, J.B. and Furuya, M. (2000) *Plant Cell* **12**, 1063–1078.

Holmes, M. and Schaefer, E. (1981) *Planta.* **153**, 267–272.

Huq, E., Al-Sady, B. and Quail, P.H. (2003) *Plant J.* **35**, 660–664.

Jang, I.C., Yang, J.Y., Seo, H.S. and Chua, N.H. (2005) *Genes Dev.* **19**, 593–602.

Kendrick, R.E. and Hillmann, W.S. (1971) *Am. J. Bot.* **58**, 424–428.

Kendrick, R.E. and Kronenberg, G.H.M. (1994) *Photomorphogenesis in Plants*, Kluwer, Dordrecht, The Netherlands.

Kim, L. (2002) PhD thesis at the Institute of Biology II/Botany, Albert-Ludwig University, Freiburg, Germany.

Kim, L., Kircher, S., Toth, R., Adam, E., Schaefer, E. and Nagy, F. (2000) *Plant J.* **22**, 125–133.

Kircher, S., Gil, P., Kozma-Bognar, L., Fejes, E., Speth, V., Husselstein-Muller, T., Bauer, D., Adam, E., Schäfer, E. and Nagy, F. (2002) *Plant Cell* **14**, 1541–1555.

Kircher, S., Kozma-Bognar, L., Kim, L., Adam, E., Harter, K., Schaefer, E. and Nagy, F. (1999) *Plant Cell* **11**, 1445–1456.

Kraml, M. (1994) In: *Photomorphogenesis in Plants,* 3rd edn (eds R.E. Kendrick and G.H.M. Kronenberg), pp. 417—445, Kluwer, Dordrecht, The Netherlands.

Lagarias, J.C. and Lagarias, D.M. (1989) *Proc. Natl. Acad. Sci. USA* **86**, 5778–5780.

Lissemore J.L. and Quail, P.H. (1988) *Mol. Cell Biol.* **8**, 4840–4850.

MacKenzie, J.M.J., Coleman, R.A., Briggs, W.R. and Pratt, L.H. (1975) *Proc. Natl. Acad. Sci. USA* **72**, 799–803.

Marme, D., Marchal, B. and Schaefer, E. (1971) *Planta.* **100**, 331–336.

Marme, D. and Schaefer, E. (1972) *Z. Pflanzenphysiol.* **67**, 192–194.

Marrocco, K., Zhou, Y., Bury, E., Dieterle, M., Funk, M., Genschik, P., Krenz, M., Stolpe, T. and Kretsch, T. (2006) *Plant J.* **45**, 423–438.

Mathews, S. and Sharrock, R.A. (1997) *Plant Cell Environ.* **20**, 666–671.

Matsushita, T., Mochizuki, N. and Nagatani, A. (2003) *Nature* **424**, 571–574.

McCurdy, D.W. and Pratt, L.H. (1986) *J. Cell Biol.* **103**, 2541–2550.

Moesinger, E., Batschauer, A., Schaefer, E. and Apel, K. (1985) *Eur. J. Biochem.* **147**, 137–142.

Moesinger, E. and Schaefer, E. (1984) *Planta* **161**, 444–450.

Nagy, F. and Schaefer, E. (2000) *Curr. Opin. Plant Biol.* **3**, 450–454.

Nagy, F. and Schaefer, E. (2002) *Annu. Rev. Plant Biol.* **53**, 329–355.

Ni, M., Tepperman, J.M. and Quail, P.H. (1998) *Cell* **95**, 657–667.

Otto, V., Mösinger, E., Sauter, M. and Schaefer, E. (1983) *Photochem. Photobiol.*, **38**, 693–700.

Otto, V., Schaefer, E., Nagatani, A., Yamamoto, K.T. and Furuya, M. (1985) *Plant Cell Physiol.* **25**, 1579–1584.

Parker, M.W., Hendricks, S.B., Borthwick, H.A. and Scully, N.J. (1945) *Science* **102**, 152–155.

Quail, P.H., Marme, D. and Schafer, E. (1973a) *Nat. New Biol.* **245**, 189–191.

Quail, P.H., Schaefer, E. and Marme, D. (1973b) *Plant Physiol.* **52**, 128–131.

Quail, P.H., Schaefer, E. and Marmè, D. (1973c) *Plant Physiol.* **52**, 124–127.

Ryu, J.S., Kim, J.I., Kunkel, T., Kim, B.C., Cho, D.S., Hong, S.H., Kim, S.H., Fernandez, A.P., Kim, Y., Alonso, J.M., Ecker, J.R., Nagy, F.,Lim, P.O., Song, P.S., Schaefer, E. and Nam, H.G. (2005) *Cell* **120**, 395–406.

Sakamoto, K. and Nagatani, A. (1996) *Plant J.* **10**, 859–68.

Schaefer, E. (1975) *Photochem. Photobiol.* **21**, 189–91.

Schaefer, E., Lassig, T.U. and Schopfer, P. (1976) *Photochem. Photobiol.* **24**, 267–573.

Schaefer, E., Marchal, B. and Marmé, D. (1972) *Photochem. Photobiol.* **15**, 457–464.

Schaefer, E., Marchal, B. and Marmé, D. (1971) *Planta* **101**, 265–276.

Schmidt, W., Marmé, D., Quail, P.H. and Schaefer, E. (1973) *Planta* **111**, 329–336.

Seo, H.S., Watanabe, E., Tokutomi, S., Nagatani, A. and Chua, N.H. (2004) *Genes Dev.* **18**, 617–622.

Seo, H.S., Yang, J.Y., Ishikawa, M., Bolle, C., Ballesteros, M.L. and Chua, N.H. (2003) *Nature* **423**, 995–999.

Sharrock, R.A. and Clack, T. (2002) *Plant Physiol.* **130**, 442–456.

Sharrock, R.A. and Clack, T. (2004) *Proc. Natl. Acad. Sci. USA* **101**, 11500–11505.

Sharrock, R.A. and Quail, P.H. (1989) *Genes Dev.* **3**, 1745–1757.

Shinomura, T., Nagatani, A., Hanzawa, H., Kubota, M., Watanabe, M. and Furuya, M. (1996) *Proc. Natl. Acad. Sci. USA* **93**, 8129–8133.

Siegelman, H.W. and Firer, E.M. (1964) *Biochemistry* **3**, 418–423.

Speth, V., Otto, V. and Schaefer, E. (1986) *Planta* **168**, 299–304.

Speth, V., Otto, V. and Schaefer, E. (1987) *Planta* **171**, 332–338.

Stone, H.J. and Pratt, L.H. (1979) *Plant Physiol.* **63**, 680–682.

Sweere, U., Eichenberg, K., Lohrmann, J., Mira-Rodado, V., Baurle, I., Kudla, J., Nagy, F., Schafer, E. and Harter, K. (2001) *Science* **294**, 1108–1111.

Toole, E.H., Borthwick, H.A., Hendricks, S.B. and Toole, V.K. (1953) *Proc. Int. Seed Test Assoc.* **18**, 267–276.

van der Woude, W.J. (1987) In: *Phytochrome and Photoregulation in Plants. Proceedings of the XVI Yamada Conference* (ed, M. Furuya), pp. 249–258. Academic Press, New York, USA.

Yamaguchi, R., Nakamura, M., Mochizuki, N., Kay, S.A. and Nagatani, A. (1999) *J. Cell Biol.* **145**, 437–445.

Yeh, K.C. and Lagarias, J.C. (1998) *Proc. Natl. Acad. Sci. USA* **95**, 13976–13981.

2 Cryptochromes

Alfred Batschauer, Roopa Banerjee and Richard Pokorny

2.1 Introduction

Cryptochromes (cry) are sensory photoreceptors operating in the UV-A and blue light regions of the electromagnetic spectrum. They were first discovered in the plants, *Arabidopsis thaliana* and *Sinapis alba*, in 1993. Subsequently they have been identified in many other plant species, bacteria, fungi, animals and humans. Therefore, cryptochromes can be considered as the most widely distributed among the photoreceptor families. Cryptochromes are related in their sequence to DNA repair enzymes, the DNA photolyases, and they share the same chromophore cofactors. Since their discovery, a large quantity of data on the biological functions, signalling mechanisms, biochemistry and structure of cryptochromes has been accumulated and the reader is referred to several recent reviews and chapters on these topics (Banerjee and Batschauer, 2005; Batschauer, 2005; Cashmore, 2005; Lin and Shalitin, 2003; Partch and Sancar, 2005; van Gelder and Sancar, 2005). This chapter covers the biological function, the spectroscopic, biochemical and structural properties of plant cryptochromes, and examines aspects of their signalling mechanism. The role of cryptochromes in the photocontrol of flowering is presented in more detail in Chapter 8, and Chapter 5 outlines the effects of phosphorylation and dephosphorylation on cryptochrome function.

2.2 Cryptochrome genes and their evolution

Two different strategies led to the molecular cloning of cryptochrome genes in 1993. Margaret Ahmad and Anthony Cashmore (Ahmad and Cashmore, 1993) screened for T-DNA-tagged *Arabidopsis* mutants with the same phenotype as *hy4*, a mutant isolated by Maarten Koornneef and coworkers in 1980 (Koornneef *et al.*, 1980). The *hy4* mutant, in contrast to the wild type, had a long hypocotyl when the seedlings were grown under white or blue light. When grown in darkness, red or far-red light *hy4* hypocotyl growth inhibition was essentially normal (Ahmad and Cashmore, 1993; Jackson and Jenkins, 1995). These results indicated that a gene encoding either a blue light photoreceptor or a component in blue light signalling is affected in *hy4*. The insertion of a T-DNA facilitated the molecular cloning of the *HY4* gene (Ahmad and Cashmore, 1993). It turned out that HY4 has striking sequence similarity to class I CPD photolyases. These enzymes use the energy of photons in the UV-A/blue region of the spectrum to catalyse the repair of cyclobutane pyrimidine dimers (CPDs), caused by the exposure of DNA to UV-B (see Section 2.3.3.1 and Sancar, 2003, for review). The fact that HY4 had homology with photolyases but lacked photolyase activity (Lin *et al.*, 1995a; Malhotra *et al.*, 1995), combined with additional supportive findings (described below), led to the conclusion that *HY4* encodes a UV-A/blue light receptor rather than a component in blue light signalling. Therefore, HY4 was renamed cryptochrome 1 (Lin *et al.*, 1995a), a term used earlier for unknown blue light receptors (Gressel, 1979; Senger, 1984).

A second approach simultaneously led to the isolation of a cryptochrome gene from white mustard (*Sinapis alba* L.) (Batschauer, 1993). Here, polymerase chain reaction was used to amplify DNA fragments from a white mustard cDNA library using degenerate oligonucleotides that resembled conserved regions in some of the class I CPD photolyase known at that time. However, this gene was originally considered to be a DNA photolyase as the lack of mutant alleles for this gene prevented confirmation of its role *in planta*. Later studies demonstrated that the white mustard gene did not encode a photolyase and thus it was most likely a *bonafide* cryptochrome (Malhotra *et al.*, 1995). The similarity between cryptochromes and DNA photolyase was not completely unexpected because DNA photolyase and its chromophores were discussed as models for blue light receptors before the cryptochromes had been identified at the molecular level (Galland and Senger, 1988, 1991; Lipson and Horwitz, 1991).

After cryptochromes were identified in *Arabidopsis* and *Sinapis*, they were found in many other plant species, animals, fungi and bacteria either by using heterologous probes to screen for cryptochrome genes or by identifying such sequences in the growing databases of genome or EST projects.

Most plants seem to possess more than one cryptochrome. *Arabidopsis* contains two well-characterized cryptochromes (Ahmad and Cashmore, 1993; Hoffman *et al.*, 1996; Lin *et al.*, 1996b) and a third (cry3 or A.t.cryDASH) for which the biological function is not yet well defined (Brudler *et al.*, 2003; Kleine *et al.*, 2003). Tomato has three cryptochromes, CRY1a, CRY1b, CRY2 (Perrotta *et al.*, 2000; Perrotta *et al.*, 2001), and a putative cryDASH (G. Giuliano, personal communication), rice has three, CRY1a, CRY1b and CRY2 (Matsumto *et al.*, 2003), *Adiantum* has at least five cryptochromes (Kanegae and Wada, 1998; Imaizumi *et al.*, 2000) and the moss *Physocomitrella patens* has at least two (Imaizumi *et al.*, 1999; Imaizumi *et al.*, 2002).

Plant cryptochromes do not group together with animal cryptochromes in phylogenetic trees (see Figure 2.1). They are more closely related to class I DNA photolyases that repair cyclobutane pyrimidine dimers and are mostly found in microbial organisms including the yeast *Saccharomyces cerevisiae* and other fungi. In contrast, animal cryptochromes group with (6–4) photolyases that repair another type of DNA photoproduct, and are found exclusively in eukaryotes (for review see Sancar, 2003). One hypothesis for the evolution of the cryptochrome/photolyase family is that several gene duplication events gave rise to the present-day photolyases and cryptochromes (Kanai *et al.*, 1997; Todo 1999). The most ancestral gene (possibly a CPD photolyase) duplicated to give rise to class I CPD photolyases and to class II CPD photolyases that are now present in metazoans and plants

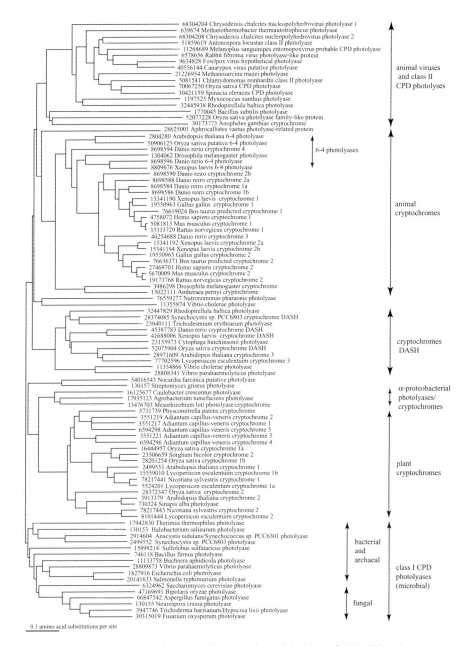

Figure 2.1 Unrooted phylogenic tree of the cryptochrome/photolyase family. Selected sequences covering all kingdoms of life were included for the tree construction. Each sequence is indicated by its NCBI GI number followed by the name of source organism and the original classification. The tree shows that photolyases surprisingly present in animal viruses group together with class II CPD photolyases, animal cryptochromes group with 6–4 photolyases whereas plant cryptochromes are more closely related to the class I CPD photolyases from microbes including archaea, bacteria and fungi, especially to those of α-proteobacteria. DASH cryptochromes form a separate group. Thus, plants may have received their cryptochrome genes by a dual horizontal transfer from former endosymbionts that gave rise to mitochondria (α-proteobacteria) and chloroplasts (cyanobacteria), respectively (see text). The tree was calculated using ClustalX and displayed using TreeView software, respectively.

(Ahmad *et al.*, 1997; Kato *et al.*, 1994; Petersen *et al.*, 1999; Taylor *et al.*, 1996; Todo *et al.*, 1994; Yasuhira and Yasui, 1992; Yasui *et al.*, 1994), as well as in some bacteria (O'Connor *et al.*, 1996; Yasui *et al.*, 1994) and animal viruses (Afonso *et al.*, 1999; Sancar, 2000; Srinivasan *et al.*, 2001; Todo, 1999). The class I CPD photolyase duplicated again to give rise to the more recent class I CPD photolyases and the progenitor of cryptochromes and (6–4) photolyases. The latter duplicated again to give rise to the present-day cryptochromes in plants and the (6–4) photolyases and the cryptochromes in animals. More recent duplications led to an increase in the number of cryptochrome genes in animals and plants, for example, *CRY1* and *CRY2* in *Arabidopsis*.

Recent work led to the discovery of a third group of cryptochromes, the cry-DASH family (Brudler *et al.*, 2003). DASH stands for *Drosophila*, *Arabidopsis*, *Synechocystis* and *Homo sapiens*. Members of the cryDASH family have been found so far in cyanobacteria (Brudler *et al.*, 2003; Hitomi *et al.*, 2000; Ng and Pakrasi, 2001), plants (Brudler *et al.*, 2003; Kleine *et al.*, 2003), marine bacteria (Daiyasu *et al.*, 2004), *Neurospora crassa* (Daiyasu *et al.*, 2004), *Vibrio* sp. (Worthington *et al.*, 2003) and the vertebrates zebrafish and *Xenopus* (Daiyasu *et al.*, 2004).

2.3   Cryptochrome domains, chromophores and structure

2.3.1   Domain structure of the cryptochromes

As mentioned above, plant cryptochromes have significant sequence similarity with class I CPD photolyases. For example, a stretch of 500 amino acids within the N-terminal region of the cryptochromes CRY1 and CRY2 from *Arabidopsis* show about 30% sequence identity at the protein level with *E. coli* photolyase (Figure 2.2). This domain has been shown to bind the chromophores (see below) and is therefore required for light sensing. In contrast to photolyases, most of the plant and animal cryptochromes carry extensions of varying length at their C-terminus (cryptochrome C-terminus or CCT domain). This is schematically shown in Figure 2.2. The longest extension found is 367 amino acids for *Chlamydomonas* cryptochrome (total length 867 amino acids) (Small *et al.*, 1995), whereas AcCRY5 of *Adiantum capillus veneris* lacks such an extension (Imaizumi *et al.*, 2000; Kanegae and Wada, 1998) as does white mustard CRY2 (Batschauer, 1993). Although the sequences of plant cryptochromes are mostly conserved in the cofactor binding, N-terminal region of approximately 500 amino acids (the so-called photolyase homology or PHR domain), there is some conservation in the C-terminal extensions as well. The first motif described to be conserved between *Arabidopsis* CRY1 and CRY2 in this region is the so-called STAES (Ser-Thr-Ala-Glu-Ser$_n$) motif (n stands for 4 or 5 Ser residues in cry1 and cry2, respectively; Hoffman *et al.*, 1996). Further upstream are two other conserved regions in plant cryptochromes, one of which contains a varying number of acidic residues (Asp and Glu, A motif), and DQXVP (D motif) which is close to the start of the C-terminal extension. Together these motifs are named the DAS domain (Lin, 2002). The conservation of the DAS domain in cryptochromes of moss, fern and seed plants indicates that it must have been present already in early

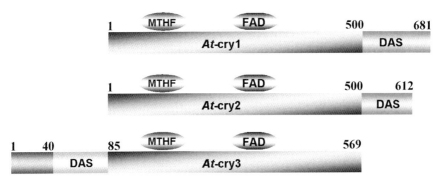

Figure 2.2 Domain structure of plant cryptochromes. The three cryptochromes identified in *Arabidopsis* and their cofactors are shown schematically. The highest conservation among the protein sequences is found in the photolyase homology region (PHR, central part). This region is about 500 amino acids long and binds the FAD and MTHF cofactors non-covalently. Cry1 and cry2 carry an additional domain at the C-terminal end (CCT) that varies in length and sequence. However, the CCTs contain three motifs conserved in all land plant cryptochromes and named together the DAS domain. For cry2 it was demonstrated that its CCT is required for nuclear import and contains a bipartite nuclear localization signal. In addition, the CCTs of cry1 and cry2 mediate signalling and interact with various proteins (see Figure 5.1). In contrast to cry1 and cry2, cry3 carries an extension at the N-terminal end, part of which (amino acids 1–40) is required for the import of cry3 into chloroplasts and mitochondria. Most of the DAS domain is also conserved in the further part of the cry3 N-terminal extension. For references see text.

land plants. As mutations in *CRY1* that introduce stop codons or amino acid changes directly before or within the D motif (*hy4-3* and *hy4-9* alleles) cause phenotypic changes (Ahmad and Cashmore, 1993, Ahmad *et al.*, 1995), this region must be essential for the biological function of the molecule. The importance of the C-terminal extensions has also been shown by domain-switch experiments where the C-terminal extension of *Arabidopsis* CRY1 was fused to the PHR domain of CRY2 and vice versa (Ahmad *et al.*, 1998a). Both combinations of fused photoreceptors were biologically active. Surprisingly, the DAS domain is more or less conserved in *Arabidopsis* cry3 (D-motif sequence NDHIHRVP compared to EDQMVP in cry1 and NDQQVP in cry2, A-motif sequence EEEID compared to EEDEE in cry1 and EEEEE in cry2 and a conserved tandem of four Ser residues in S-motif sequence), which does not carry a C-terminal extension, instead it carries an extension at the N-terminus (see Figure 2.2). Considering the fact that cry2 and cry1 are nuclear proteins and cry3 is targeted to chloroplasts and mitochondria (see Section 2.5.2) the DAS-domain could, in addition to its putative role in signalling, also have other functions. The role of the extensions in cryptochrome signalling, subcellular localization and biochemistry will be further discussed below.

2.3.2 *Cryptochrome chromophores*

Since cryptochromes are not highly abundant proteins in plants and because of technical difficulties, they have not yet been purified in amounts sufficient to allow determination of their associated cofactors. However, heterologous expression of

Arabidopsis cry1 and cry2 and white mustard cryptochrome in *E. coli* or insect cells have shown that plant cryptochromes bind FAD non-covalently and in 1:1 stoichiometry with the protein (Lin *et al.*, 1995b; Malhotra *et al.*, 1995). This finding was not unexpected because the flavin-binding pocket of *E. coli* DNA photolyase is well conserved in the plant cryptochromes. In addition, *E. coli*-expressed *Arabidopsis* cry1 and *Sinapis* cry2 contained non-covalently bound methenyltetrahydrofolate (MTHF) (Malhotra *et al.*, 1995). This was surprising as the photolyase amino acids that make contact with this cofactor are not well conserved in these cryptochromes. However, it was analyzed that although the *Sinapis* cry2 protein lacked the C-terminal extension, it bound both cofactors (Malhotra *et al.*, 1995); this work demonstrated that the photolyase-related N-terminal domain is sufficient for chromophore binding. *Arabidopsis* cry3 expressed in *E. coli* was also found to bind FAD and MTHF in stoichiometric amounts (Pokorny *et al.*, 2005). Therefore, it seems that the chromophore composition of plant cryptochromes is the same irrespective of their family grouping (Figure 2.1) and subcellular localization (see below). However, purification of the holoproteins from plant tissue is still required to test the binding and function of these chromophores *in planta*.

2.3.3   Photolyase and cryptochrome structure

The structure of cryptochrome, deduced thus far, is very similar to that of *E. coli* photolyase with respect to the α-carbon backbone (overall) structure. The common PHR domain comprises two structurally different domains, the N-terminal α/β-domain, which adopts a dinucleotide-binding fold (five parallel β-sheets surrounded by four α-helices and one 3_{10}-helix) and the C-terminal α-domain to which the cofactors are bound. These two domains are separated by a connector region that exhibits only limited regular secondary structure (see Plate 2.1) and bridges equivalent secondary structures from N- and C-terminal domains in cryptochromes and *E. coli* photolyase, respectively.

2.3.3.1   Photolyase structure and reaction mechanism

The structure and reaction mechanism of class I DNA photolyases are well studied. Considering the importance of the photolyase molecule for comparative studies with cryptochrome, the current knowledge of photolyase structure and function will be summarized briefly. The reader is also referred to recent reviews for greater detail (e.g. Sancar, 2003). The protein structures of three class I CPD photolyases (*Escherichia coli*, *Anacystis nidulans*, *Thermus thermophilus*) have been solved (Komori *et al.*, 2001; Park *et al.*, 1995; Tamada *et al.*, 1997). These, in common with other photolyases, contain the catalytic cofactor flavin FAD, which adopts a U-shaped conformation, where the isoalloxazine ring is in close proximity to the adenine ring. FAD is essential for catalysis and only active in its two-electron reduced deprotonated form (FADH$^-$). In addition, class I CPD photolyases contain a second cofactor that absorbs light and transfers the energy to FADH$^-$ or to FADH$^\circ$. Thus, the second cofactor is a chromophore that acts as an antenna. Although this cofactor is not required for catalysis, it increases the rate of repair under limiting light conditions (Kim *et al.*, 1991). In most species, the second chromophore is the pterin 5,10-methenyltetrahydropteroylpolyglutamate (methenyltetrahydrofolate, MTHF). In other species, such as *Anacystis nidulans*, the second cofactor is the deazaflavin-type chromophore 8-hydroxy-7,8-didemethyl-5-deazariboflavin (8-HDF). However, both FAD and the second cofactor are present in 1:1 stoichiometric amounts in photolyases and neither of them is covalently bound to the apoenzyme.

The CPD photolyase reaction mechanism consists of light-driven electron transfer from the fully reduced flavin (FADH$^-$) to the CPD creating an unstable CPD radical anion and a neutral flavin radical (FADH$^\circ$). The CPD radical causes the spontaneous cleavage of the carbon bonds within the cyclobutane ring which transfers the electron back to FADH$^\circ$, thus completing the reaction cycle. For *E. coli* photolyase, the efficiencies of energy transfer between the antenna and FADH$^-$ and of the electron transfer between FADH$^-$ and CPD are high with values of 62% and 89%, respectively. The energy transfer between the antenna and FADH$^\circ$ in *E. coli* photolyase is even faster with higher efficiency (92%) (for review see Sancar, 2003).

Energy transfer from MTHF to FADH$^-$ and to the fully oxidized FAD was recently demonstrated also for bacterial cryptochrome 1 from *Vibrio cholerae*. This was the first cryptochrome that was shown to contain both cofactors in significant and nearly stoichiometric amounts when purified in its native form (Saxena *et al.*, 2005). It was found that the MTHF to FAD energy transfer in this cryptochrome occurs with a sixfold higher rate than the transfer to FADH$^-$. Nevertheless, the second energy transfer was still found to be more than four times faster, with a twofold increase in MTHF fluorescence lifetime, when compared to *E. coli* photolyase. These results could suggest different binding interactions and local structures of MTHF in photolyases and cryptochromes, respectively, though this has yet to be proven. However, these experiments also suggest mechanistic similarities between photolyases that repair damaged DNA and cryptochromes that mediate blue light signalling.

Photolyase has high affinity for its substrate, but a reciprocal low affinity for undamaged DNA. For example, the binding constant of *E. coli* photolyase to the thymine dimer in DNA is about 10^{-9} M, whereas the binding constant for undamaged DNA is about 10^{-4} M (Husain and Sancar, 1987; Sancar, 2003). *In vitro* binding and enzyme assays have shown that plant cryptochromes have neither detectable photolyase activity (Hoffman *et al.*, 1996; Kleine *et al.*, 2003; Lin *et al.*, 1995b; Malhotra *et al.*, 1995) nor show significant binding to pyrimidine dimer-containing DNA (Malhotra *et al.*, 1995). Recent data have demonstrated, however, that some cryptochromes including *Arabidopsis* cry3 may well have DNA-binding activity (see Section 2.4.3).

2.3.3.2 Cryptochrome structure

The production of full-length plant cryptochromes in large quantities was hampered for a long time as expression in *E. coli* or yeast cells resulted in aggregated protein preparations that could not be reconstituted with cofactors. Consequently, efforts to crystallize plant cryptochromes were not successful, and even spectroscopic and biochemical studies were limited. However, more recent efforts to produce full-length plant cry1 and cry2 in Sf9 or Sf21 insect cells have been more effective (Bouly *et al.*, 2003; Giovani *et al.*, 2003; Lin *et al.*, 1995b; Shalitin *et al.*, 2003). This work has allowed spectrosocopic and biochemical characterization of cryptochromes that is described in Section 2.4.

In contrast to most plant and animal cryptochromes, the crystallization of *Synechocystis* cryptochromes was successful and its structure was solved at the atomic level (Brudler *et al.*, 2003). This cryptochrome (cryDASH) contains only FAD but no second cofactor. Its overall structure is very similar to the *E. coli* photolyase with differences that explain its lack of photolyase activity. In particular, the pocket that binds the pyrimidine dimer in photolyase is wider and flatter in cryDASH because of the replacement of two amino acids and the rotation of a Tyr residue out of the pocket. One of the amino-acid changes affects the electronic structure of FAD and thus probably alters its ability to transfer an electron. The second change in the pocket (Trp to Tyr) along with other changes on the protein surface are also important for substrate binding, probably these changes reduce the binding affinity for the photolyase substrate (Brudler *et al.*, 2003).

The first successful crystallization of plant cryptochromes was made in 2004 by the group of Johann Deisenhofer. They expressed the PHR domain of *Arabidopsis* cry1 (residues 1–509) as a soluble protein in *E. coli*, and crystallized this protein and solved its structure at 2.6 Å resolution (Brautigam *et al.*, 2004) (see Plate 2.1). They found that FAD was the only cofactor present in this cryptochrome structure, although in earlier experiments MTHF was found to be associated with cry1 (Malhotra *et al.*, 1995). However, the cry1 PHR region that corresponds to the MTHF binding pocket in *E. coli* photolyase was found to be largely filled with amino acid side chains, making the binding mode of MTHF to this protein unclear. Again, the overall structure is very similar to that of *E. coli* photolyase and *Synechocystis* cryDASH with differences that account for its lack of photolyase activity. The surface of cry1-PHR is predominantly negatively charged with a small concentration of positive charge near the FAD-access cavity. This contrasts with the photolyases and cry-DASH which have a positively charged groove on their surfaces near to the FAD-access cavity. The two Trp residues that are important for specific thymine-dimer and DNA binding in *E. coli* photolyase are changed in cry1 PHR to Leu and Tyr, respectively. Some other differences in this region result in a larger FAD-access cavity with a unique chemical environment when compared with the cavities of other members of the photolyase/cryptochrome superfamily. All these above-mentioned features together could effectively account for the lack of cry1 PHR photolyase activity, and also DNA-binding activity, a property of cryDASH (Brautigam *et al.*, 2004). Another unique feature of cry1 PHR is the presence of a disulfide bond between Cys residues in the N-terminal α/β domain and the connector region. It is not currently known whether this bond exists also *in vivo* and what is its role in the signalling mechanism (Brautigam *et al.*, 2004).

Recently, our group has expressed *Arabidopsis* cryptochrome 3, a cryDASH subfamily member, as a soluble protein in *E. coli* and in collaboration with the group of Lars-Oliver Essen we have crystallized this protein and solved its structure at 1.9 Å resolution (Pokorny *et al.*, 2005). As this protein contains both FAD and

MTHF cofactors in a 1:1 ratio, after purification, analysis of its structure should provide the structural support that is missing so far for cross-talk between both cofactors in cryptochromes (e.g. energy transfer; Section 2.4.4). Another interesting feature observed upon crystallization of cry3, in contrast to the other cryptochromes and photolyases that have been crystallized so far, is a presence of its dimer in crystals. Because cryptochrome dimerization might be functionally important for the mediation of light-triggered conformational changes, as recently exemplified for cry1 (Sang *et al.*, 2005; see Section 2.6.1), our cry3 structure could shed light on this aspect of cryptochrome function.

The fact that none of the solved cryptochrome structures contains the C-terminal extension highlights the need for crystallization of at least one more full-length cryptochrome or a CCT domain alone. It would be very illuminating to see such a structure as it would inform on interaction between the PHR and CCT domains and thus further elucidate the mechanism of cryptochrome signalling. A partial answer could come from our cry3 structure as this protein contains the putative DAS domain in its N-terminal extension, a feature contained within CCTs of cry1 and cry2 (Section 2.1). However, due to its different location within the receptor, it could differ substantially from the structures of true CCT domains.

2.4 Cryptochrome biochemistry and spectroscopy

2.4.1 Phosphorylation

Plant cryptochromes are phosphorylated *in vivo* and *in vitro* upon blue light exposure and phosphorylation probably affects both their activity and stability (Ahmad *et al.*, 1998c; Shalitin *et al.*, 2002; Shalitin *et al.*, 2003; Bouly *et al.*, 2003). As Chapter 5 deals with the role of phosphorylation in light signalling, the reader is referred to this chapter for further details on cryptochrome in this context.

2.4.2 Nucleotide-binding and kinase activity

In vitro studies with recombinant *Arabidopsis* cry1 expressed in insect cells and purified to apparent homogeneity have given more insight into the molecular mechanism of nucleotide binding and phosphorylation of cryptochromes (Bouly *et al.*, 2003; Shalitin *et al.*, 2003). These studies showed that cry1 autophosphorylates and that autophosphorylation is blue light dependent. In the work by Bouly *et al.* (2003), this was analyzed in detail. It was shown that autophosphorylation depends not only on blue light but also on the presence of the FAD cofactor, and the flavin antagonists such as KI and oxidizing agents abolish the blue light-induced phosphorylation. Since cryptochromes do not have homology to known protein kinases, one concern has been that *in vitro* phosphorylation is caused by a co-purified kinase and not by autophosphorylation. However, recombinant cry1, as well as cry1 purified from plant cells, binds to ATP-agarose. Furthermore, the binding affinity of cry1 for ATP ($K_d = 20$ μM) is in the same range as described for other ATP-binding proteins with high and specific affinity for ATP. The stoichiometry of ATP bound to recombinant cry1 was determined to be 0.4, indicating that cry1 contains one binding site for ATP. The recently solved crystal structure of the photolyase-related domain of *Arabidopsis* cry1 (see Section 2.3.3.2) indeed shows binding of the non-cleavable ATP analog AMP-PNP close to the FAD binding pocket with a distance of 4.8 Å to the closest FAD atom (Brautigam *et al.*, 2004). However, based on this structure the mechanism by which cryptochromes could mediate a phospho-group transfer is still an open question.

The *in vitro* (Bouly *et al.*, 2003) and *in vivo* (Shalitin *et al.*, 2003) kinetic studies of cry1 phosphorylation give similar results with saturation being reached between 30 and 60 min after the onset of blue light. Interestingly, preillumination of cry1 with blue light, in the absence of the ATP substrate, followed by the addition of ATP in the absence of blue light, nevertheless, leads to the phosphorylation of cry1 (Bouly *et al.*, 2003). This indicates that cry1 remains activated, at least for some time, after it is transferred to darkness. The identity of the amino acids phosphorylated in cry1 *in vitro* was determined and only serine was identified (Bouly *et al.*, 2003). Since the autophosphorylated cry1 does not show the same shift in mobility on SDS-PAGE as the cry1 isolated from plant material, one may assume that *in planta* autophosphorylates one upon blue light treatment but is also phosphorylated by other kinases. The blue light dependency of additional phosphorylation could be caused by a conformational change of cry after it has absorbed light, thus giving access to a kinase. Another explanation could be that blue light activates a kinase, which then phosphorylates the cryptochromes. In any case, further characterization of cry1 and cry2 *in vivo* and *in vitro* will be needed to fully determine the molecular mechanism of cry phosphorylation.

2.4.3   DNA-binding activity

As outlined in Section 2.3.3.1, photolyase has a high binding affinity for its substrate and a lower affinity for undamaged DNA, which is not sequence specific. Interestingly, it seems that the ability of photolyase to bind with low affinity to undamaged DNA is conserved in some cryptochromes, for example, cryDASH from the cyanobacterium *Synechocystis* (Brudler *et al.*, 2003). Since the structure of *Synechocystis* cryDASH has been solved (see Section 2.3.3.2), this allowed comparison with already known structures from microbial (class I) CPD photolyases. Importantly, the structure of a DNA photolyase together with its substrate has been solved (Mees *et al.*, 2004). This revealed the residues involved in CPD-binding and those that make contact with the DNA backbone. Five Arg residues (Arg226, Arg278, Arg342, Arg344, Arg397, numbering according to *E. coli* photolyase) on the protein surface that contribute to a positive electrostatic potential and are situated close to the substrate binding pocket, were considered to be important for DNA binding (Park *et al.*, 1995). Interestingly, all of these Arg residues are conserved in *Synechocystis* cryDASH for which binding to undamaged DNA has been demonstrated with an equilibrium dissociation constant of around 2 μM (Brudler *et al.*, 2003), similar to that described for *E. coli* photolyase. Indeed, the substrate co-crystal structure of *Anacystis nidulans* photolyase shows that the DNA makes several inter-actions on the protein surface with basic amino acid residues (Mees *et al.*, 2004). In *Arabidopsis* cry3, all of the above mentioned Arg residues are conserved (Brudler *et al.*, 2003; Kleine *et al.*, 2003) and it has been shown that cry3 also binds to DNA (Kleine *et al.*, 2003).

For all the other plant cryptochromes there is no direct proof for, or against, DNA binding. From random fusions of GFP with *Arabidopsis* cDNAs, a fusion protein was identified, which bound to chromatin. The fusion protein carried the C-terminal part of cry2 (Cutler *et al.*, 2000). It is not clear from this study, however, whether the chromatin association was mediated by the interaction of the cry2 C-terminus with other proteins or by direct binding of cry2 to DNA.

The question remains whether DNA binding of cryptochromes is regulated by light and what function this binding might have on the cellular response to light. In the case of *Synechocystis* cryDASH, it was concluded from a comparison of the gene-expression profiles from wild type and the *cry* mutant that cryDASH could act as a repressor of transcription (Brudler *et al.*, 2003).

2.4.4 Electron transfer

As described in Section 2.3.3.1, photolyases use light-driven electron transfer from the reduced flavin cofactor $FADH^-$ to the substrate for catalysis. In photolyase, only the fully reduced flavin is catalytically active, not the semireduced or fully oxidized form (for review see Sancar, 2003; Partch and Sancar 2005). Photolyase containing semireduced or oxidized FAD can be transformed to the catalytically active form by photo-excitation of the FAD in the presence of reducing agents in the medium. This photoreduction involves conserved tryptophans, and in some photolyases tyrosine residues, which transfer electrons to the excited FAD. Owing to the similarities between photolyase and cryptochromes in amino acid sequence and cofactor composition it was speculated that cryptochromes might use light-driven electron transfer for signalling (Cashmore *et al.*, 1999; Malhotra *et al.*, 1995).

Indeed, it has been shown for *Arabidopsis* cry1 that electron transfer could be involved in cryptochrome signalling (Giovani *et al.*, 2003). The photoreceptor used in this study was expressed and purified from baculovirus-transfected insect cells and contained fully oxidized FAD. After ns-laser flash excitation of the FAD, transient absorbance changes were monitored and the recovery kinetics indicated three components with half-lives of about 1 ms, 5 ms and >100 ms. From the kinetics and the observed spectral changes, it was concluded that upon excitation the semireduced radical $FADH^\circ$ is formed concomitantly with a neutral tryptophan radical. There was further evidence for electron transfer from a tyrosine to the tryptophan radical (Trp°), as was demonstrated for *Anacystis nidulans* photolyase. Addition of β-mercaptoethanol as an external electron donor led to the reduction of the tyrosine radical and to accumulation of $FADH^\circ$. Based on these studies one can conclude that the FAD cofactor can be photoreduced in cryptochromes, as in photolyase, involving Trp and Tyr radicals. In principle, all of these internal radicals as well as external electron donors that reduce the Tyr° or electron acceptors, which can be reduced by FADH°, could mediate the signalling. However, physiological electron donors or acceptors of cryptochromes have not yet been identified.

Recently, there have been more insights into how electron transfer reactions control their signalling from studies using cry1 tryptophan mutants (Zeugner *et al.*, 2005). Based on homology studies with *E. coli* photolyase, redox inactive Phe was substituted for Trp400 and Trp324, the predicted electron donors proximal to the FAD and to the exposed cry1 surface, respectively. These substitutions indeed led to impaired cry1 dependent blue light responses *in vitro* and *in vivo*. The insect cell expressed mutants showed a marked decrease in photoreduction of FAD in the presence of ß-mercaptoethanol compared to the wild-type cry1. Transient flash laser absorption spectroscopy in the absence of the external electron donor showed a rapid concomitant formation of the neutral semireduced flavin radical and a neutral tryptophanyl radical, as well as a polyphasic decay in the wild-type cry1. By comparison the absorbance changes in the tryptophan mutants were five times weaker. Moreover, a strongly enhanced fluorescence in the Trp400 mutant compared to the wild type and the Trp324 mutant was observed suggesting that Trp400 is primary electron donor to the FAD and that it quenches the fluorescence of the fully oxidized FAD due to fast electron transfer. Thus, the two tryptophan residues appear to be indispensable for efficient electron transfer to the FAD. Further, the mutants were also impaired in blue light stimulated *in vitro* autophosphorylation, implying a functional relevance of the intramolecular electron transfer. When the mutant proteins were expressed in *Arabidopsis* in a cryptochrome deficient background, there was both reduced anthocyanin accumulation and reduced hypocotyl growth inhibition under blue light compared to seedlings containing wild-type cry1, thus suggesting the importance of the electron transfer reaction for *in vivo* photoreceptor function. Hence, a light-dependent intramolecular electron transfer to the FAD could be the primary step to trigger the plant cryptochromes to undergo a conformation change or other biochemical changes, thereby initiating their signalling pathway.

2.5 Expression and biological activity of cryptochromes

2.5.1 Expression and light regulation of cryptochromes in planta

The transcription of *Arabidopsis CRY1* and *CRY2* is under photoperiodic and circadian clock control (Bognár *et al.*, 1999; El-Assal *et al.*, 2003; Harmer *et al.*, 2000; Toth *et al.*, 2001). *CRY3* transcript levels are transiently upregulated in etiolated *Arabidopsis* seedlings by continuous far-red light mediated through phyA (S. Meier, A. Batschauer, unpublished data), and light effects on *CRY* transcript levels were also described for the fern *Adiantum* (Imaizumi *et al.*, 2000). How the differential expression of cryptochromes could affect their biological function is not well understood and it is not clear if the fluctuations in transcript levels are reflected in corresponding changes of the cry protein levels. What is known, however, is that light does impact on *Arabidopsis* cry2 protein levels (Ahmad *et al.*, 1998a; Lin *et al.*, 1998). Exposure of etiolated *Arabidopsis* seedlings to blue light leads to a rapid decrease in the amount of cry2. This effect is fluence rate and wavelength dependent. When treated with low fluence rate (1 μmol m^{-2} s^{-1}) blue light for 6 h, cry2 protein levels were comparable to those in dark grown seedlings. However, 1 h of high fluence rate blue light was effective in significantly reducing cry2 protein, and after 24 h of exposure cry2 was undetectable.

Although UV-A light also led to a decrease in the amount of cry2 protein, red light had no effect even when seedlings were exposed for prolonged periods and at fluence rates (Lin *et al.*, 1998). The fact that the amount of cry2 protein is only affected by light with wavelengths below 500 nm and that this process is very similar for wild type and *cry1* mutant plants (Ahmad *et al.*, 1998a) suggests that cry2 could regulate its own degradation. However, the involvement of other blue light receptors in this process has not been tested rigorously so far.

The rapid downregulation of the cry2 protein in blue light together with the observation that its transcript level is not reduced by light suggests that blue light either induces degradation of cry2 or blocks translation of its mRNA. To distinguish between these possibilities, dark-grown seedlings were incubated with the protein-synthesis inhibitor cycloheximide and then treated with blue light. Since no difference in the disappearance of cry2 was observed between inhibitor-treated and control plants, it is very likely that blue light induces the degradation of cry2 (Ahmad *et al.*, 1998a).

In order to define the region of cry2 involved in its degradation, domain-switch experiments were performed in which different regions of *Arabidopsis CRY1* and *CRY2* were exchanged and the chimeric genes expressed under control of the CaMV 35S promoter in the *cry1* mutant of *Arabidopsis* (Ahmad *et al.*, 1998a). The fusion proteins that contained either the C-terminal extension (amino acids 506–611) or the N-terminal region (amino acids 1–505) of cry2 were biologically active and showed significantly lower levels in blue than in red light, indicating that both domains of cry2 can mediate degradation. Since chimeric proteins of the GFP or GUS reporters with either the N-terminal or the C-terminal domain of cry2 are not reported to be light labile (Guo *et al.*, 1999; Kleiner *et al.*, 1999), one may conclude that both cry2 domains are required to mediate degradation. This result also supports the view that cry2 is likely to induce its own degradation.

Studies comparing two naturally occurring CRY2 alleles in the Cape Verde Islands (Cvi) and Landsberg *erecta* (Ler) *Arabidopsis* accessions have provided additional insights into CRY2 function. In Cvi *CRY2* methionine substitutes for valine at position 367 and this appears to cause the early flowering in short days (SDs) and day-length insensitivity of this accession (El-Assal *et al.*, 2001). Interestingly, when dark-grown Cvi and Ler seedlings were transferred to blue light at 40 μmol m^{-2} s^{-1}, the depletion of Cvi-cry2 and Ler-cry2 proteins was very similar. However, when plants were grown under SD conditions during the photoperiod Cvi-cry2 was degraded much more slowly than Ler-cry2 and it reaccumulated much faster in the following dark period. The same authors have also shown that the levels of Cvi-cry2 and Ler-cry2 in plants kept under LD conditions are very similar and do not oscillate significantly. Taken together, these data show that the level of cry2 protein is under photoperiodic control and that a single amino acid substitution within cry2 leads to some stabilization in light when plants are kept in SDs. This extended stability of Cvi-cry2 in SD is most likely the direct cause for the early-flowering phenotype of this accession under these conditions. The molecular mechanism of how the amino acid substitution at position 367 affects cry2 stability has not yet been investigated.

The photoperiodic effect on cry2 stability was also addressed by Chentao Lin and coworkers (Mockler *et al.*, 2003). As Koornneef and coworkers found (El-Assal *et al.*, 2003), the cry2 protein oscillates strongly under SD conditions with high levels at the end of the dark phase and low levels during the light phase. In LD conditions, the oscillation was very weak and cry2 levels are constitutively low throughout the cycle. When white light was replaced by monochromatic light, the same oscillation in the cry2 level was observed under SD conditions with blue light but not with red light. Surprisingly, the cry2 level was very low in SD red light conditions. When plants were transferred from blue light SDs to continuous red light cry2 levels were constitutively high, whilst a transfer from red light SDs resulted in low cry2 levels. After transfer from SD blue light to continuous blue the cry2 level remained low but showed some increase in the subjective night phase. Taking these results together the following can be concluded: (1) the cry2 protein level strongly oscillates under SD but not under LD conditions; (2) oscillation of the cry2 protein level is mainly controlled by protein degradation and not by the circadian expression of the *CRY2* gene; (3) blue light induces oscillation and cry2 degradation; (4) red light partially antagonizes blue light control of cry2.

The degradation of cry2 is reminiscent of the degradation of phyA, which is also rapidly broken down upon light treatment. In the case of phyA, ubiquitination has been shown to occur upon light treatment (Clough *et al.*, 1999). For cry2, ubiquitination has not been demonstrated; nevertheless, there are some indications that cry2 could be degraded by the proteasome pathway. COP1 (see Chapter 6 for further details) is a putative subunit of the E3 ubiquitin ligase complex, mediating the proteolytic degradation of the bZIP transcription factor HY5 in darkness. In light, HY5 is not degraded and activates the transcription of genes, such as *CHS* (Ang *et al.*, 1998; Hardtke *et al.*, 2000; Osterlund *et al.*, 2000), which have been shown to be active in the light but not in the dark. In *Arabidopsis* seedlings carrying the weak *cop1-6* allele, the degradation of cry2 in blue light is impaired and the ratio between phosphorylated and unphosphorylated cry2 is increased (Shalitin *et al.*, 2002) (see Section 2.4.1). This suggests that phosphorylated cry2 is the substrate for degradation, and requires functional COP1 for the process to be efficient. In support of this conclusion are the results from yeast-two-hybrid interaction studies that show physical interaction of cry2 with COP1 (Wang *et al.*, 2001). Also cry1, which is not degraded in light, interacts with COP1 (Yang *et al.*, 2001) and from this association, it was concluded that the interaction between COP1 and cry1 is involved in cry signalling (Wang *et al.*, 2001; Yang *et al.*, 2001) (see Section 2.6.2.1). A problem in assuming COP1 involvement in cry2 degradation is the fact that COP1 is transported out of the nucleus in light (Osterlund and Deng, 1998; von Arnim and Deng, 1994), whereas cry2 seems to be located in the nucleus independent of the light conditions (see Section 2.5.2). However, since the degradation of cry2 seems to be much faster than the translocation of COP1, there could be enough COP1 present in the nucleus after dark-light transition to initiate cry2 degradation. Thus,

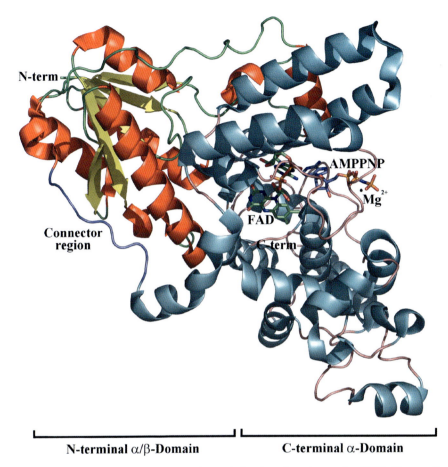

N-term

AMPPNP

Mg^{2+}

FAD

C-term

Connector region

| N-terminal α/β-Domain | C-terminal α-Domain |

Plate 2.1   Structure of *A. thaliana* cryptochrome 1 photolyase homology region (PHR). The overall structure is very similar to that of *E. coli* photolyase. In different colours are shown the N-terminal α/β-domain (red helices, yellow β-sheets, green coils) and the C-terminal α-domain (cyan helices, magenta coils) joined by the connector loop (blue). The positions of the N- and C-terminus are indicated (N- and C-term, respectively). The flavine adenine dinucleotide (FAD) cofactor is bound to the C-terminal α-domain in an U-shaped conformation with its adenine ring being in close proximity to the isoalloxazine ring as found in photolyases. No second cofactor is present in the structure. The non-hydrolysable ATP-analog adenosine 5′-(β,γ-imido)triphosphate (AMPPNP) is bound to the C-terminal α-domain close to the FAD cofactor (see text) while its triphosphate moiety is complexed with Mg^{2+} ion. The coordinates were taken from public protein structure database and the structure was displayed using PyMOL software. This structure was solved by Brautigam *et al.* (2004).

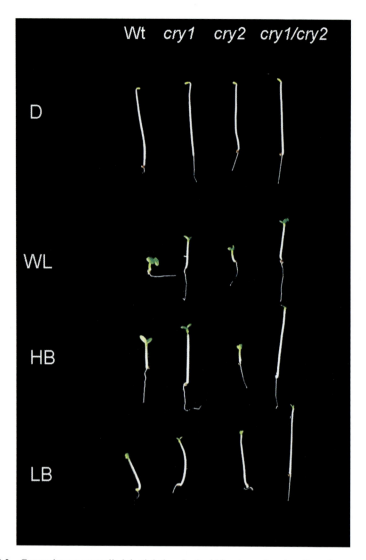

Plate 2.2 Cryptochrome controlled deetiolation. In *Arabidopsis*, cry1 and cry2 regulate most of the blue light-specific developmental programs. Shown are only phenotypic effects during deetiolation caused by mutations in *cry1* or *cry2*. *Arabidopsis* wild type (WT) and photoreceptor mutants grown for 3 days in darkness (D), white light (WL) or continuous blue light of 30 μmol m^{-2} s^{-1} (high blue, HB) or 1 μmol m^{-2} s^{-1} (low blue, LB), respectively given from the top. In wild-type seedlings, hypocotyl growth is inhibited by blue light whereas opening of the hypocotyl hook and cotyledon opening and expansion is promoted by blue light. HB is more efficient than LB. The lack of cry1 (*cry1*) is most evident under HB conditions, whereas the lack of cry2 (*cry2*) becomes more evident under LB conditions, which is seen in particular for the double mutant (*cry1/cry2*).

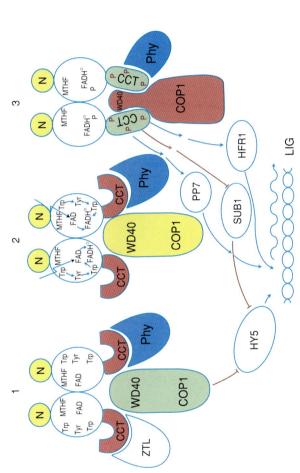

Plate 2.3 Model of cryptochrome signalling. This model suggests that cryptochromes form homodimers irrespective of the light conditions, and that the N-terminal photolyase homology region (PHR) mediates homodimerization. The α-helical domain of PHR binds the two cofactors, FAD and MTHF and carries the amino acid residues (Tyr, Trp) involved in transfer of an electron to FAD after it has been excited by light. (**1**) Ground state of the receptor with fully oxidized FAD. Cry1 and cry2 bind already in the ground state via their C-terminal domain (CCT) the E3 ubiquitin ligase COP1 through the WD40 domain of COP1. The cry–COP1 interaction in darkness has no effect on the ability of COP1 to recruit the transcription factor HY5 for ubiquitination and proteasomal degradation. Further interaction of the CCT of cry1 with the C-terminus of phyA and the Zeitlupe (ZTL) protein was described. However, it is not known whether these interactions are light-controlled in the context of full-length receptor protein. In case of cry2 its interaction with phyB seems to be light-controlled as indicated by the colocalization of cry2 and phyB in nuclear speckles only after light-treatment. (**2**) First step in light-driven activation of cry. Excitation of fully oxidized FAD leads to the formation of flavin semiquinone (FADH°). The electrons required for the reduction of FAD are provided by a tryptophan residue (Trp) proximal to FAD, which itself is reduced by more distally located Trp and Tyr residues. Energy transfer from excited MTHF to FAD has only been demonstrated yet for a cryptochrome from *Vibrio cholerae*. (**3**) Signalling state of cryptochrome. After FADH° formation, several serine residues of cry become phosphorylated (P) within the CCT but most likely also within the PHR domain. CCT phosphorylation is essential for signal transduction. It is not clear yet whether cross-phosphorylation occurs between the two cry molecules within the dimer. The transition from the ground state to the signalling state seems to release the CCT from a blocked state to its active state probably by conformational changes occurring in the CCT. The active state of CCT represses COP1 function and causes translocation of COP1 out of the nucleus. As a consequence, HY5 accumulates in the nucleus and switches on the transcription of light-induced genes (LIG). In addition, further downstream components are regulated by the signalling state of cry. SUB1, a calcium-binding cytosolic protein that operates upstream of HY5 as a repressor of cry signalling, is inhibited. The positive factor PP7, a nuclear-localized phosphatase, is activated by cry as is the putative transcription factor HFR1. As for most models the lifespan of this one cannot be predicted. For further details and references see text.

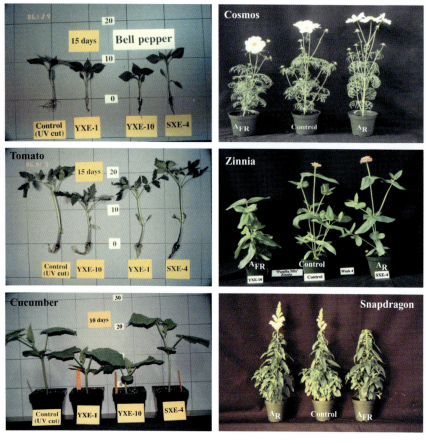

Plate 12.1   Plant response under red and far-red light absorbing photoselective greenhouse films. In the left column, YXE-1 and YXE-10 are photoselective films with different far-red light absorbing dyes. SXE-4 is a photoselective film with a red light absorbing dye. In the right column, A_R and A_{FR} are red and far-red light absorbing films, respectively.

further research is needed to elucidate the molecular events in blue light-induced cry2 degradation.

2.5.2 Cellular localization

The localization of cry2 was studied in detail using GUS and GFP as reporters, and also by immunological methods (Guo *et al.*, 1999; Kleiner *et al.*, 1999). All approaches showed consistently that cry2 is localized in the nucleus. In contrast to phytochromes (see Chapter 1), there seems to be no light effect on nuclear targeting of cry2. The CCT of cry2 is required and sufficient for translocation into the nucleus, and this region contains a bipartite nuclear localization signal (NLS).

Protoplasts transfected with cry2-GFP or cry2-RFP produced a homogenous signal within the nucleus when the cells were kept in darkness or red light. However, blue light treatment caused the rapid formation of the so-called nuclear speckles in tobacco (Más *et al.*, 2000), as well as in *Arabidopsis* and parsley protoplasts (M. Müller and A. Batschauer, unpublished data). These cry2 speckles co-localize with phyB (Más *et al.*, 2000); this further supports the observation that cry2 and phyB interact (Más *et al.*, 2000) (see also Chapter 1 and Section 2.6.2.3). The localization of *Arabidopsis* cry1 was studied in onion epidermal cells by bombarding them with cry1-GFP constructs (Cashmore *et al.*, 1999) and in transgenic *Arabidopsis* plants as GUS fusions (Yang *et al.*, 2000). The cry1-fusion proteins were also found in the nucleus. In contrast to cry2, there is a light effect described for the localization of cry1. The CCT1 fused to GUS was found to be enriched in the nucleus in dark-grown plants and to be cytosolic in light-treated plants (Yang *et al.*, 2000). Since the CCTs do not bind chromophores, the observed light effect on the localization of the fusion protein is not self-mediated. However, this does not rule out that endogenous cryptochromes are involved in this process. Some of the cryptochromes of the fern *Adiantum capillus-veneris* (see Wada, 2003), as well as the cryptochromes in animals (for review see Partch and Sancar, 2005), are also transported to the nucleus.

The more recently discovered cry3 (Kleine *et al.*, 2003) carries an extension at the N-terminus, and the most extreme N-terminal region has significant similarities with targeting signals for import into chloroplasts and mitochondria. Using cry3-GFP fusion proteins and *in vitro* import studies it was shown that cry3 is indeed transported into both organelles. Since the N-terminus of cry3 is necessary and sufficient for the import into chloroplasts and mitochondria, it must contain a dual targeting signal (Kleine *et al.*, 2003). The function that cry3 fulfils in these organelles remains to be investigated.

2.5.3 Growth responses controlled by cryptochromes

As already mentioned, *cry1* was identified in a screen for *Arabidopsis* mutants with reduced hypocotyl growth inhibition in white and blue light. Because of the above-mentioned redundancy of photoreceptor action in white light, the effects of mutation of *CRY1* are less pronounced in white than in blue light (Ahmad and Cashmore, 1993; Jackson and Jenkins, 1995). In contrast to blue light, the lack of cry1 and of cry2 seems to have no effects in darkness (Ahmad and Cashmore, 1993; Lin *et al.*, 1998;

Jackson and Jenkins, 1995) (see Plate 2.2). However, there are reports describing red-light effects, which are affected in the *cry1* mutant. This could be explained by a role of the flavosemiquinone radical form of the photoreceptor, which absorbs even above 500 nm or that cryptochrome acts as a component in phytochrome signalling (Devlin and Kay, 2000; Chapter 8).

Although the lack of cry1 has a strong effect on hypocotyl growth inhibition under high fluence rates of blue light, the *cry2* mutant shows essentially no difference from wild type under these conditions (see Plate 2.2). However, under lower fluence rates (1 μmol m^{-2} s^{-1} and less) the lack of cry2 becomes clearly visible as its hypocotyls are much longer than those of the wild type under these conditions (Lin *et al.*, 1998) (see Plate 2.2). Even more pronounced is this effect in the *cry1/cry2* double mutant, revealing a redundancy of action in deetiolation where cry1 operates primarily under high light and cry2 under low light conditions. Interestingly, a similar situation was also observed for phytochromes (see Chapter 9). The fact that cry2 does not operate under high fluence rates of blue light during seedling development is explained by the observation that under such conditions the cry2 protein is rapidly degraded (Lin *et al.*, 1998). Cry2 degradation is further discussed in Section 2.5.1.

Detailed physiological analysis has revealed that the underlying molecular mechanisms that modulate hypocotyl growth are complex and involve many different mechanisms. Growth inhibition can be observed after just 30 s of blue light exposure. Within 30 min the growth rate decreases to almost zero and then reaches a steady rate for several days that is much lower than that observed for dark-grown seedlings (Folta and Spalding, 2001; Parks *et al.*, 1998; Parks *et al.*, 2001). By analyzing *Arabidopsis* mutants deficient in *cry1*, *cry2* and *phototropin1* (*phot1*), it was shown that the early response (within 30 min) was similar to wild type in *cry* single and *cry1/cry2* double mutants but strongly reduced in the *phot1* mutant, demonstrating that phot1, and not the crys, is responsible for this early growth inhibition. Although cry1 and cry2 are not required for the early response, they mediate a very fast membrane depolarization caused by the activation of anion channels, which precedes the early inhibition response (Parks *et al.*, 2001). Blocking the blue light-regulated anion channels chemically has no effect on the phot1-regulated growth inhibition, but does affect the second phase (30–120 min of blue light) that is controlled by cry1 and cry2. Afterwards, growth inhibition in blue light (under high fluence rates) seems to be controlled only by cry1, and this is independent of anion channel activity. The cry-regulated growth inhibition phase is, however, delayed in the *phot1* mutant, indicating that phot1 affects cry signalling (for review see Parks *et al.*, 2001).

Besides hypocotyl growth inhibition, cry mutant alleles also impair other processes during deetiolation, such as cotyledon opening (Lin *et al.*, 1998), cotyledon expansion (Jackson and Jenkins, 1995; Weller *et al.*, 2001), inhibition of petiole elongation (Jackson and Jenkins, 1995), anthocyanin formation (Ahmad *et al.*, 1995; Jackson and Jenkins, 1995; Ninu *et al.*, 1999; Weller *et al.*, 2001) and alter gene expression, which is discussed in Section 2.5.4.

The expression of either CRY1 or CRY2 under control of the constitutive and strong cauliflower mosaic virus 35S promoter in *Arabidopsis* or tobacco leads essentially to phenotypes that are opposite to *cry1* or *cry2* loss of function mutants and result in exaggerated inhibition of hypocotyl and petiole growth, and enhanced cotyledon opening and anthocyanin production (Ahmad *et al.*, 1998a; Lin *et al.*, 1995a,, 1996a, 1998). However, detailed inspection of expression profiles of blue light-regulated genes showed that overexpression of cry1 did not always have effects opposite to what was observed in the *cry1/cry2* double mutant (Ma *et al.*, 2001) (see Section 2.5.4).

The cry1 overexpressor showed enhanced sensitivity for UV-A and blue light, as expected, but also to green light (Lin *et al.*, 1995a, 1995b, 1996a). The effect of green light could be due to the presence of a semi-reduced FAD chromophore in the cryptochrome, which has however not been demonstrated yet *in planta*. Surprisingly, there is no direct correlation between the amount of cry1 photoreceptor in *Arabidopsis* seedlings and their sensitivity to light. A comparison of the action spectra and threshold values for hypocotyl growth inhibition (end-point measurements) of wild type, *cry1* single and *cry1/cry2* double mutants and cry1 overexpressor showed that an increase in the amount of cry1 by a factor of 10 leads to a shift in the threshold values of less than a factor of 3. In addition, the shape of the action spectrum was altered by cry1 overexpression (Ahmad *et al.*, 2002). Although limiting levels of signalling components or adverse effects causing 'light-stress' could explain the first observation, the effects on the shape of the action spectrum by cry1 overexpression are not understood.

Besides *Arabidopsis*, the function of cryptochromes has also been studied in moss and fern plants (Wada, 2003) and to some extent in tomato. As mentioned above, tomato has three cryptochrome genes (CRY1a, CRY1b, CRY2) and a putative cryDASH. The characterization of the function of cry1 was done by expression of antisense constructs (Ninu *et al.*, 1999) and by mutant analysis (Weller *et al.*, 2001). As with *Arabidopsis,* cry1 regulates the inhibition of hypocotyl growth and the induction of anthocyanin formation in tomato. However, phenotypic changes of the tomato *cry1* mutant and antisense plants were also found that have not been observed in *Arabidopsis*. These characteristics include reduced chlorophyll content in seedlings and effects on stem elongation, apical dominance, and chlorophyll content in leaves and fruits of adult plants. In addition to the growth and differentiation processes regulated by cryptochromes as described above, there are some reports of cryptochromes having a role in phototropism and stomatal opening (Ahmad *et al.*, 1998b; Mao *et al.*, 2005; Whippo and Hangarter, 2003) normally associated with phototropins (Chapter 3).

2.5.4   *Regulation of gene expression through cryptochromes*

As outlined above, cryptochromes regulate many physiological and developmental processes in plants and probably most of these processes involve differential gene expression, at least in part. When the complete sequence of the *Arabidopsis* genome became available (*Arabidopsis* Genome Initiative, 2000), genome-wide expression profiling became feasible and was used to analyze light effects on gene expression in *Arabidopsis*. In one of these studies (Ma *et al.*, 2001) long-term effects of light treatment on gene expression were analyzed by growing the seedlings for 6 days in continuous darkness or in white, blue, red or far-red light. In addition, seedlings grown for 4.5 days in darkness were treated with light for 36 h. Besides wild-type plants, mutants lacking phyA, phyB, or cry1/cry2, as well as overexpressors of phyA, phyB and cry1, were included in this study. It was shown that of the 9216 analyzed ESTs (representing about 6120 unique genes) 32% showed of at least twofold differential expression in white light. Under monochromatic light conditions, 73%, 57% and 40% of the genes expressed differentially in white light were affected by red, blue and far-red light, respectively. These numbers already show that the expression of most of these genes is affected by different wavelengths of light.

Although only a few genes had been previously identified as being downregulated by light, expression profiling showed that, of the differentially expressed genes, about 40% are repressed. This value is more or less the same in all light qualities. These data showed that at least 26 pathways seem to be coordinately upregulated or downregulated by white light. The 11 pathways downregulated by white light include those for the mobilization of stored lipids, enzymes which are probably no longer needed under these conditions; for ethylene and brassinonoid synthesis, hormones known to be involved in repressing photomorphogenesis (see below); and for cell wall degradation and water transport across the plasma membrane and the tonoplast, which are probably involved in enhancing hypocotyl elongation growth in darkness.

Establishing the number and specific groups of genes that are affected by cryptochrome function provides valuable information of how these photoreceptors signal. A genomics study that addressed this question was undertaken by Ma *et al.* (2001) using the *cry1/cry2* double mutant and the cry1 overexpressor. Most of the genes that were up- or downregulated in the wild type by blue light were not differentially expressed in the *cry1/cry2* double mutant under the same light conditions, demonstrating that cryptochromes are the major photoreceptors for regulation of gene expression in blue light and that other photoreceptors such as phototropins and phytochromes seem to play only a minor role under these conditions. However, overexpression of cry1 under control of the constitutive and strong 35S promoter of cauliflower mosaic virus resulted in reduced expression of 18% of the genes that are induced in wild type and 7% of the genes that were not upregulated in wild type under blue light showed upregulation in the cry1 overexpressor. This result shows that the enhanced level of the cry1 photoreceptor leads to both quantitative and qualitative effects on gene expression and explain, at least in part, why the increase in photoreceptor concentration does not result in a corresponding shift in the threshold response curve for hypocotyl growth inhibition (Ahmad *et al.*, 2002) (see Section 2.5.3). However, driving the expression of the photoreceptor with a promoter, which causes ectopic expression, could also result in side effects resulting from the presence of the photoreceptor in cells where it is normally absent. Studies on the tissue- and cell-specific expression of *CRY1* and *CRY2* using promoter–reporter fusions in transgenic *Arabidopsis* plants have, however, shown that both genes seem to be expressed in all organs and tissues (Lin, 2002; Toth *et al.*, 2001).

In another DNA microarray study using Affimetrix gene chips, differences between wild type and *cry1* gene expression under blue light were analyzed (Folta *et al.*, 2003). Although effects of blue light treatment for at least 36 h were analyzed in the study of Ma *et al.* (2001), Folta *et al.* (2003) screened for differential effects 45 min after the onset of blue light, a time point when hypocotyl growth inhibition is already under cryptochrome control (Parks *et al.*, 2001) (see Section 2.5.3). They found that 420 (5%) of the 8298 analyzed transcripts were differentially expressed in the *cry1* mutant, about half of them with higher and half of them with lower transcript levels compared to wild type. A possible explanation for the downregulation of transcripts in blue light in the *cry1* mutant, which are upregulated in wild type, is that their expression is regulated at the level of transcription and RNA stability, both of which are positively affected by cry1.

Among the pathways where gene expression is differentially affected soon after the onset of blue light are the cell cycle, auxin and gibberellin synthesis or signalling, and cell wall metabolism. All of the cell cycle genes are upregulated in the *cry1* mutant, indicating that cry1 suppresses cell division at this developmental stage, although hypocotyl growth inhibition is caused by reduced cell elongation. Most of the differentially expressed genes for auxin and gibberellin synthesis or signalling were upregulated in the *cry1* mutant, indicating that blue light represses these pathways via cryptochromes and this represses cell elongation. About half of the differentially expressed cell-wall genes are upregulated and the other half downregulated in the *cry1* mutant. Inspecting the known or putative functions of the encoded proteins, one can conclude that cry1 suppresses the expression of genes involved in cell wall loosening, but enhances the expression of genes involved in cell wall strengthening (Folta *et al.*, 2003; Ma *et al.*, 2001).

In conclusion, the gene expression profiling studies show the following: (1) Most of the blue light effects on gene expression are mediated by cryptochromes. (2) Cryptochromes have short- (minutes) and long-term (days) effects on gene expression. (3) Cryptochromes affect hormone biosynthesis and signalling by repressing auxin and gibberellin pathways at early stages and the brassinosteroid pathway in a later stage of development. (4) Genes involved in extension growth through cell-wall relaxation and increasing water transport through the plasma membrane and the tonoplast are suppressed by blue light via the cryptochromes. (5) Overexpression of cryptochrome 1 has both quantitative and qualitative effects on the gene expression pattern. (6) Although cryptochromes seem to affect transcription rates in most cases, there is also evidence for effects of cryptochromes on the stability of some transcripts. (7) Many of the genes that are regulated by cryptochromes are also controlled by phytochromes.

2.6   Cryptochrome signalling

The initial signalling mechanism of cryptochromes is still not well understood. However, the combined results from studies on their post-translational modifications, in particular phosphorylation, intramolecular electron transfer and the identification of interacting proteins, have provided informative insights into how cryptochromes work. Whereas the two former aspects are already discussed in Sections 2.4.1 and 2.4.4, we focus here on the output domain, interacting proteins and further downstream signalling components of cryptochromes.

2.6.1   Dimerization and output domains

Cryptochromes are distinguished from photolyases not only because they lack photolyase activity but also because they possess regions that are located C-terminally (cry1, cry2) or N-terminally (cry3) that extend from the photolyase homology domain (see Figure 2.2). As outlined in Section 2.1, these extensions differ in size and amino acid sequence but contain a motif (the DAS motif) that is well conserved in most plant cryptochromes. Ahmad *et al.* (1995) have already shown that mutations in the C-terminal domain of cry1 (CCT1) can cause loss-of-function of photoreceptor activity. This indicated that the C-terminal domain is specifically involved in cryptochrome signalling. However, since no spectroscopic studies have been performed on the mutant variants, one cannot exclude the possibility that the CCT has influence on the spectral integrity of this photoreceptor. Anthony Cashmore and coworkers addressed the question of how CRY1-CCT and CRY2-CCT mediated their effects on growth and development in transgenic *Arabidopsis* (Yang *et al.*, 2000). When not fused to another protein, both CCTs were unstable. Consequently, the CCTs of CRY1 and CRY2 were fused with β-glucuronidase (GUS), a reporter protein known to have a very high stability in plant cells. Indeed, very high levels of the GUS–CCT protein fusions could be achieved with expression under control of the CaMV 35S promoter, higher than expression of full-length cry1 under control of the same promoter. Interestingly, most of the transgenic lines overexpressing the CCT of CRY1 or CRY2 showed a phenotype similar to the *cop* (*constitutive photomorphogenic*) mutants (see Chapter 6). These mutants exhibit dark phenotypes that resemble those of light-grown seedlings. Common traits include short hypocotyls, enhanced anthocyanin production, initiation of chloroplast development and expression of genes normally induced by light. In addition, the overexpressors of CCT1 or CCT2 flowered earlier than the wild type under SD conditions. The observed effects of overexpression of the CCTs were specific as all control plants, expressing GUS alone or fusions of GUS with the CCTs of human or Drosophila cryptochromes, or with the C-termini of *Arabidopsis* phyA or phyB or the fusion of GUS with the N-terminal domain of *Arabidopsis* cry1, showed no phenotypic changes compared to wild type. Most importantly, seedlings overexpressing GUS-CCT1s carrying point mutations shown to inactivate full-length cry1, did not exhibit a *cop* phenotype. The effect of overexpression of the CCTs was seen not only in darkness but also under various light conditions where the overexpressors had significantly shorter hypocotyls than the wild-type controls. The action of the CCT1 seems to be independent of the other photoreceptors since its overexpression in the *cry1, phyA, phyB* or *hy1* (*hy1*: phytochromobilin synthesis mutant) backgrounds had nearly the same effect as when expressed in the wild type. Based on these findings, it was suggested (Yang *et al.*, 2000) that the CCTs are the signalling domains of cryptochromes that, in the intact receptors, are repressed by the photolyase-related sensory domain in the dark state. Upon light excitation, the CCT is altered either chemically and/or structurally to allow the propagation of the light signal to a downstream partner. The light-induced changes of the CCT are probably initiated by electron transfer processes occurring in the N-terminal domain as outlined above (see Plate 2.3). Since the fusion of GUS with the CCTs mimics a constitutively active receptor, it is unlikely that electrons are transferred from the sensory domain to the output domain, because the GUS fusion cannot perform photochemistry. Instead, it seems more likely that the photochemical processes within the sensory domain induce conformational changes in the CCTs that allow signal transduction. If so, the fusion with GUS would push the CCT to its signalling state. Alternatively, the subcellular localization of the cryptochromes could be altered by light that permits interaction with a positive element in the light, or a negative element in darkness. However, the available data on the subcellular distribution of cryptochromes do not indicate that this is the case. As outlined in Section 2.5.2, full-length cry2 and fusions of GUS or GFP with the CCT of cry2 seem to be localized in the nucleus independent of the light conditions. The fusion of GUS with the CCT1 was found in the nucleus of dark-grown seedlings and to be excluded from the nucleus after transfer to light (Yang *et al.*, 2000). Therefore, overexpressed CCTs of cry1 and cry2 are localized in different cellular compartments, although causing similar phenotypes.

An indication that the *Arabidopsis* cryptochromes may need a dimeric structure for activity *in vivo* also came from the observation that transgenic seedlings expressing the CCTs fused to GUS display a COP phenotype (Yang *et al.*, 2000). This suggested that in etiolated seedlings, the action of the C-terminal domain in native cry could be suppressed by the N-terminal domain. Indeed, this has been demonstrated in transgenic seedlings overexpressing the cry1 N-terminal domain fused to the myc epitope in the wild-type background. These seedlings that express both the endogenous cry1 and the N-terminal domain fusion have a mutant phenotype. Overexpression of this fusion in the *cry1* mutant background had a dominant negative effect (Sang *et al.*, 2005). One explanation for these results is that crys directly interact or dimerize *in vivo*. Yeast-two-hybrid assays by the same group have demonstrated cry1–cry1, CNT1–CNT1 (cry1–N-terminal domains), CNT1–cry1, CNT2–CNT2 (cry2 N-terminal domains) and CNT2–cry2 interactions and shown that they are light independent. However, no evidence has been provided for interaction between the carboxy- and the amino-terminal domain. Chemical cross-linking, size exclusion and co-immunoprecipitation studies of *Arabidopsis* expressing myc or TAP- tagged cry1 also suggest that cry1 homodimerization is light independent. Mutations in the N-terminal domain that compromise cry1 phosphorylation and activity (Shalitin *et al.*, 2003) also abolish CNT1–cry1 dimerization, indicating that indeed dimerization is a requisite for cry activity. Further, overexpression of the mutant CNT1 fused to the C-terminal domain in a *cry1* mutant background failed to show enhanced blue light responses. Likewise, mutant CNT1 overexpression in the wild-type background did not show a dominant negative phenotype. Since GUS itself oligomerizes, GUS-cry1/2 and GUS-CCT expressors show a COP phenotype, while GUS-CNT expressors do not, it is now clear that the GUS-CCT COP phenotype results from GUS activity. The GUS oligomer functions as the light-modified N-terminal domain dimers of cry1, resulting in the conformational change of the C-terminal domains and eventually allowing it to interact with partners such as COP1. *In vitro* pull-down assays using yeast expressed cry1 and cry2 with radioactively labelled CNT, CCT and full-length cry (by *in vitro* transcription/translation) showed light-independent cry–cry and cry–CNT interaction but not cry–CCT interaction. However, such interactions depend to a large extent on the conformation of proteins, and *in vitro* translated proteins, also owing to the absence of cofactors, may not assume the appropriate structures. *In vitro* intermolecular cry interactions involving the C-terminal domain have been observed in our group using insect cell expressed cry2 with *E. coli* expressed CCT of cry2 by binding studies (Banerjee and Batschauer, unpublished data). Further, the C-terminal domain of crys has flexible structures with substantial intrinsic disorder and attain stability only on interaction with the N-terminal domains as shown by CD spectroscopy, NMR and partial proteolysis (Partch *et al.*, 2005). An *in vitro* light-dependent increase in instability of the CCT of cry1 has been observed. This was based on its susceptibility to proteolysis, which may result from the disruption of the intramolecular interaction of the C-terminal and the N-terminal domains. It is conceivable that blue light modifies the properties of the dimer resulting in a change in the CCT. The CCT structurally has all the properties of a signal transducing domain that could recognize diverse interacting partners. Induced folding on binding has been shown to enhance specific reactions at low affinity, suitable for signalling. However, this has not been demonstrated for the CCT. Indeed, the structural nature of the dimerization and the mechanism of how the dimers affect signalling is yet to be unravelled.

Notable is the fact that cry1 mutants of the N-terminal domain that fail to show phosphorylation and biological activity also do not dimerize (Sang *et al.*, 2005). Based on *in vitro* studies in our group, we observed that autophosphorylation of cry2 results in quantitative increase of the equilibrium forms from an oligomeric to a more of monomeric state that is stimulated by blue light. This could imply that this shift in equilibrium could be a switch to control its activity (Banerjee and Batschauer, unpublished). How this is significant *in vivo* is to be studied. In the case of cry2, monomerization could also be the trigger for its degradation.

2.6.2   *Cryptochrome partners*

The mechanism(s) through which cryptochromes transduce light signals is still intensively studied. Based on the similarity of cryptochromes in their amino acid sequence and cofactor composition with DNA photolyase, it was speculated that cryptochromes could use a similar signalling mechanism to photolyases where catalysis is driven by light-mediated electron transfer (Cashmore, 1999; Malhotra *et al.*, 1995). Based on recent studies on light-induced absorbance changes (see Section 2.4.4), it is very likely that intramolecular electron transfer after photoexcitation of the flavin is the primary event in cryptochrome signalling. In addition, several proteins were identified that physically interact with cryptochromes, and by genetic approaches further components were found that are more downstream in the cryptochrome signalling chain.

2.6.2.1 Interaction with COP1

The interaction of cry1 and cry2 with COP1 is, with respect to its biological significance, the best understood interaction of cryptochromes with downstream partners. COP1 (see Chapter 6 for further details and references on COP proteins) functions as an E3 ubiquitin ligase that targets the basic leucine zipper (bZIP) transcription factor HY5 for degradation via the proteasome or the COP9 signalosome (Hardtke et al., 2000; Osterlund et al., 2000). In light, COP1 is transported out of the nucleus (Osterlund and Deng, 1998; von Arnim and Deng, 1994), which rescues HY5 from degradation. The depletion of COP1 from the nucleus can be seen as a mechanism for its long-term inactivation. HY5 acts as a positive element in light signalling by binding to regulatory sequences known to be present in many light-induced genes. Independent research of Xing-Wang Deng and Anthony Cashmore with their coworkers has shown that cry2 (Wang et al. 2001) and cry1 (Yang et al., 2001) bind to COP1. Among the three conserved motifs (ring finger domain, coiled-coil region, WD40 repeat) of COP1 it seems to be the C-terminal WD40 repeat that is required for binding to the CCTs of both cryptochromes. The design of the constructs used for these yeast-two-hybrid interaction studies suggests that the CCTs can only bind to the WD40 domain of COP1 when present as dimers (CCT2 fused to GUS or CCT1 fused to the LexA DNA binding domain). Yeast-two-hybrid studies using full-length cry1, co-immunoprecipitation studies with plant extracts for cry1 and cry2 and studies on the subcellular distribution of GFP-CCT1 provided support for this notion. Notably, none of these results indicates that the interaction of COP1 with cryptochromes is regulated by light. Therefore, one must assume that the light signal perceived by the sensory domain of cry transduces a signal to COP1 leading to inhibition of its activity and/or translocation out of the nucleus. As mentioned above, it is the WD40 domain of COP1 that is required for interaction with CCT. The WD40 domain is also known to be essential for binding of COP1 to HY5. However, based on mutation studies within this domain, which have different effects on the interaction with either HY5 or CCT1, it was concluded that different binding modes of COP1 for HY5 and CCT1 might exist (Yang et al., 2001). Therefore, it is possible that cry and HY5 bind, at least in darkness, simultaneously to COP1. In light, cry could compete with HY5 for COP1 binding. Interestingly, Yang et al. (2001) found that besides CCT1 also the C-terminal domain of phyB, but not of phyA, interacts with COP1. The domain of COP1 required for phyB interaction has not been defined, but it is unlikely that the phyB–COP1 interaction has the same consequence as the cry–COP1 interaction because to our knowledge overexpression of the C-terminal domain of phyB causes no cop phenotype.

2.6.2.2 Interaction with zeitlupe/ADAGIO1

Zeitlupe (ZTL), also named ADAGIO1, belongs to small family of closely related PAS/LOV domain proteins consisting of three members in Arabidopsis (see Chapters

3 and 8 and references therein for further details). In continuous white light, the *ztl* knock-out mutant has a long period phenotype for multiple clock outputs indicating that ZTL could act in the light-input to the clock or be an integral component of the clock (Jarillo *et al.*, 2001). All members of the Zeitlupe family bind FMN (Imaizumi *et al.*, 2003); thus, they most likely function as photoreceptors (Imaizumi *et al.*, 2003). Interestingly, Jarillo *et al.* (2001) have also shown by yeast-two-hybrid and pull-down studies with *in vitro* transcribed/translated proteins that the C-termini of CRY1 and PHYB interact with ZTL. It is not clear yet how these interactions could modulate the activity of one or both partners. In the case of the ZTL–cry interaction it is, however, still tempting to speculate that light-driven electron transfer could occur between the two flavo proteins, although intramolecular electron transfer within cryptochrome is at present a more fashionable model (see Plate 2.3).

2.6.2.3   Interaction with phytochromes

Most light-regulated processes in plants are controlled by several photoreceptors, and many examples for such photoreceptor coactions have been described (for review see Mohr, 1986, Casal, 2000). In principle, one can imagine that these coactions could occur at several levels, for example, through shared signalling partners (see Section 2.6.3) or direct physical interaction between photoreceptors. There is a large body of evidence that different photoreceptors operate through common signalling components. However, there is also evidence for a direct interaction between cry1 and phyA (Ahmad *et al.*, 1998c) and between cry2 and phyB (Más *et al.*, 2000). With the yeast-two-hybrid system Ahmad *et al.* (1998c) have shown that the C-terminal domain of CRY1 (CCT1) interacts with the C-terminal domain of *Arabidopsis* PHYA spanning the region of amino acids 624–1100, whereas a shorter PHYA fragment (amino acids 689–939) lacking a domain thought to be involved in PHYA dimerization did not interact with the CCT1. Interestingly, all tested CCT1s carrying point mutations leading to amino acid changes did not interfere with binding to the C-terminal domain of PHYA. One of these mutant CCT1 versions (*hy4–19*) was also used in the study of Yang *et al.* (2000) to check for *cop* phenotypes when over-expressed in *Arabidopsis* as a GUS-fusion. Since the overexpressed *hy4–19* CCT1 did not cause a *cop* phenotype and showed only weak interaction with COP1 (Yang *et al.*, 2001), one must assume that the mechanism of interaction of CCT1 with COP1 and PHYA is different.

Steve Kay and coworkers have shown (Más *et al.*, 2000) that cry2 and phyB interact *in planta*. Evidence for this interaction is based on co-immunoprecipitation of cry2 with phyB-specific antibodies using protein extracts isolated from cry2 overexpressing *Arabidopsis* plants. In addition, the same authors have shown that cry2 fused to the red fluorescence protein (RFP) forms nuclear speckles when the cells are irradiated with blue light. PhyB fused to the green fluorescent protein also formed nuclear speckles when treated with red light. Cells co-transformed with cry2-RFP and phyB-GFP that had already formed phyB speckles under red light and afterwards treated with blue light showed co-localization of the RFP and GFP signals in the same nuclear speckles. Fluorescence resonance energy transfer (FRET) microscopy of these nuclear speckles showed emission of RFP fluorescence when

GFP was excited demonstrating a direct molecular interaction of cry2 with phyB in the nuclear speckles. Interestingly, nuclear speckle formation has not been seen for the overexpressed fusion of full-length cry1 with RFP (Más *et al.*, 2000), although the GFP–CCT1 fusion forms such speckles when co-expressed with COP1 (Wang *et al.*, 2001). In summary, results from the interaction and localization studies suggest that the nuclear speckles contain phyB, COP1, cry2 and cry1. The sensitivities of cry2 and cry1 for their recruitment into the nuclear speckles may depend on the COP1 and phyB concentration or ratios within these speckles.

2.6.3 Further downstream components

SUB1 (short under blue light) was identified in a screen for mutants that were short under low fluence rate ($3 \, \mu mol \, m^{-2} \, s^{-1}$) blue light, but normal under a similar fluence rate of red light (Guo *et al.*, 2001). The *sub1* mutant short hypocotyl phenotype was also observed under far-red light, but not in darkness. This suggested a specific role of SUB1 in both cry and phyA signalling. The mutation in *sub1* is caused by a T-DNA insertion in the 3′ UTR of the *SUB1* gene, resulting in decreased *SUB1* transcript and protein levels. Accordingly, SUB1 is a repressor of light signalling. Since the phenotype of *sub1* is more pronounced under low fluence rates than under high fluence rates of blue light, SUB1 seems to operate preliminary under low light conditions. Analysis of double mutants defective in *sub1* and one of the photoreceptors (*cry1, cry2, phyA*) showed that the combinations of *sub1* with *cry* mutants had phenotypes similar to the *sub1* parent, whereas the *sub1/phyA* double mutant showed the *phyA* phenotype. This indicates that *sub1* is epistatic to *cry2* and operates downstream of cry, whereas *phyA* is epistatic to *sub1*. As outlined above, both cry2 and phyA are light labile. cry2 and phyA proteins levels were similar in the *sub1* mutant and the wild type, indicating that SUB1 did not operate by moderating photoreceptor levels. Thus SUB1 may act as a branch point regulating cryptochrome and phyA signal transduction. Since the *sub1* mutant has no phenotype in darkness, SUB1 must operate upstream of COP1 and HY5. Indeed, compared to wild-type plants no increase in the HY5 protein level was found in dark-grown *sub1* mutants, but stronger accumulation of HY5 after transfer from darkness to light (Guo *et al.* 2001). *SUB1* belongs to a small family consisting of three members in *Arabidopsis* that encode proteins of about 550 amino acid residues with a conserved EF-hand such as Ca^{2+}-binding motif. Ca^{2+}-binding could be confirmed through filter binding studies showing significant binding activity, which is, however, much lower than that of calmodulin. The Ca^{2+}-binding activity of a component in cryptochrome and phyA signalling is reminiscent of earlier findings that suggested a role for Ca^{2+} in phytochrome and cryptochrome signalling (Long and Jenkins, 1998, Neuhaus *et al.*, 1993). As outlined above, cry1 and cry2 were found to be located in the nucleus. Localization studies performed with GUS–SUB1 fusion expressed in onion cells indicate that SUB1 is cytosolic and enriched at the nuclear envelope or ER (Guo *et al.*, 2001). Therefore, a signalling mechanism involving direct interaction of cry1 with SUB1 is more difficult to imagine. However, a small cytosolic fraction of cry2 may have been overlooked and, in the case of cry1, there is evidence for a light-driven nuclear export that could allow direct physical contact with cytosolic proteins.

Although SUB1 acts as a negative element in cryptochrome signalling, the Ser/Thr protein phosphatase 7 (AtPP7) was found to be a positive element specific for cryptochrome signalling (Moller *et al.*, 2003). Transgenic lines expressing *PP7* antisense RNA have elongated hypocotyls under constant white and blue light, but show no difference to wild-type seedlings when grown under continuous red or far-red light. The most severe phenotype of *PP7* antisense plants is similar to the *hy4* (*cry1*) knockout mutant indicating that PP7 is indeed an important component for cryptochrome signalling. Besides hypocotyl inhibition, *PP7* repression also affects cotyledon expansion, cotyledon opening and blue light-induced gene expression in a similar way as seen for the null *cry1* mutant, indicating that all these responses are regulated by a common signalling chain. However, some differences between the *hy4* mutant and the *PP7* antisense lines were observed, for example, in expression of the *CHS* and *CAB* genes, indicating that PP7 is not an immediate upstream component of the cryptochrome signalling pathway. Similar to SUB1, PP7 bears (two) putative Ca^{2+}-binding EF hands. Whether the phosphatase activity of PP7 is stimulated by Ca^{2+} or Mn^{2+} is, to our knowledge, still a matter of debate. It has also been shown that PP7 is a nuclear protein (Andreeva and Kutuzov, 2001; Moller *et al.*, 2003), which does not physically interact with cryptochromes (Moller *et al.*, 2003). A direct interaction with cryptochrome of a positively acting component of the cry signalling pathway that has phosphatase activity is also not expected in view of the fact that phosphorylation seems to be essential for cryptochrome function (see Section 2.4.1 and Chapter 5).

Besides PP7, the putative bHLH transcription factor long hypocotyl in far-red (HFR1) is a positively acting component in cryptochrome signalling (Duek and Fankhauser, 2003). As its name suggests, HFR1 was first identified as a component involved in most but not all far-red light responses that are mediated by phyA (see Chapter 4 and references therein for further details). Since the phyA mutant is also affected in the hypocotyl growth inhibition under blue light (Whitelam *et al.* 1993), it is more complicated to separate phyA and cryptochrome responses under these wavelengths. Duek and Fankhauser (2003) addressed the question that asked whether, in addition to its clear role in phyA signalling, HFR1 could also have a role in cry1 signal transduction. To do this, they analyzed *hfr1* single and double mutants in combination with null alleles of *phyA*, *cry1* and *cry2* under different fluence rates of blue light. Under high intensity blue the *cry1/hfr1* double mutant was phenotypically similar to the *cry1* single mutant. This demonstrated that *cry1* is epistatic to *hfr1* under these conditions and suggested that HFR1 acts in the cry1 signalling chain. At lower blue fluence rates, *cry1/hfr1* had a more pronounced phentotype than the *cry1* single mutant, indicating that under these conditions HFR1 functions also in another pathway, most likely that of phyA. Protein levels of cry1 are the same in wild type and the *hfr1* mutant under all tested light conditions demonstrating that HFR1 does not affect levels of the photoreceptor. Interestingly, the transcript levels of *HFR1* appear to be negatively regulated by red light, suggesting another level of control in this pathway. HFR1 is therefore an example of a positive signalling component acting in the cry1 and phyA pathways. The suppression of *HFR1* expression through red light indicates crosstalk with light-stable phytochromes resulting in a reduced signalling of cry1 and phyA.

2.7 Summary

Since the discovery of cryptochromes in 1993, our knowledge about these photoreceptors has increased dramatically. Genetic studies have revealed the extent of cryptochrome function as they regulate nearly all growth and differentiation processes in plants. Recent progress in understanding cryptochrome function at the atomic level was achieved by solving cryptochrome structures made possible by crystallization of the protein. This opens the door to future work that may include modification, or redesign of receptor activity, and provides a means to examine spectroscopic effects on photoreceptor function. In contrast to most other photoreceptors, knowledge of the cryptochrome photocycle is still incomplete and bridging this gap will be a major challenge for future cryptochrome research.

Acknowledgements

We thank Birte Dohle and Stefan Meier for preparing the figures, the Deutsche Forschungsgemeinschaft for the support of our research, and Karen J. Halliday for editing this chapter.

References

Afonso, C.L., Tulman, E.R., Lu, Z., Oma, E., Kutish, G.F. and Rock, D.L. (1999) The genome of *Melanoplus sanguinipes* entomopoxvirus. *J.Virol.*, **73**, 533–552.
Ahmad, M., Grancher, N., Heil, M., Black, R.C., Giovani, B., Galland, P. and Lardemer, D. (2002) Action spectrum for cryptochrome-dependent hypocotyl growth inhibition in *Arabidopsis. Plant Physiol.*, **129**, 774–785.
Ahmad, M., Jarillo, J. and Cashmore, A.R. (1998a) Chimeric proteins between cry1 and cry2 Arabidopsis blue light photoreceptors indicate overlapping functions and varying protein stability. *Plant Cell.*, **10**, 197–207.
Ahmad, M., Jarillo, J.A., Klimczak, L.J., Landry, L.G., Peng, T., Last, R.L. and Cashmore, A.R. (1997) An enzyme similar to animal type II photolyasees mediates photoreactivation in *Arabidopsis. Plant Cell*, **9**, 199–207.
Ahmad, M., Jarillo, J.A., Smirnova, O. and Cashmore, A.R. (1998b) Cryptochrome blue-light photoreceptors of *Arabidopsis* implicated in phototropism. *Nature*, **392**, 720–723.
Ahmad, M., Jarillo, J.A., Smirnova, O. and Cashmore, A.R. (1998c) The CRY1 blue light photoreceptor of *Arabidopsis* interacts with phytochrome A in vitro. *Mol. Cell*, **1**, 939–948.
Ahmad, M., Lin, C. and Cashmore, A.R. (1995). Mutations throughout an *Arabidopsis* blue-light phoptoreceptor impair blue-light responsive anthocyanin accumulation and inhibition of hypocotyl elongation. *Plant J.*, **8**, 653–658.
Ahmad, M. and Cashmore, A.R. (1993) *HY4* gene of *A. thaliana* encodes a protein with characteristics of a blue-light photoreceptor. *Nature*, **366**, 162–166.
Andreeva, A.V. and Kutuzov, M.A. (2001) Nuclear localization of the plant Ser/Thr phosphatase PP7. *Mol. Cell. Biol. Res. Commun.*, **4**, 345–352.

Ang, L.-H., Chattopadhyay, S., Wei, N., Oyama, T., Okada, K., Batschauer, A. and Deng, X.-W. (1998) Molecular interaction between COP1 and HY5 defines a regulatory switch for light control of Arabidopsis development. *Mol. Cell*, **1**, 213–222.

Banerjee, R. and Batschauer, A. (2005) Plant blue light receptors. *Planta*, **220**, 498–502.

Batschauer, A. (1993) A plant gene for photolyase: an enzyme catalyzing the repair of UV-light-induced DNA damage. *Plant J.*, **4**, 705–709.

Batschauer, A. (2005) Plant cryptochromes: Their genes, biochemistry, and physiological roles. In: *Handbook of Photosensory Receptors* (eds J.L. Spudich and W.R. Briggs), Wiley-VCH, Weinheim, Germany, pp. 211–246.

Bognár, L.K., Hall, A., Adam, E., Thain, S.C., Nagy, F. and Millar, A.J. (1999) The circadian clock controls the expression pattern of the circadian input photoreceptor, phytochrome B. *Proc. Natl. Acad. Sci. USA*, **96**, 14652–14657.

Bouly, J.-P., Giovani, B., Djamei, A., Mueller, M., Zeugner, A., Dudkin, E.A., Batschauer, A. and Ahmad, M. (2003) Novel ATP-binding and autophosphorylation activity associated with Arabidopsis and human cryptochrome 1. *Eur. J. Biochem.*, **270**, 2921–2928.

Brautigam, C.A., Smith, B.S., Ma, Z., Palnitkar, M., Tomchick, D.R., Machius, M. and Deisenhofer, J. (2004) Structure of the photolyase-like domain of cryptochrome 1 from *Arabidopsis thaliana*. *Proc. Natl. Acad. Sci USA*, **101**, 12142–1247.

Brudler, R., Hitomi, K., Daiyasu, H., Toh, H., Kucho, K., Ishiura, M., Kanehisa, M., Roberts, V.A., Todo, T., Trainer, A. and Getzoff, E.D. (2003) Identification of a new cryptochrome class: structure, function, and evolution. *Mol. Cell*, **11**, 59–67.

Casal, J.J. (2000) Phytochromes, cryptochromes, phototropin: Photoreceptor interactions in plants. *Photochem. Photobiol.*, **71**, 1–11.

Cashmore, A.R., Jarillo, J.A., Wu, Y.J. and Liu, D. (1999) Cryptochromes: Blue light receptors for plants and animals. *Science*, **284**, 760–765.

Cashmore, A.R. (2005). Plant cryptochrome signaling. In: *Handbook of Photosensory Receptors* (eds J.L. Spudich and W.R. Briggs), Wiley-VCH, Weinheim, Germany, pp. 247–258.

Clough, R.C., Jordan-Beebe, E.T., Lohman, K.N., Marita, J.M., Walker, J.M., Gatz, C. and Vierstra, R.D. (1999) Sequences within both the N- and C-terminal domains of phytochrome A are required for Pfr ubiquitination and degradation. *Plant J.*, **17**, 155–167.

Cutler, S.R., Ehrhardt, D.W., Griffits, J.S. and Somerville, C.R. (2000) Random GFP::cDNA fusions enable visualisation of subcellular structures in cells of *Arabidopsis* at a high frequency. *Proc. Natl. Acad. Sci. USA*, **97**, 3718–3723.

Daiyasu, H., Ishikawa, T., Kuma, K., Iwai, S., Todo, T. and Toh, H. (2004) Identification of cryptochrome DASH from vertebrates. *Genes Cells*, **9**, 479–495.

Devlin, P.F. and Kay, S.A. (2000) Cryptochromes are required for phytochrome signaling to the circadian clock but not for rhythmicity. *Plant Cell*, **12**, 2499–2509.

Duek, P.D. and Fankhauser, C. (2003) HFR1, a putative bHLH transcription factor, mediates both phytochrome A and cryptochrome signalling. *Plant J.*, **34**, 827–836.

El-Assal, S.E.-D., Alonso-Blanco, C., Peeters, A.J.M., Wagemaker, C., Weller, J.L. and Koornneef, M. (2003) The role of cryptochrome 2 in flowering in *Arabidopsis*. *Plant Physiol.*, **133**, 1–13.

El-Assal, S.E.-D., Alonso-Blanco, C., Peeters, A.J.M., Raz, V. and Koornneef, M. (2001) A QTL for flowering time in *Arabidopsis* reveals a novel allele of *CRY2*. *Nat. Genet.*, **29**, 435–440.

Folta, K.M., Pontin, M.A., Karlin-Neumann, G., Bottini, R. and Spalding, E.P. (2003) Genomic and physiological studies of early cryptochrome 1 action demonstrate roles for auxin and gebberellin in the control of hypocotyl growth by blue light. *Plant J.*, **36**, 203–214.

Folta, K.M. and Spalding, E.P. (2001) Opposing roles of phytochrome A and phytochrome B in early cyptochrome-mediated growth inhibition. *Plant J.*, **28**, 333–340.

Galland, P. and Senger, H. (1988) The role of pterins in the photoperception and metabolism of plants. *Photochem. Photobiol*, **48**, 811–820.

Galland, P. and Senger, H. (1991) Flavins as possible blue-light photoreceptors. In: *Photoreceptor Evolution and Function* (ed M.G. Holmes MG). Academic Press, London, pp. 65–124.

Giovani, B., Byrdin, M., Ahmad, M. and Brettel, K. (2003) Light-induced electron transfer in a cryprochrome blue-light photoreceptor. *Nat. Struct. Biol.*, **10**, 489–490.

Gressel, J. (1979) Blue light photoreception. *Photochem. Photobiol.*, **30**, 749–754.

Guo, H., Duong, H., Ma, N. and Lin, C. (1999) The *Arabidopsis* blue light receptor cryptochrome 2 is a nuclear protein regulated by a blue light-dependent post-transcriptional mechanism. *Plant J.*, **19**, 279–287.

Guo, H., Mockler, T., Duong, H. and Lin, C. (2001) SUB1, an *Arabidopsis* Ca^{2+}-binding protein involved in cryptochrome and phytochrome coaction. *Science*, **291**, 487–490.

Hardtke, C.S., Gohda, K., Osterlund, M.T., Oyama, T., Okada, K. and Deng, X.-W. (2000) HY5 stability and activity in *Arabidopsis* is regulated by phosphorylation in its COP1 binding domain. *EMBO J.*, **19**, 4997–5006.

Harmer, S.L., Hogenesch, J.B., Straume, M., Chang, H.S., Han, B., Zhu, T., Wang, X., Kreps, J.A. and Kay, S.A. (2000) Orchestrated transcription of key pathways in *Arabidopsis* by the circadian clock. *Science*, **290**, 2110–2113.

Hitomi, K., Okamoto, K., Daiyasu, H., Miyashita, H., Iwai, S., Toh, H., Ishiura, M. and Todo, T. (2000) Bacterial cryptochrome and photolyase: characterization of two photolyase-like genes of *Synechocystis* sp. PCC6803. *Nucleic Acids Res.*, **28**, 2353–2362.

Hoffman, P.D., Batschauer, A. and Hays, J.B. (1996) *AT-PHH1*, a novel gene from *Arabidopsis thaliana* related to microbial photolyases and plant blue light photoreceptors. *Mol. Gen. Genet.*, **253**, 259–265.

Husain, I. and Sancar A. (1987) Binding of *E. coli* DNA photolyase to a defined substrate containing a single T◇T dimer. *Nucleic Acids Res.*, **15**, 1109–1120.

Imaizumi, T., Kadota, A., Hasebe, M. and Wada, M. (2002) Cryptochrome light signals control development to supress auxin sensitivity in the moss *Physcomitrella patens*. *Plant Cell*, **12**, 81–95.

Imaizumi, T., Kanegae, T. and Wada, M. (2000) Cryptochrome nucleocytoplasmic distribution and gene expression are regulated by light quality in the fern *Adiantum capillus-veneris*. *Plant Cell*, **12**, 81–96.

Imaizumi, T., Kiyosue, T., Kanegae, T. and Wada, M. (1999) Cloning of the cDNA Encoding the Blue-light Photoreceptor (Cryptochrome) from the Moss *Physcomitrella patens* (Accession No. AB027528). *Plant Phys.*, **120**, 1205.

Imaizumi, T., Tran, H.G., Swartz, T.E., Briggs, W.R. and Kay, S.A. (2003) FKF1 is essential for photoperiodic-specific light signalling in *Arabidopsis*. *Nature*, **426**, 302–306.

Jackson, J.A. and Jenkins, G.I. (1995) Extension-growth responses and expression of flavonoid biosynthesis genes in the *Arabidopsis hy4* mutant. *Planta*, **197**, 233–239.

Jarillo, J.A., Capel, J., Tang, R.H., Yang, H.Q., Alonso, J.M., Ecker, J.R. and Cashmore, A.R. (2001) An Arabidopsis circadian clock component interacts with both CRY1 and phyB. *Nature*, **410**, 487–490.

Kanai, S., Kikuno, R., Toh, H., Ryo, H. and Todo, T. (1997) Molecular evolution of the photolyase-blue-light photoreceptor family. *J. Mol. Evol.*, **45**, 535–548.

Kanegae, T. and Wada, M. (1998) Isolation and characterization of homologues of plant blue-light photoreceptor (cryptochrome) genes from the fern *Adiantum capillus-veneris*. *Mol. Gen. Genet.*, **259**, 345–353.

Kato, T.Jr., Todo, T., Ayaki, H., Ishizaki, K., Morita, T., Mitra, S. and Ikenaga, M. (1994) Cloning of a marsupial DNA photolyase gene and the lack of related nucleotide sequences in placental mammals. *Nucleic Acids Res.*, **22**, 4119–4124.

Kim, S.-T., Heelis, P. F., Okamura, T., Hirata, Y., Mataga, N. and Sancar, A. (1991) Determination of rates and yields of interchromophore (folate → flavin) energy transfer and intermolecular (flavin → DNA) electron transfer in *Escherichia coli* photolyase by time-resolved fluorescence and absorption spectroscopy. *Biochemistry*, **30**, 11262–11270.

Kleine, T. Lockhart, P. and Batschauer, A. (2003) An *Arabidopsis* protein closely related to *Synechocystis* cryptochrome is targeted to organelles. *Plant J.*, **35**, 93–103.

Kleiner, O., Kircher, S., Harter, K. and Batschauer, A. (1999) Nuclear localization of the Arabidopsis blue light receptor cryptochrome 2. *Plant J.*, **19**, 289–296.

Komori, H., Masui, R., Kuramitsu, S., Yokoyama, S., Shibata, T., Inoue, Y. and Miki, K. (2001) Crystal structure of thermostable DNA photolyase: pyrimidine-dimer recocognition mechanism. *Proc. Natl. Acad. Sci. USA*, **98**, 13560–13565.

Koornneef, M., Rolf, E. and Spruit, C.J.P. (1980) Genetic control of light-inhibited hypocotyl elongation in *Arabidopsis thaliana* (L). HEYNH. *Z. Pflanzenphysiol.*, **100**, 147–160.

Lin, C., Ahmad, M. and Cashmore, A.R. (1996a) *Arabidopsis* cryptochrome 1 is a soluble protein mediating blue light-dependent regulation of plant growth and development. *Plant J.*, **10**, 893–902.

Lin, C., Ahmad, M., Chan, J. and Cashmore, A.R. (1996b) CRY2: a second member of the Arabidopsis cryptochrome gene family. *Plant Physiol.*, **110**, 1047.

Lin, C., Ahmad, M., Gordon, D. and Cashmore, A.R. (1995a) Expression of an Arabidopsis cryptochrome gene in transgenic tobacco results in hypersensitivity to blue, UV-A, and green light. *Proc. Natl. Acad. Sci. USA*, **92**, 8423–8427.

Lin, C., Robertson, D.E., Ahmad, M., Raibekas, A.A., Schuman Jornes, M., Dutton, P.L. and Cashmore, A. (1995b) Association of flavin adenine dinucleotide with the *Arabidopsis* blue light receptor CRY1. *Science*, **269**, 968–970.

Lin, C. and Shalitin, D. (2003) Cryptochrome Structure and Signal Transduction. *Annu. Rev. Plant Biol.*, **54**, 469–496.

Lin, C., Yang, H., Guo, H., Mockler, T., Chen, J. and Cashmore, A.R. (1998) Enhancement of blue-light sensitivity of Arabidopsis seedlings by a blue light receptor cryptochrome 2. *Proc. Natl. Acad. Sci. USA*, **95**, 2686–2690.

Lin, C. (2002) Blue light receptors and signal transduction. *Plant Cell*, **14** (Suppl.), S207–S225.

Lipson, E.D. and Horwitz, B.A. (1991) Photosensory reception and transduction In: *Sensory Receptors and Signal Transduction* (eds J.L. Spudich and B.H. Satir), Wiley, New York, p. 64.

Long, J.C. and Jenkins, G.I. (1998) Involvement of plasma membrane redox activity and calcium homeostasis in the UV-B and UV-A/blue light induction of gene expression in *Arabidopsis. Plant Cell*, **10**, 2077–2086.

Ma, L., Li, J., Qu, L., Hager, J., Chen, Z., Zhao, H. and Deng, X.-W. (2001) Light control of *Arabidopsis* development entails coordinated regulation of genome expression and cellular pathways. *Plant Cell*, **13**, 2589–2607.

Malhotra, K., Kim, S.-T., Batschauer, A., Dawut, L. and Sancar, A. (1995) Putative blue-light photoreceptors from *Arabidopsis thaliana* and *Sinapis alba* with high degree of sequence homology to DNA photolyase contain the two photolyase cofactors but lack DNA repair activity. *Biochemistry*, **34**, 6892–6899.

Mao, J., Zhang, Y.-C., Sang, Y., Li, Q.-H. and Yang, H.-Q. (2005) A role for *Arabidopsis* cryptochromes and COP1 in the regulation of stomatal opening. *Proc. Natl. Acad. Sci. USA*, **102**, 12270–12275.

Más, P., Devlin, P.F., Panda, S. and Kay, S.A. (2000) Functional interaction of phytochrome B and cryptochrome 2. *Nature*, **408**, 207–211.

Matsumoto, N., Hirano, T., Iwasaki, T. and Yamamoto, N. (2003) Functional analysis and intercellular localization of rice cryptochromes. *Plant Physiol.*, **133**, 1494–1503.

Mees, A., Klar, T., Gnau, P., Hennecke, U., Eker, A.P., Carell, T. and Essen, L.O. (2004) Crystal structure of a photolyase bound to a CPD-like DNA lesion after in situ repair. *Science*, **30**, 1789–1793.

Mockler, T.C., Yang, H., Yu, X., Parikh, D., Cheng, Y., Dolan, S. and Lin C. (2003) Regulation of photoperiodic flowering by *Arabidopsis* photoreceptors. *Proc. Natl. Acad. Sci. USA*, **100**, 2140–2145.

Moller, S.G., Kim, Y.S., Kunkel, T. and Chua, N.-H. (2003) PP7 is a positive regulator of blue light signalling in *Arabidopsis. Plant Cell*, **15**, 1111–1119.

Mohr, H. (1986). Coaction between pigment systems. In: *Photomorphogenesis in Plants* (eds R.E. Kendrick and G.H.M. Kronenberg), Martinus-Nijhoff, The Netherlands, pp. 547–564.

Neuhaus, G., Bowler, C., Kern, R. and Chua, N.-H. (1993) Calcium/calmodulin-dependent and -independent phytochrome signal transduction pathways. *Cell*, **73**, 937–952.

Ng, W.O. and Pakrasi, H.B. (2001) DNA photolyase homologs are the major UV resistance factors in the cyanobacterium *Synechocystis* sp. PCC 6803. *Mol. Gen. Genet.*, **264**, 924–930.

Ninu, L., Ahmad, M., Miarelli, C., Cashmore, A.R. and Giuliano, G. (1999) Cryptochrome 1 controls tomato development in response to blue light. *Plant J.*, **18**, 551–556.

O'Connor, K.A., McBridge, M.J., West, M., Yu, H., Trinh, L., Yuan, K., Lee, T. and Zusman, D.R. (1996) Photolyase of *Myxococcus xanthus*, a Gram-negative eubacterium, is more similar to photolyases found in Archea and 'higher' eukaryotes than to photolyase of other eubacteria. *J. Biol. Chem.*, **271**, 6252–6359.

Osterlund, M.T. and Deng, X.-W. (1998) Multiple photoreceptors mediate the light-induced reduction of GUS-COP1 from *Arabidopsis* hypocotyl nuclei. *Plant J.*, **16**, 201–208.

Osterlund, M.T., Hardtke, C.S., Wei, N. and Deng, X.-W. (2000) Targeted destabilization of HY5 during light-regulated development of *Arabidopsis*. *Nature*, **405**, 462–466.

Park, H.W., Kim, S.-T., Sancar, A. and Deisenhofer, J. (1995) Crystal structure of DNA photolyase from *Escherichia coli*. *Science*, **268**, 1866–1872.

Parks, B.M., Cho, M.H. and Spalding, E.P. (1998) Two genetically separable phases of growth inhibition induced by blue light in *Arabidopsis* seedlings. *Plant Physiol.*, **118**, 609–615.

Parks, B.M., Folta, K.M. and Spalding, E.P. (2001) Photocontrol of stem growth. *Curr. Opin. Plant Biol.*, **4**, 436–440.

Partch, C.L., Clarkson, M.W., Özgür, S., Lee, A.L. and Sancar, A. (2005) Role of structural plasticity in signal transduction by the cryptochrome blue-light photoreceptor. *Biochemistry*, **44**, 3795–3805.

Partch, C. and Sancar, A. (2005) Photochemistry and photobiology of cryptochrome blue-light photopigments: The search for a photocycle. *Photochem. Photobiol.*, **81**, 1291–1304.

Perotta, G., Yahoubyan, G., Nebuloso, E., Renzi, L. and Giuliano, G. (2001) Tomato and barley contain duplicated copies of cryptochrome 1. *Plant Cell Environ.*, **24**, 991–997.

Perrotta, G., Ninu, L., Flamma, F., Weller, J.L., Kendrick, R.E., Nebuloso, E. and Giuliano, G. (2000) Tomato contains homologues of Arabidopsis cryptochromes 1 and 2. *Plant Mol. Biol.*, **42**, 765–773.

Petersen, J.L., Lang, D.W. and Small G.D. (1999) Cloning and characterization of a class II DNA photolyase from *Chlamydomonas*. *Plant Mol. Biol.*, **40**, 1063–1071.

Pokorny, R., Klar, T., Essen, L.-O. and Batschauer, A. (2005) Crystallization and preliminary X-ray analysis of cryptochrome 3 from *Arabidopsis thaliana*. *Acta Crys.*, **F61**, 935–938.

Sancar, A. (2003) Structure an function of DNA photolyase and cryptochrome blue-light photoreceptors. *Chem. Rev.*, **103**, 2203–2237.

Sancar, G.B. (2000) Enzymatic photoreactivation: 50 years and counting. *Mutat. Res.*, **451**, 25–37.

Sang, Y., Li, Q. H., Rubio, V., Zhang, Y. C., Mao, J., Deng, X. W. and Yang, H. Q. (2005) N-terminal domain-mediated homodimerization is required for photoreceptor activity of cryptochrome 1. *Plant Cell*, **17**, 1569–1584.

Saxena, C., Wang, H., Kavakli, I.H., Sancar, A. and Zhong, D. (2005) Ultrafast dynamics of resonance energy transfer in cryptochrome. *J. Am. Chem. Soc.*, **127**, 7984–7985.

Senger, H. (1984). Cryptochrome, some terminological thoughts. In: *Blue Light Effets in Biological Systems* (ed H. Senger), Springer Verlag, Berlin, p. 72.

Shalitin, D., Yang, H., Mockler, T. C., Maymon, M., Guo, H., Whitelam, G.C. and Lin, C. (2002) Regulation of Arabidopsis cryptochrome 2 by blue-light-dependent phosphorylation. *Nature*, **417**, 763–67.

Shalitin, D., Yu, X., Maymon, M., Mockler, T. and Lin, C. (2003) Blue light-dependent *in vivo* and *in vitro* phosphorylation of Arabidopsis cryptochrome 1. *Plant Cell*, **15**, 2421–2429.

Small, G.D., Min, B.Y. and Lefebvre, P.A. (1995) Characterization of a *Chlamydomonas reinhardtii* gene encoding a protein of the DNA photolyase blue light receptor family. *Plant Mol. Biol.*, **28**, 443–454.

Srinivasan, V., Schnitzlein, W.M. and Tripathy, D.N. (2001) Fowlpox virus encodes a novel DNA repair enzyme, CPD-photolyase, that restores infectivity of UV light-damaged virus. *J. Virol.*, **75**, 1681–1688.

Tamada, T., Kitadokora, K., Higuchi, Y., Inaka, K., Yasui, A., de Ruiter, P.E., Eker, A.P.M. and Miki, K. (1997) Crystal structure of DNA photolyase from *Anacystis nidulans*. *Nat. Struct. Biol.*, **4**, 887–891.

Taylor, R., Tobin, A.K. and Bray, C.M. (1996) Nucleotide sequence of an Arabidopsis thaliana cDNA with high homology to the class II CPD photolyase present in higher eukaryotes. *Plant Physiol.*, **112**, 862.

The Arabidopsis Genome Initiative (2000) Analysis of the genome sequence of the flowering plant *Arabidopsis thaliana. Nature*, **408**, 796–815.

Todo, T. (1999) Functional diversity of the DNA photolyase/blue light receptor family. *Mutat. Res.*, **434**, 89–97.

Todo, T., Ryo, H., Takemori, H., Toh, H., Nomura, T. and Kondo, S. (1994) High-level expression of the photorepair gene in Drosophila and its evolutionary implications. *Mutat. Res. DNA Repair*, **315**, 213–228.

Toth, R., Kevei, E., Hall, A., Millar, A.J., Nagy, F. and Kozma-Bognar, L. (2001) Circadian clock-regulated expression of phytochrome and cryptochrome genes in *Arabidopsis. Plant Physiol.*, **127**, 1607–1616.

Van Gelder, R.N. and Sancar, A. (2005) Animal cryptochromes. In: *Handbook of Photosensory Receptors* (eds J.L. Spudich and W.R. Briggs), Wiley-VCH, Weinheim, Germany, pp. 259–276.

von Arnim, A.G. and Deng, X.-W. (1994) Light inactivation of *Arabidopsis* photomorphogenic repressor COP1 involves a cell-specific regulation of its nucleo-cytoplasmatic partitioning. *Cell*, **79**, 1035–1045.

Wada, M. (2003) Blue light receptors in fern and moss. In: *Photoreceptors and Light Signalling* (ed A. Batschauer), The Royal Society of Chemistry, Cambridge, UK, pp. 328–342.

Wang, H., Ma, L.-G., Li, J.-M., Zhao, H.-Y. and Deng, X.-W. (2001) Direct interaction of Arabidopsis cryptochromes with COP1 in light control development. *Science*, **294**, 154–158.

Weller, J.L., Perrotta, G., Schreuder, M.E., van Tuinen, A., Koornneef, M., Giuliano, G. and Kendrick, R.E. (2001) Genetic dissection of blue-light sensing in tomato using mutants deficient in cryptochrome 1 and phytochromes A, B1 and B2. *Plant J.*, **25**, 427–440.

Whippo, C.W. and Hangarter, R.P. (2003) Second positive phototropism results from coordinated co-action of the phototropins and cryptochromes. *Plant Physiol.*, **132**, 1499–1507.

Whitelam, G.C, Johnson, E., Peng, J., Carol, P., Anderson, M.L., Cowl, J.S. and Harberd, N.P. (1993) Phytochrome A null mutants of *Arabidopsis* display a wild-type phenotype in white light. *Plant Cell*, **5**, 757–768.

Worthington, E.N., Kavakli, I.H., Berrocal-Tito, G., Bondo, B.E. and Sancar, A. (2003) Purification and characterization of three members of the photolyase/cryptochrome familiy glue-light photoreceptors from *Vibrio cholerae. Biol. Chem.,* **278**, 39143–39154.

Yang, H.Q., Tang, R.H. and Cashmore, A.R. (2001) The signaling mechanism of *Arabidopsis* CRY1 involves direct interaction with COP1. *Plant Cell*, **13**, 2573–2587.

Yang, H.-Q., Wu, Y.-J., Tang, R.-H., Liu, D., Liu, Y. and Cashmore, A.R. (2000) The C termini of *Arabidopsis* cyptochromes mediate a constitutive light response. *Cell*, **103**, 815–827.

Yasuhira, S. and Yasui, A. (1992) Visible light-inducible photolyase gene from goldfish *Carassius auratus. J. Biol. Chem.*, **267**, 25644–25647.

Yasui, A., Eker, A., Yasuhira, S., Yajima, H., Kobayashi, T., Takao, M. and Oikawa, A. (1994) A new class of DNA photolyases present in various organisms including aplacental mammals. *EMBO J.*, **13**, 6143–6151.

Zeugner, A., Byrdin, M., Bouly, J.-P., Bakrim, N., Giovani, B., Brettel, K. and Ahmad, M. (2005) Light-induced electron transfer in *Arabidopsis* cryptochrome-1 correlates with *in-vivo* function. *J. Biol. Chem.*, **280**. 19437–19440.

3 Phototropins and other LOV-containing proteins

John M. Christie

3.1 Introduction

Given their apparent sedentary lifestyle, plants have evolved an array of sophisticated processes to detect and respond to changes in their surrounding environment. Some of these processes involve movement. For instance, environmental stimuli such as light can trigger a range of movement responses that serve to optimize the photosynthetic efficiency of plants. These include phototropism, light-induced stomatal opening and chloroplast relocation in response to changes in light intensity (Briggs and Christie, 2002; Kagawa, 2003; Celaya and Liscum, 2005).

The above-mentioned responses are activated by blue light and are under the control of specific photoreceptors known as the phototropins (Briggs *et al.*, 2001a; Briggs and Christie, 2002). Although the effects of blue light on plant physiology have been studied for well over a century, much of our knowledge regarding phototropin receptor function has emerged over the past decade using the plant genetic model, *Arabidopsis thaliana*. In addition, major advances in our understanding of how these receptors are activated by blue light has come from detailed biochemical and photochemical analyses of the proteins themselves. Such studies have led to the discovery of the LOV domain, a flavin-binding motif within the phototropin molecule, that functions as a blue light sensor for the receptor. Light-sensitive LOV domains have also been identified in proteins besides the phototropins. Such novel LOV-containing proteins are present in fungi, bacteria and archaeabacteria, as well as plants, demonstrating that this regulatory light switch is used widely throughout nature.

This chapter will highlight some of the recent advances relating to the functional roles of the phototropins and the mechanisms associated with phototropin receptor activation. A summary of the current understanding of the signaling events coupling receptor activation to specific phototropin-mediated responses will also be discussed as will the function and biochemical properties of newly identified LOV-containing proteins. For further information, readers are directed to a number of recently published book chapters on the phototropins (Briggs *et al.*, 2005; Christie and Briggs, 2005; Crosson, 2005; Swartz and Bogomolni, 2005; Suetsugu and Wada, 2005) and other LOV-containing proteins (Crosson, 2005; Dunlap, 2005; Dunlap and Loros, 2005; Schultz, 2005). These chapters together with their listed citations will provide a valuable resource for the interested reader.

3.2 Phototropins and their biological functions

3.2.1 Physiological roles in higher plants

Phototropins are ubiquitous in higher plants and have been identified in several plant species including rice, maize, oat, pea and *Arabidopsis* (Briggs *et al.*, 2001b). As mentioned already, genetic analysis using *Arabidopsis* has been instrumental in identifying the molecular nature of the phototropins, and establishing their roles as blue light receptors. *Arabidopsis* contains two phototropins designated phot1 and phot2 (Christie and Briggs, 2001; Briggs and Christie, 2002). Genetic analysis of phot-deficient mutants has revealed that phot1 and phot2 exhibit partially overlapping roles in regulating phototropism, after which they are named (Christie *et al.*, 1999; Briggs *et al.*, 2001a). Both phot1 and phot2 act to regulate hypocotyl phototropism in *Arabidopsis* in response to high intensities of unilateral blue light (Sakai *et al.*, 2001). In contrast, hypocotyl phototropism under low light conditions is solely mediated by phot1 (Liscum and Briggs, 1995; Sakai *et al.*, 2000; Sakai *et al.*, 2001). Thus, phot1 represents the primary phototropic receptor in *Arabidopsis* by acting over a broad range of blue light intensities. The functional activity of phot1 and phot2 most likely results from differential gene expression. In dark-grown seedlings *PHOT2* transcript levels are induced by exposure to light (Jarillo *et al.*, 2001b; Kagawa *et al.*, 2001) through the activation of the red/far-red light receptor phytochrome A (Tepperman *et al.*, 2001). Long-term exposure of dark-grown seedlings to light results in a decrease in *PHOT1* transcript levels (Kanagae *et al.*, 2000; Sakamoto and Briggs, 2002), which is also dependent upon phytochrome photoactivation (Elliot *et al.*, 2004).

Phytochrome action has been known for some time to enhance phototropic curvature in *Arabidopsis* (Janoudi *et al.*, 1997; Stowe-Evans *et al.*, 2001). Recent genetic analysis has revealed that this enhancement is achieved through a phyA-mediated suppression of the gravitropic response pathway (Lariguet and Fankhauser, 2004). This mechanism may be unique to *Arabidopsis* as it has yet to be demonstrated for other plant species (Iino, 2006). Cryptochromes, members of a second blue light photoreceptor family, also serve to modulate the degree of phototropic curvature in *Arabidopsis* (Ahmad *et al.*, 1998; Lascève *et al.*, 1999; Whippo and Hangarter, 2003). Hence, optimal hypocotyl growth reorientation towards light requires the co-action of three different photoreceptor families. It should be noted, however, that whilst phytochrome and cryptochrome photoreceptors influence the magnitude of the phototropic hypocotyl response in *Arabidopsis*, only the phototropins act as directional light sensors.

Further genetic analysis of phot-deficient mutants has revealed that phot1 and phot2 act to control other processes, in addition to phototropism, that serve to fine-tune the photosynthetic status of the plant. The opening of stomata (pores in the epidermis) in response to blue light allows plants to regulate CO_2 uptake for photosynthesis and water loss through transpiration. This response is controlled redundantly by phot1 and phot2, where, in contrast to phototropism, both receptors contribute equally, acting across the same light intensity range (Kinoshita *et al.*, 2001). Recently, a role for cryptochromes in regulating blue light-induced stomatal opening has been reported (Mao *et al.*, 2005). Combined mutant analysis indicates that cryptochromes function additively with the phototropins to mediate this blue light response. Indeed, light-induced stomatal opening in *Arabidopsis* appears to be complex as this process is also regulated by two other photodetection systems, one that is blue–green reversible (Talbott *et al.*, 2003) and another that is responsive to UV-B (Eisinger *et al.*, 2003). The photoreceptors responsible for mediating the latter two effects have yet to be identified.

In higher plants, chloroplasts display two types of movement within the cell depending on the external light conditions (Wada *et al.*, 2003): an accumulation movement to low light intensities, which maximizes light capture for photosynthesis and an avoidance movement that prevents photodamage to the photosynthetic apparatus in excess light (Kasahara *et al.*, 2002a). Phot1 and phot2 overlap in function to control the chloroplast accumulation response (Sakai *et al.*, 2001), whereas the avoidance response is controlled exclusively by phot2 (Jarillo *et al.*, 2001b; Kagawa *et al.*, 2001). Incidentally, phot2 was originally identified as a photoreceptor for light-induced chloroplast movement from a genetic screen for mutants impaired in chloroplast avoidance movement (Kagawa *et al.*, 2001). Phot1 appears to be more sensitive than phot2 in activating chloroplast accumulation movement because phot2 is reported to require a higher light threshold to mediate this response (Kagawa and Wada, 2000; Sakai *et al.*, 2001).

Phototropins are now associated with controlling other extension-growth responses besides phototropism. These include cotyledon expansion (Ohgishi *et al.*, 2004) and leaf expansion (Sakamoto and Briggs, 2002). In addition, the rapid inhibition of hypocotyl elongation upon transfer of dark-grown seedlings to blue light appears to be controlled exclusively by phot1 (Folta and Spalding, 2001). More recently, Takemiya *et al.* (2005) have shown that the phototropins are responsible for promoting growth of *Arabidopsis* under weak light conditions. Plants grown under red light supplemented with very low levels of blue display a threefold increase in fresh weight compared to those grown under red light alone. Blue light-induced growth enhancement is absent in mutants lacking both phot1 and phot2. Moreover, analysis of *phot1* and *phot2* single mutants demonstrates that phot1, as found for chloroplast accumulation movement, is more sensitive than phot2 in promoting growth in response to blue light. The growth enhancement mediated by the phototropins most likely results from an increase in photosynthetic performance due to changes in chloroplast movement, stomatal opening and leaf expansion. It is unlikely that this response involves changes in growth-related gene expression since microarray analysis indicates that the phototropins play a minor role in blue light-induced transcriptional regulation (Ohgishi *et al.*, 2004). Despite these findings, phot1 activity has been found to be essential for the destabilization of specific nuclear and chloroplast transcripts in response to high intensity blue light (Folta and Kaufman, 2003).

Several studies employing different experimental approaches have shown that phototropin activation leads to an increase in cytosolic Ca^{2+} concentrations (Baum *et al.*, 1999; Babourina *et al.*, 2002; Harada *et al.*, 2003; Stoelzle *et al.*, 2003). Folta *et al.* (2003) using the calcium-specific chelator BAPTA were able to demonstrate that the rapid, blue light-induced increase in cytosolic Ca^{2+} observed in dark-grown seedlings is associated with the phot1-mediated inhibition of hypocotyl growth. Evidently, phot2 plays no role in the rapid inhibition of hypocotyl growth as *phot1* mutants completely lack this response (Folta and Spalding, 2001). Yet, Harada *et al.* (2003) have reported that phot2 and phot1 mediate a rapid blue light-dependent increase in cytosolic Ca^{2+} in *Arabidopsis* leaves from different subcellular compartments. Therefore, Ca^{2+} may act as signal messenger in processes other than hypocotyl growth inhibition. Intriguingly, as found for chloroplast accumulation movement and the promotion of growth, phot2 is less sensitive than phot1 in mediating blue light-induced calcium fluxes (Harada *et al.*, 2003), suggesting that phot1 and phot2 may exhibit different photochemical properties. This possibility is discussed in more detail later in the chapter in relation to photochemical and biochemical characterization of the photoreceptor proteins.

3.2.2   Physiological roles in lower plants

Blue light responses have been studied extensively in ferns, mosses and green algae owing to their simplified cell architecture (Suetsugu and Wada, 2005). As in higher plants, the fern *Adiantum capillus-veneris* has two phototropins (Nozue *et al.*, 2000; Kagawa *et al.*, 2004) that likely mediate phototropism and light-induced chloroplast movements in this organism. Genetic analysis indicates that phot2 alone, like its higher plant counterpart, is responsible for mediating chloroplast avoidance movement in *Adiantum* (Kagawa *et al.*, 2004). Besides the phototropins, *Adiantum* contains a novel photoreceptor phy3, which has recently been assigned the name neochrome (Suetsugu *et al.*, 2005). *Adiantum* neochrome (neo) is a chimeric protein comprising a phytochrome photosensory domain fused to the N-terminus of an entire phototropin receptor (Nozue *et al.*, 1998; Nozue *et al.*, 2000; Suetsugu *et al.*, 2005). Genetic studies have shown that neo is required for phototropism and chloroplast relocation in *Adiantum* (Kawai *et al.*, 2003), both of which are regulated by red and blue light in this organism.

Four phototropins have been identified in the moss *Physcomitrella patens* (Kasahara *et al.*, 2004). Chloroplast movement in *Physcomitrella*, as in ferns, is induced by red light as well as blue light (Kadota *et al.*, 2000). Gene disruption using homologous recombination has been used to probe the functional roles of *Physcomitrella* phototropins. Interestingly, a loss of phototropin activity has been shown to affect both blue and red light-induced chloroplast movements in this organism (Kasahara *et al.*, 2004), implying that phototropins may act downstream of phytochrome in mediating red light-induced chloroplast relocation in *Physcomitrella*. Although no neo-type photoreceptor has been identified in *Physcomitrella*, two *NEO* genes have been identified in the filamentous green alga *Mougeotia scalaris*, in addition to two *PHOT* genes (Suetsugu *et al.*, 2005). Both *Mougeotia NEO* genes can rescue red light-induced chloroplast movement in a neo-deficient *Adiantum* mutant, indicating that these algal proteins also function as photoreceptors for chloroplast photorelocation movements. Comparison of the algal and fern *NEO* genes suggests that they have arisen independently, providing an intriguing example of convergent evolution.

Only one phototropin is known in the biflagellate unicellular green alga *Chlamydomonas reinhardtii* (Huang *et al.*, 2002; Kasahara *et al.*, 2002b) where it appears to have a unique function. RNA interference studies have shown that reduced levels of phototropin impair various stages in the cycle of sexual reproduction in *Chlamydomonas* (Huang and Beck, 2003), which is dependent on blue light. Even though the physiological function of *Chlamydomonas* phot is quite diverse to that of higher plant phototropins, the gene encoding *Chlamydomonas* phototropin has been shown to restore phot1- and phot2-mediated responses when introduced into the *phot1phot2* double mutant of *Arabidopsis* (Onodera *et al.*, 2005), implying that the mechanism of action of higher and lower plant phototropins is highly conserved.

3.3   Phototropin structure, localization and activity

3.3.1   *Phototropin structure and localization*

Phototropins are flavoprotein photoreceptors whose structure can be divided into two segments: a photosensory domain at the N-terminus and a serine/threonine kinase domain at the C-terminus (Figure 3.1). The phototropins belong to the AGC family of kinases (named after cAMP-dependent protein kinase, cGMP-dependent protein kinase G and phospholipid-dependent protein kinase C) and are members of the AGC-VIIIb subfamily (Bögre *et al.*, 2003). Members of this subfamily contain a DFD motif in subdomain VII instead of the DFG motif typically found in AGC kinase family members (Watson, 2000; Bögre *et al.*, 2003). The aspartate residue of the DFG motif is required for chelating Mg^{2+}, an ion necessary for phosphate transfer and is essential for phot1 activity (Christie *et al.*, 2002) and phot1 function (Celaya and Liscum, 2005) in *Arabidopsis*.

The N-terminal photosensory domain of the phototropins contains a repeated domain of approximately 110 amino acids called LOV1 and LOV2. LOV domains are members of the large and diverse superfamily of PAS (Per, Arnt, Sim) domains associated with cofactor binding and mediating protein interactions (Taylor and Zhulin, 1999). The LOV domains, however, are more closely related to a subset of proteins within the PAS domain superfamily that are regulated by external signals such as light, oxygen or voltage, hence the acronym LOV (Huala *et al.*, 1997). Phototropin LOV domains bind the cofactor flavin mononulceotide (FMN) that allows the photoreceptor molecule to detect light (Christie *et al.*, 1999; Salomon *et al.*, 2000). Indeed, the LOV domain has become a signature motif used to identify the presence of photosensory proteins in plants and other organisms (Crosson, 2005). For instance, LOV domains have been found in various proteins from plants, fungi and bacteria and, as will be discussed later, represent novel blue light receptors.

Both phot1 and phot2 are hydrophilic proteins but have been shown to localize to, and co-purify with, the plasma membrane in *Arabidopsis* (Sakamoto and Briggs, 2002; Kong *et al.*, 2006) and other plant species (Briggs *et al.*, 2001b). The nature of their association with the plasma membrane remains unknown, but may involve some post-translational modification or binding of a protein membrane anchor. Blue

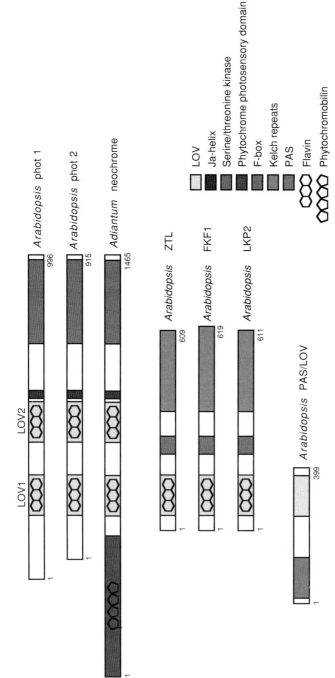

Figure 3.1 Protein structures of LOV-containing proteins found in higher plants. Three classes of LOV-containing proteins found in *Arabidopsis* are shown: the phototropin blue light receptors (phot1 and phot2), the Zeitlupe/Adagio family (ZTL, FKF1, LKP2) and the PAS/LOV protein. The protein structure of neochrome, a novel chimeric photoreceptor from the fern *Adiantum* is also shown which consists of a phytochrome photosensory domain, with covalently bound phytochromobilin as a chromophore, fused to the N-terminus of an entire phototropin receptor. Each of these proteins contains one or two LOV domains that function as binding sites for the blue light absorbing flavin chromophore FMN. As yet, it is unknown whether the LOV domain of the *Arabidopsis* PAS/LOV protein functions as a FMN-binding site. Additional domain structures within these proteins are indicated.

light irradiation has been shown to cause a rapid internalization (within minutes) of phot1 from the plasma membrane (Sakamoto and Briggs, 2002; Knieb *et al.*, 2004). Similarly, Kong *et al.* (2006) have found that a fraction of phot2 re-localizes to the Golgi apparatus upon blue light irradiation. Although the functional consequence of this partial redistribution is currently not known, the kinase domain of phot2 appears to be essential for Golgi localization. It will now be important to establish whether phot1 is also relocalized to the Golgi upon blue light excitation and whether this internalization phenomenon plays a role in photoreceptor signaling or desensitization.

3.3.2 Phototropin autophosphorylation

Insights into the biochemical properties of the phototropins were obtained prior to the isolation of the first phototropin gene back in 1997 (Huala *et al.*, 1997). Briggs and colleagues were the first to identify a plasma membrane protein from dark-grown pea epicotyls that became phosphorylated upon irradiation with blue light (Gallagher *et al.*, 1988). Extensive photochemical and biochemical characterization of the light-induced phosphorylation reaction and its correlation with phototropism indicated that the unknown phosphoprotein was a candidate phototropic receptor that undergoes autophosphorylation in response to blue light treatment (Briggs *et al.*, 2001b). This hypothesis was substantiated several years later when it was observed that mutants of *Arabidopsis* deficient in phot1 lacked the blue light-induced phosphorylation reaction (Reymond *et al.*, 1992b; Liscum and Briggs, 1995).

Isolation of the *Arabidopsis PHOT1* gene and characterization of its encoded protein almost 10 years after the initial discovery of the light-dependent phosphorylation reaction demonstrated that it was indeed a photoreceptor for phototropism. When expressed in insect cells, phot1 undergoes autophosphorylation in response to blue light irradiation in the absence of any other plant proteins, implying that recombinant phot1 is a functional photoreceptor kinase (Christie *et al.*, 1998). Mutation of an essential aspartate residue within the phot1 kinase domain results in a loss of phot1 autophosphorylation, demonstrating that light-induced phosphorylation is mediated by phot1 itself and not some other kinase present in the insect cell extracts (Christie *et al.*, 2002). Furthermore, recombinant phot1 non-covalently binds the chromophore FMN and displays spectral characteristics that match the action spectrum for phototropism and other phototropin-mediated responses (Christie *et al.*, 1998, 1999; Briggs and Christie, 2002). Phot2 displays similar spectral properties and autophosphorylation activity to phot1 when expressed in insect cells (Sakai *et al.*, 2001; Christie *et al.*, 2002). However, it is not known whether phototropin autophosphorylation occurs intra- or inter-molecularly. For phot1, biochemical evidence suggests that such a process might involve intermolecular communication between distinct phototropin molecules (Reymond *et al.*, 1992a). Given that the LOV1 domain of oat phot1 has been shown to dimerize *in vitro* (Salomon *et al.*, 2004), it seems likely that the full-length receptors themselves form dimers. Whether receptor dimerization is necessary for phototropin autophosphorylation requires further investigation.

Autophosphorylation, at least for phot1, has been shown to occur on multiple serine residues (Palmer *et al.*, 1993; Short *et al.*, 1994; Salomon *et al.*, 1996). Phot1 from several plant species has been reported to show reduced electrophoretic mobility after blue light irradiation, consistent with autophosphorylation on multiple sites (Short *et al.*, 1993; Liscum and Briggs, 1995; Knieb *et al.*, 2005). Autophosphorylation of oat phot1 is accompanied by a loss in immunoreactivity with an antibody raised against the N-terminal region of *Arabidopsis* phot1 (Salomon *et al.*, 2003). Curiously, Knieb *et al.* (2005) have found that UV-C (280 nm) irradiation induces an electrophoretic mobility shift for oat phot1 without any change in immunoreactivity, suggesting that distinct serine residues may be phosphorylated in response to different light qualities. A recent study by Salomon *et al.* (2003) has identified eight serine residues within oat phot1 that become phosphorylated upon illumination. Two of these sites (Ser27, Ser30) are located before LOV1, near the N-terminus of the protein. The remaining six sites (Ser274, Ser300, Ser317, Ser325, Ser332, and Ser349) are located in the peptide region between LOV1 and LOV2. Salomon *et al.* (2003) also demonstrated that phot1 autophosphorylation *in vivo* is fluence dependent; the two serine residues situated at the extreme N-terminus are phosphorylated in response to low fluences of blue light, whereas the remaining sites are phosphorylated either at intermediate or high fluences. The authors therefore proposed that the hierarchical pattern of autophosphorylation might result in different biochemical consequences; the low fluence phosphorylation of phot1 may initiate receptor signaling by modifying the interaction status between the receptor and a specific signaling partner, whereas the high-fluence-activated phosphorylation sites may play a role in receptor desensitization. Support for this hypothesis has recently come from Kinoshita *et al.* (2003) who have shown that phot1 from *Vicia faba* (broad bean) guard cells binds a 14-3-3 protein upon autophosphorylation. 14-3-3 proteins belong to a highly conserved protein family that typically bind to phosphorylated target proteins and regulate signaling in eukaryotic cells (Ferl, 2004). Specifically, 14-3-3 binding to *Vicia* phot1 requires phosphorylation of Ser358 situated between LOV1 and LOV2 which is equivalent to Ser325 of oat phot1 that is phosphorylated in response to intermediate fluences of blue light (Salomon *et al.*, 2003). Thus, consistent with the proposed mechanism, autophosphorylation of phot1 in response to low and/or intermediate fluence rates of blue light may initiate signaling by binding a 14-3-3 protein. Further work is now needed to clarify the role of 14-3-3 proteins in phototropin signaling especially since this phenomenon is not restricted to stomatal guard cells (Kinoshita *et al.*, 2003). Interestingly, *Chlamydomonas* phot lacks the N-terminal extension preceding LOV1 present in higher plant phototropins, but is still able to restore phot1- and phot2-mediated responses when introduced into the *phot1phot2* double mutant of *Arabidopsis* (Onodera *et al.*, 2005), implying that the N-terminal phosphorylation sites of higher plant phototropins are not essential for phototropin signaling.

Autophosphorylation of phot1 *in vivo* has been shown to return to its inactive state in darkness following a saturating pulse of blue light (Short and Briggs 1990; Hager and Brich, 1993; Salomon *et al.*, 1997a; Kinoshita *et al.*, 2003). Moreover, the recovered photoreceptor system can be rephosphorylated in response to a second blue light pulse (Hager *et al.*, 1993; Salomon *et al.*, 1997a; Kinoshita *et al.*, 2003). These findings therefore demonstrate that the phot1, and most likely phot2, possess the ability to regenerate back to their non-phosphorylated form. The mechanisms associated with this recovery process are still not known but it is tempting to speculate on the involvement of an as yet unidentified protein phosphatase.

3.4   Light sensing by the LOV domains

3.4.1   LOV-domain photochemistry

Much work over the last five years has focused on understanding the underlying processes by which the phototropins detect blue light and use this information accordingly to activate specific physiological responses. As alluded to earlier, the LOV domains along with their associated FMN cofactors function as the 'eyes' of the receptor protein enabling the phototropins to detect the presence of light. Purification of sufficient quantities of LOV-domain proteins expressed in *Escherichia coli* has greatly facilitated the spectral and structural analysis of these light-sensing motifs. Purified LOV domains are highly fluorescent owing to their bound chromophore FMN (Figure 3.2A). Moreover, each phototropin LOV domain binds one molecule of FMN (Christie *et al.*, 1999). In addition, the spectral properties of recombinant LOV1 and LOV2 fusion proteins (Christie *et al.*, 1999; Salomon *et al.*, 2000) are very similar to those of the full-length photoreceptor proteins expressed in insect cells (Christie *et al.*, 1998; Kasahara *et al.*, 2002b), showing absorption in the blue/UV-A regions of the spectrum that closely match the action spectrum for phototropin-mediated responses (Briggs and Christie, 2002).

Spectral analysis of LOV domain fusion proteins has uncovered a unique mode of photochemistry underlying the primary mechanisms associated with light sensing by the phototropins (Figure 3.2B). In darkness, the FMN chromophore is non-covalently associated within the LOV domain forming a spectral species, designated LOV_{447} that absorbs maximally at 447 nm (Swartz *et al.*, 2001). Absorption of blue light by the FMN chromophore results in the formation of an excited singlet state, which subsequently decays into a flavin triplet state species (LOV_{660}) by intersystem crossing, absorbing maximally in the red region of the spectrum (Swartz *et al.*, 2001; Kennis *et al.*, 2003; Kottke *et al.*, 2003). The FMN triplet state is the primary photoproduct of the LOV-domain photocycle occurring within nanoseconds after the absorption of light (Kennis *et al.*, 2003). Decay of the flavin to the triplet state involves an electronic redistribution within the FMN chromophore that increases the basicity of the N5 nitrogen. It is then proposed that the N5 nitrogen of the FMN triplet state is stabilized by the abstraction of a proton from a thiol group of a conserved cysteine residue within the LOV domain (Crosson and Moffat, 2001; Kennis *et al.*, 2003). Protonation of N5, in turn, increases the electrophilicity of the C(4a) carbon of the flavin isoalloxazine ring and promotes nucleophilic attack by the thiol anion, resulting in the formation of a covalent adduct between the FMN chromophore and the active site cysteine. FMN-cysteinyl adduct formation occurs

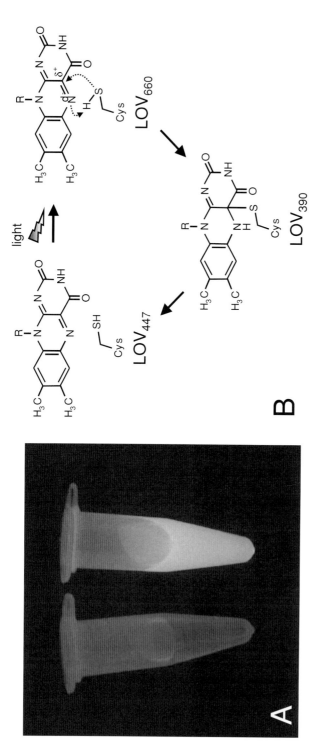

Figure 3.2 LOV-domain photochemistry. (A) Phototropin LOV domains expressed and purified from *E. coli* bind FMN and exhibit green fluorescence when illuminated with ultraviolet light. The fluorescence emitted from a purified preparation of phot1 LOV2 is shown (right) compared to that of a control protein preparation that does not bind FMN (left). (B) Schematic representation of LOV-domain photochemistry. The photocycle scheme is adapted from Crosson and Moffat (2001) and Kennis *et al.* (2003). Phototropin LOV domains contain a conserved cysteine that is required for their photochemical reactivity. In darkness, the FMN chromophore is non-covalently bound within the LOV domain forming a species that absorbs maximally at 447 nm (LOV$_{447}$). Light drives the production of a highly reactive triplet state flavin (LOV$_{660}$) that leads to formation a covalent bond between the C(4a) carbon of the FMN chromophore and a conserved cysteine residue within the LOV domain (LOV$_{390}$). At least for phototropin LOV domains, the photoreaction process is self-contained and fully reversible in darkness. Further details of the photocycle are described within the main text.

in the order of microseconds producing a species (LOV$_{390}$) that absorbs maximally at 390 nm.

Formation of LOV$_{390}$ within the phototropin LOV domains is fully reversible in darkness returning the LOV domain back to its initial ground state, LOV$_{447}$ (Salomon *et al.*, 2000; Swartz *et al.*, 2001). To date, little information is known as to how the covalent C–S bond formed upon illumination is able to spontaneously break and return to LOV$_{447}$ in subsequent darkness. Yet, dark-recovery of LOV$_{390}$ to LOV$_{447}$ is significantly slower in LOV domains that have been lyophilized and resuspended in D$_2$O compared to those rehydrated with H$_2$O, indicating that dark-recovery may be limited by proton transfer events (Swartz *et al.*, 2001; Corchnoy *et al.*, 2003). Consistent with light-induced adduct formation, replacement of the active site cysteine with either serine or alanine results in a complete loss of photochemical reactivity (Salomon *et al.*, 2000; Swartz *et al.*, 2001). Substitution of the conserved cysteine with methionine results in the formation of a unique photoproduct species absorbing in the red region of the spectrum (Kottke *et al.*, 2003). This species is stable both under aerobic and denaturing conditions and consists of a covalent adduct between the introduced methionine and the N5 nitrogen of the FMN chromophore (Federov *et al.*, 2003).

Although there is still some debate as to the primary mechanisms associated with adduct formation (Swartz *et al.*, 2001; Kottke *et al.*, 2003; Kay *et al.*, 2003; Schleicher *et al.*, 2004; Richter *et al.*, 2005), it is generally accepted that the FMN-cysteinyl adduct species LOV$_{390}$ represents the active signaling state that leads to photoreceptor activation. Indeed, light-induced adduct formation has been shown for several phototropin LOV domains using a number of biophysical approaches (Salomon *et al.*, 2001; Crosson and Moffat, 2002; Holzer *et al.*, 2002; Swartz *et al.* 2002; Ataka *et al.*, 2003; Iwata *et al.*, 2003, Federov *et al.*, 2003). LOV domains can therefore cycle between active (LOV$_{390}$) and inactive (LOV$_{447}$) states. Formation of LOV$_{390}$ results in a loss of absorption (in the blue region of the spectrum) and fluorescence, both of which are recoverable in darkness (Salomon *et al.*, 2000; Kasahara *et al.*, 2002b). Hence, the photocycle of the phototropin LOV domains can be readily monitored by means of absorbance or fluorescence spectroscopy. Kennis *et al.* (2004) using ultra-fast spectroscopy have shown that formation of LOV$_{390}$ can be reversed to its initial dark state upon illumination with near UV-light. The biological significance of this photoreversibility is not immediately obvious and requires further investigation.

3.4.2   LOV-domain structure

Crystal structures of LOV1 and LOV2 from different phototropins have been solved and show a close resemblance in overall structure to other PAS domains (Crosson and Moffat, 2001; Federov *et al.*, 2003). The structures of LOV1 and LOV2 are almost identical and comprise five antiparallel β-strands interconnected by two α-helices, similar to that which had been determined previously by molecular modeling (Salomon *et al.*, 2000). The FMN chromophore is held tightly within a central cavity by hydrogen bonding and Van der Waals forces via 11 conserved amino acids

(Crosson and Moffat, 2001; Federov *et al.*, 2003). The constraints imposed by the protein environment surrounding the flavin chromophore account for the vibronic fine structure observed in the absorbance spectrum of the LOV domain, which is not observed for free flavins in solution (Salomon *et al.*, 2000; Swartz *et al.*, 2001).

Importantly, crystal structures of LOV1 and LOV2 have been solved in both the dark and illuminated states (Crosson and Moffat, 2002; Federov *et al.*, 2003). In darkness, the sulfur atom of the conserved cysteine within the LOV domain is located several angstroms from the C(4a) carbon of the FMN isoalloxazine ring. Structures of LOV1 and LOV2 in their illuminated states reveal movements of the conserved cysteine side chain and the flavin ring structure that are required to bring about formation of the FMN-cysteinyl adduct (Crosson and Moffat, 2002; Federov *et al.*, 2003). These findings are consistent with circular dichroism (CD) measurements obtained for LOV2 indicating that adduct formation brings about a major structural change in the flavin moiety (Salomon *et al.*, 2000; Corchnoy *et al.*, 2003).

Recent chromophore exchange analysis provides additional information with regard to LOV-domain photochemistry and structure. Durr *et al.* (2005) used hydrophobic-interaction chromatography to successfully replace the FMN chromophore of LOV2 from oat phot1 with FAD, riboflavin and other flavin derivatives. Replacement of FMN with either FAD or riboflavin had little effect on the absorption properties and photocycle of LOV2 except that the kinetics for dark recovery became significantly faster in the presence of riboflavin. Evidently, the ribityl phosphate side chain of FMN is not essential for light-driven adduct formation, since riboflavin can function efficiently as a chromophore. Moreover, the adenosine moiety of the FAD bound to LOV2 could be readily cleaved with phosphodiesterase, indicating that this part of the cofactor extruded from the protein and could not be accommodated within the central chromophore pocket. More importantly, LOV2 was found to have a higher affinity for FMN over FAD, suggesting that the natural chromophore for phototropins in plants is also likely to be FMN. Then again, the identity of the chromophore bound *in vivo* will remain inconclusive until direct analysis of the chromophore bound to purified plant phototropins can be carried out.

3.4.3 Functional roles of LOV1 and LOV2

The phototropins are the only proteins identified to date that possess two LOV domains (Briggs *et al.*, 2001a). As yet, the significance of two chromophore-binding sites within the phototropin molecule is not completely understood. However, recent studies have uncovered important insights into the functional roles of LOV1 and LOV2. Detailed photochemical analysis has shown that LOV1 and LOV2 exhibit different quantum efficiencies and reaction kinetics (Salomon *et al.*, 2000; Kasahara *et al.*, 2002b), implying that these domains may have different light-sensing roles in regulating phototropin activity. Support for this hypothesis has come from recent structure–function studies. Christie *et al.* (2002) have used the cysteine to alanine mutation described earlier, which blocks LOV-domain photochemistry to ascertain the roles of LOV1 and LOV2 in regulating phototropin function. Photochemical reactivity of LOV2 is required for phot1 kinase activity and to elicit phot1-mediated hypocotyl phototropism in *Arabidopsis* to low intensities of unilateral blue light. LOV1, on the other hand, plays at most a minor light-sensing role and is not sufficient to elicit phot1-induced phototropic curvature. Thus, at least for phototropism, LOV2 is essential for phot1 function in *Arabidopsis*. Consistent with these findings, Kagawa *et al.* (2004) have reported that the LOV2 domain of phot2 plays a dominant role in regulating the chloroplast avoidance response since a truncated version of phot2 comprising only the LOV2 domain and the C-terminal kinase domain is able to complement the chloroplast avoidance movement in a *phot2* mutant of *Adiantum*. Peptide sequences C-terminal to the kinase domain of phot2 were also found to be necessary for biological activity.

Although the above findings clearly demonstrate that LOV2 plays an important role in regulating phototropin activity, the exact role of LOV1 remains unclear. The LOV1 domain of oat phot1 has been reported to self-dimerize, whereas the LOV2 does not (Salomon *et al.*, 2004). LOV1 may therefore play a role in receptor dimerization. If so, receptor dimerization may be affected by light, and in turn, control the sensitivity of a phototropin receptor complex. Yet, irradiation has no effect on the status of LOV1 dimerization *in vitro* (Salomon *et al.*, 2004). Alternatively, LOV1 may be involved in regulating phototropin-activated processes other than phototropism. Another possible function for LOV1 could be to regulate the lifetime of phototropin receptor activation. Kagawa *et al.* (2004) have estimated the signal lifetime for phot2 activation required to mediate chloroplast avoidance movement in *Adiantum*. The rate of dark recovery measured for *Adiantum* phot2 LOV2 alone was too fast to account for the lifetime signal measured for phot2 activity *in vivo*. Nevertheless, the rate of dark recovery measured for a fusion protein of *Adiantum* phot2 containing both LOV1 and LOV2 corresponded closely to the lifetime signal for phot2 activity. Since the LOV1 domain of phot2 is not essential for chloroplast avoidance movement in *Adiantum*, Kagawa *et al.* (2004) hypothesized that LOV1 may serve to prolong the lifetime of phot2 receptor activation. A role for LOV1 in modulating the activity of bacterially expressed phot2 kinase has also been recently reported (Matsuoka and Tokutomi, 2005).

Studies of individual LOV domains have been valuable in elucidating the reaction mechanisms associated with LOV-domain photochemistry. However, bacterially expressed fusion proteins, containing both LOV domains exhibit photochemical properties that more closely resemble those of full-length phototropins expressed in insect cells (Kasahara *et al.*, 2002b). Hence, the behavior of these tandem LOV proteins more accurately reflects those of the photoreceptor proteins themselves. Although tandem LOV proteins for phot1 and phot2 exhibit similar relative quantum efficiencies, their times for dark recovery differ significantly in that phot1 recovers much slower than phot2 (Christie *et al.*, 2002; Kasahara *et al.*, 2002b; Kagawa *et al.*, 2004). The rapid recovery for phot2 would be expected to yield steady-state levels of photoproduct much lower than those of phot1. As a result, higher light intensities would be required to drive phot2 to the same photostationary equilibrium as phot1. As mentioned earlier, phot1 and phot2 have been reported to exhibit different photosensitivities in activating several phototropin-mediated responses in *Arabidopsis*, in which phot2 typically requires a higher light threshold for activity than phot1. The difference in dark recovery observed between phot1 and phot2 may therefore relate to their physiological photosensitivities.

3.4.4   Light-induced protein movements

Identifying LOV2 as the main light sensor for regulating phototropin kinase activity represents a significant advance in understanding the mechanisms associated with receptor activation. How then does light absorption and subsequent adduct formation in LOV2 lead to an activation of the C-terminal kinase domain? An obvious mechanism would involve light-induced conformational changes within the LOV2 apoprotein. Still, the photoexcited crystal structure of LOV2 shows only minor, light-induced protein changes within the vicinity of the FMN chromophore compared to that of its dark state (Crosson and Moffat, 2002). Nonetheless, Fourier transform infrared (FTIR) spectroscopy studies demonstrate that photoactivation of purified LOV2 in solution is accompanied by changes in the LOV domain apoprotein (Swartz et al., 2002; Iwata et al., 2003). In particular, Nozaki et al. (2004) have shown that the βE sheet of the LOV2 apoprotein exhibits a significant conformational change upon adduct formation. The βE sheet region contains a conserved glutamine residue that when mutated to leucine results in a loss of the light-induced conformational change (Nozaki et al., 2004; Iwata et al., 2005). X-ray crystallography indicates that the conserved glutamine forms hydrogen bonds with the FMN chromophore and undergoes side chain rotation upon adduct formation (Crosson and Moffat, 2001; 2002). Hence, this residue may serve to translate adduct formation within the chromophore-binding pocket to protein changes at the surface of LOV2. In contrast to LOV2, however, only minimal light-induced protein changes have been reported for LOV1 (Ataka et al., 2003; Losi et al., 2003; Iwata et al., 2005). It is worth noting that the conserved glutamine responsible for propagating the light-induced conformational change in LOV2 is also found in LOV1. Whether this residue is unable to undergo side chain rotation upon adduct formation or serves another functional role in LOV1 remains to be determined.

Movement of the βE sheet in LOV2 may invoke further conformational changes that in turn lead to an activation of the C-terminal kinase domain. Support for this hypothesis first came from Corchnoy et al. (2003) who used CD spectroscopy to probe for potential light-induced protein changes within an extended LOV2 peptide fragment. Corchnoy et al. (2003) observed a major conformational change upon illumination resulting in a 10–15% loss in α-helicity that might be attributed to protein sequences situated at the C-terminal side of the LOV2-core. Subsequent nuclear magnetic resonance (NMR) studies using a similar extended LOV2 fragment confirmed the findings of Corchnoy et al. (2003). Harper et al. (2003) have identified a conserved α-helix (designated Jα) that associates with the surface of LOV2 in the dark state. The Jα-helix is located at the C-terminus of LOV2 and is amphipathic in nature consisting of polar and apolar sides, the latter of which docks onto the β-sheet strands of the LOV2-core. Following adduct formation the interaction between the Jα-helix and LOV2 is disrupted; the Jα-helix becomes disordered and more susceptible to proteolysis. Independent studies using larger LOV2 fragments provide further support for light-induced helical movements (Iwata et al., 2005; Eitoku *et al.*, 2005) and a comparable helix unfolding mechanism has also been reported the bacterial PAS light sensor PYP (Hoff *et al.*, 1999). Sequence alignment of LOV1 and LOV2 domains from a wide range of phototropins reveals that peptide sequences forming the Jα-helix are only found associated with LOV2 (Harper *et al.*, 2004), consistent with the distinct functional roles reported for these domains (Christie *et al.*, 2002; Kagawa *et al.*, 2004). It is worth noting, however, that the LOV1 domain of phot2 is still able to mediate a small degree of light-activated autophosphorylation (Christie *et al.*, 2002). Although this is not apparent for phot1, it raises questions as to how LOV1 can mediate autophosphorylation in the absence of a Jα-helix.

Harper *et al.* (2004) subsequently extended their NMR investigations by monitoring the consequences of introducing polar residues into the apolar face of the Jα-helix in an attempt to artificially disrupt the interaction between the Jα-helix and the LOV2-core. Three of these mutations were found to cause disordering of the Jα-helix in the absence of light and were equally susceptible to proteolysis irrespective of the light condition. To ascertain whether unfolding of the Jα-helix is an important step that couples LOV2 photoexcitation to kinase activation, Harper *et al.* (2004) examined the effect of introducing these mutations into full-length phot1 expressed in insect cells. Incorporating the corresponding mutations into *Arabidopsis* phot1 increased dark levels of autophosphorylation relative to wild type controls, consistent with the conclusion that these mutations mimic the irradiated form of LOV2, resulting in kinase activity in the absence of light. LOV2 may therefore serve to repress kinase activity in the dark sate whereupon photoexcitation and light-dependent unfolding of the Jα-helix would act to relieve this repression (Figure 3.3). A similar PAS/kinase domain interaction mechanism has been proposed for regulating the activities of the bacterial oxygen sensor, FixL (Gong *et al.*, 1998), and the novel eukaryotic protein kinase, PAS kinase (Rutter *et al.*, 2001). A role for LOV2 as a dark-state inhibitor of kinase activity has also been reported for phot2. Matsuoka and Tokutomi (2005) have found that a bacterially expressed phot2 kinase from *Arabidopsis* can phosphorylate the artificial substrate casein *in vitro*. Substrate phosphorylation by phot2 kinase occurs constitutively but becomes light dependent upon the addition of purified LOV2. Addition of LOV1 to the reaction had no effect on substrate phosphorylation. Moreover, LOV2 binds the kinase domain of phot2 in darkness, but this interaction is alleviated in the presence of light. Curiously, the trans interaction observed between LOV2 and the kinase domain occurs in the absence of the Jα-helix. Whether the dispensability of the Jα-helix reflects a difference between the mechanisms of substrate phosphorylation and receptor autophosphorylation remains to be clarified.

3.5 Phototropin signaling

3.5.1 Phototropin-interacting proteins

Mutant alleles carrying single amino acid substitutions within the kinase domain have been identified for both phot1 and phot2, indicating that kinase activity is essential for receptor signaling (Huala *et al.*, 1997; Kagawa *et al.*, 2001; Celaya and

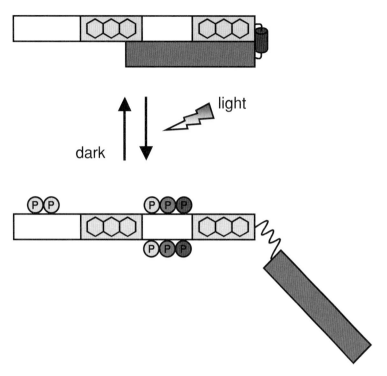

Figure 3.3 A simplified schematic overview of phototropin receptor activation. In the dark or ground state, the phototropin receptor is unphosphorylated and inactive, whereby the LOV domains have been proposed to repress receptor activity. Absorption of light by the predominant light sensor LOV2 results in a disordering of the Jα-helix and activation of the C-terminal kinase domain, which consequently leads to autophosphorylation of the photoreceptor protein. Relative positions of known phosphorylation sites are indicated and color-coded based on their hierarchical pattern of occurrence as described by Salomon *et al.* (2003): pale blue, low fluence sites; grey, intermediate fluence sites; red, high fluence sites. Further details of phototropin autophosphorylation are described within the main text.

Liscum, 2005). Yet, very little is known as to how the phototropins initiate signaling. Is autophosphorylation of the receptor molecule sufficient to elicit signaling or does the receptor phosphorylate a reaction partner in order to bring about a response? Evidence for the latter is still missing as the only substrate known for phototropin kinase activity, apart from the receptors themselves, is the artificial substrate casein (Matsuoka and Tokutomi, 2005).

Autophosphorylation, on the other hand, is likely to play a role in receptor signaling given that the phototropins are known to bind a 14-3-3 protein in stomatal guard cells in a blue light-dependent manner (Kinoshita *et al.*, 2003). Activity of the guard cell plasma membrane H^+-ATPase is essential for stomatal opening and is also regulated by phosphorylation and 14-3-3 binding in response to blue light irradiation (Kinoshita and Shimazaki, 1999, 2002). It is therefore tempting to speculate that 14-3-3 binding may facilitate a direct interaction between the phototopins and the guard cell H^+-ATPase. Yet, the fungal toxin fusicoccin has been shown to induce phosphorylation of the H^+-ATPase and subsequent 14-3-3 binding in the absence of phot1 and phot2, implying that some other protein kinase is responsible for phosphorylation of the H^+-ATPase (Kinoshita and Shimazaki, 2001; Ueno et al., 2005). Further work is now required to clarify the significance of 14-3-3 binding and how this interaction plays a role in phototropin signaling, especially since 14-3-3 binding to the phototropins has been observed in etiolated seedlings (Kinoshita et al., 2003). Recent yeast two-hybrid screening of a cDNA library derived from *Vicia faba* guard cells has identified a novel phot1-interacting protein (Emi et al., 2005). The C-terminus of the *Vicia faba* phot1-interacting protein (VfPIP) shows homology to dyneins, proteins that are associated with microtubule function in animal cells. Indeed, VfPIP localizes to cortical microtubules in *Vicia* guard cells, and may act to organize the assembly of the guard cell cytoskeleton to support stomatal opening.

The first phototropin-interacting protein to be identified was the scaffold-type protein NPH3 (non-phototropic hypocotyl 3). NPH3 is a novel protein containing several protein interaction motifs and has been shown to interact with phot1 in yeast and *in vitro* (Motchoulski and Liscum, 1999). NPH3 was identified through the isolation of phototropic mutants and is essential for phototropism in *Arabidopsis* (Liscum and Briggs, 1995; 1996; Motchoulski and Liscum, 1999; Sakai et al., 2000) and rice (Haga et al., 2005). As with the photoropins, NPH3 is associated with the plasma membrane and, although its biochemical function is unknown, most likely serves as a scaffold to assemble components of a phototropin receptor complex (Liscum and Stowe-Evans, 2000; Celaya and Liscum, 2005). NPH3 is a member of a large plant-specific gene family in *Arabidopsis* consisting of 31 members (Celaya and Liscum, 2005). A protein closely related to NPH3, designated root phototropism 2 (RPT2), was isolated from a separate genetic screen (Sakai et al., 2000). RPT2 has been shown to bind phot1 and is required for phototropism and light-induced stomatal opening (Sakai et al., 2000, 2001; Inada et al., 2004). Although the interaction between phot1 and RPT2 is unaffected by light (Inada et al., 2004), Motchoulski and Liscum (1999) have reported that the phosphorylation status of NPH3 is altered upon light exposure; NPH3 is phosphorylated in the dark and becomes dephosphorylated in response to irradiation. The functional significance of NPH3 dephosphorylation is currently unknown but it will be important to establish whether other members of the NPH3/RPT2-like (NRL) family exhibit similar properties.

3.5.2 Downstream signaling targets

To date very little is known about the signaling events occurring downstream of phototropin photoactivation. As mentioned earlier, activation of both phot1 and phot2 are known to lead to an increase in cytosolic Ca^{2+} levels in *Arabiodpsis*. Ca^{2+} therefore represents a key-signaling event downstream of receptor photoexcitation. Although pharmacological analysis has linked a role for calcium to the phot1-mediated inhibition of hypocotyl growth (Folta et al., 2003), further analysis is required to determine if calcium acts as an intracellular messenger for other phototropin-activated processes, especially since changes in intracellular calcium levels have been shown to be important for the regulation of light-induced stomatal opening (Dietrich *et al.*, 2001) and chloroplast movements (Wada *et al.*, 2003). In addition, recent electrophysiological studies indicate that phototropic bending involves changes in ion fluxes in response to blue light irradiation, including calcium (Babourina *et al.*, 2004).

Much of our knowledge regarding the downstream signaling events associated with phototropin signaling has come from genetic screens for mutants impaired in specific phototropin-mediated responses. Isolation of *Arabidopsis* mutants impaired in the chloroplast avoidance response has led to the identification of a novel F-actin-binding protein CHUP1 (Kasahara *et al.*, 2002a; Oikawa *et al.*, 2003), consistent with the evidence that chloroplast movements occur through changes in the cytoskeleton (Wada *et al.*, 2003). CHUP1 confers the ability to target GFP into the chloroplast envelope (Oikawa *et al.*, 2003), suggesting that CHUP1 may function at the periphery of the chloroplast outer membrane. Mutants lacking CHUP1 exhibit aberrant chloroplast positioning compared to wild-type plants, whereby chloroplasts are constantly gathered at the bottom of palisade cells (Oikawa *et al.*, 2003). Thus, CHUP1 most likely represents an essential component of the machinery required for chloroplast positioning and movement. A separate genetic screen designed to isolate *Arabidopsis* mutants impaired in chloroplast accumulation movement has identified the signaling component JAC1 (Suetsugu *et al.*, 2005). JAC1 is a cytosolic protein that is specifically required for chloroplast accumulation movement since *jac1* mutants display a normal avoidance response. Although the exact role of JAC1 in controlling chloroplast accumulation movement is not known, the C-terminus of JAC1 exhibits homology to auxilin, a protein that plays a role in clathrin-mediated endocytosis in animals, yeast and nematodes. The functional significance of the auxilin-like domain of JAC1 in regulating chloroplast accumulation movement awaits further characterization of the JAC1 protein.

The isolation of *Arabidopsis* mutants impaired in phototropism has provided important insights into the signaling mechanisms involved in establishing phototropic curvature. It has long been accepted that phototropic curvature in plants is mediated by an increase in growth on the shaded side of the stem resulting from an accumulation of the growth hormone auxin (Iino, 2001). The model currently favored begins with the establishment of a light gradient across the hypocotyl in response to a directional light stimulus (Iino, 2001). Indeed, Salomon *et al.* (1997b, 1997c) have shown that unilateral irradiation induces a gradient of phot1 autophosphorylation across oat coleoptiles, with a higher level of phosphorylation on the irradiated side. However, little is known about how this differential signal leads to an accumulation of auxin on the shaded side of the phototropically stimulated stem. Nonetheless, genetic analysis has demonstrated that auxin responsiveness is necessary for the development of phototropic curvature. The auxin-regulated transcription factors NPH4 and MSG2 are required for normal phototropism and gravitropism (Stowe-Evans *et al.*, 1998; Harper *et al.*, 2000; Tatematsu *et al.*, 2004), highlighting the need for auxin-regulated gene expression. The molecular identities of auxin-responsive genes involved in phototropism have recently been uncovered using a transcriptomic approach. Esmon *et al.* (2006) have meticulously monitored gene expression changes occurring across phototropically stimulated *Brassica oleracea* hypocotyls to identify auxin-responsive gene targets associated with phototropism. Gene targets of NPH4 action whose expression were found to increase on the elongating side of phototropically stimulated hypocotyls included two members of the α-expansin family, *EXPA1* and *EXPA8*. Since members of the α-expansin family are known to mediate cell wall extension, EXPA1 and EXPA8 may play important roles in the establishment of phototropic curvatures.

Genetic studies also indicate that auxin transport is required for phototropism. Mutants impaired in the localization of the putative auxin efflux carrier PIN1 exhibit altered hypocotyl phototropism (Noh *et al.*, 2003; Blakeslee *et al.*, 2004). In addition to PIN1, PIN3 a second member of the *Arabidopsis* PIN family appears to play an important role in the establishment of the lateral auxin gradient required for phototropism (Friml *et al.*, 2002). Given recent findings that phot1 photoactivation results in a change in PIN1 localization in hypocotyls cells (Blakeslee *et al.*, 2004), a regulation of auxin transporter localization may represent a major point of control in the development of phototropic curvatures. Such a mechanism is likely to be complex since PIN proteins have been shown to act in conjunction with members of a second transporter family of p-glycoproteins (PGP) to bring about active auxin transport in *Arabidopsis* (Geisler and Murphy, 2006). It will now be important to determine whether phototropin photoactivation can influence the localization of other potential auxin transporters and whether these changes in localization are mediated directly by the receptors themselves or by some other signaling mechanism.

3.6 Other LOV-containing proteins

Discovery of the phototropins and characterization of the LOV domain as a blue light-sensing motif represents a major advance in plant photomorphogenesis research. As mentioned previously, the LOV domain has been used as a signature motif to identify the presence of photosensory proteins in various organisms. As a result, novel LOV-containing proteins have been found in plants, fungi and even bacteria. These proteins differ from the phototropins in that they contain a single LOV domain, but have been shown to exhibit photochemical reactivity. The remainder of this chapter will focus on outlining the structural and biochemical properties of these novel LOV-containing proteins in addition to their known biological roles.

3.6.1 LOV-containing proteins in Arabidopsis

Besides the phototropins, several other LOV-containing proteins have been identified in *Arabidopsis*. Three of these constitute a novel family of blue light receptors known as the ZTL/ADO family that appear to localize to both the nucleus and the cytosol (Kiyosue and Wada, 2000; Yasuhara *et al.*, 2004; Fukamatsu *et al.*, 2005). The first member of the family ZTL was identified by Somers *et al.* (2000)

in a genetic screen for circadian clock mutants of *Arabidopsis*. Mutations at the *ZTL* locus exhibit a lengthened circadian period hence the German name Zeitlupe, which roughly translated means 'slow motion'. Concomitantly, ZTL was identified independently by a number of groups (Kiyosue and Wada, 2000; Nelson *et al.*, 2000; Somers *et al.*, 2000; Jarillo *et al.*, 2001a) and therefore goes by several names including ADO, which refers to the musical term adagio meaning 'slowly' (Jarillo *et al.*, 2001a). Owing to a lengthened circadian period, *ztl* mutants are altered in a number of circadian processes including clock-regulated gene expression, leaf movements and the onset of flowering (Somers *et al.*, 2000; Jarillo *et al.*, 2001a). Similarly, overexpression of *ZTL* leads to aberrant circadian clock function (Kiyosue and Wada, 2000; Somers *et al.*, 2004). Importantly, the long period phenotype of *ztl* mutants is more prominent under low light intensities than under high light, indicating a possible light-dependent role for ZTL in regulating circadian clock function (Somers *et al.*, 2000). However, *ztl* mutants have recently been reported to exhibit a long-period phenotype in the absence of light (Somers *et al.*, 2004), indicating that ZTL may play a more central role in regulating the circadian clock besides mediating light input.

The second member of the ZTL/ADO family, FKF1 (Flavin-binding, Kelch repeat, F-box 1) was identified alongside ZTL (Nelson *et al.*, 2000) and functions to regulate flowering time in response to day length by controlling the expression and activity of CONSTANS (CO), a key factor required for the photoperiodic control of flowering (Imaizumi *et al.*, 2003). The third member, LKP2 (LOV, Kelch, Protein 2) was identified from a search of the *Arabidopsis* genome for novel proteins containing a LOV domain (Schultz *et al.*, 2001). Overexpression of *LKP2* results in arrhythmic phenotypes for several circadian responses and also impairs the photoperiodic control of flowering (Schultz *et al.*, 2001). Thus, each member of the ZTL/ADO family appears to play an important role in regulating circadian clock function.

ZTL, FKF1 and LKP2 proteins share three characteristic domains: a phototropin-like LOV domain at the N-terminus followed by an F-box motif and six kelch repeats at the C-terminus (Figure 3.1). The F-box motif is typically found in E3 ubiquitin ligases which function to target proteins for degradation via the ubiquitin-proteosome system (Smalle and Vierstra, 2004). It is therefore considered that ZTL and its homologues mediate their effects on circadian control by regulating the turnover of clock-associated components. Indeed, ZTL has been reported to modulate circadian clock function by targeting TOC1, a key component of the circadian oscillator, for degradation (Más *et al.*, 2003). Likewise, FKF1 has recently been shown to control *CO* expression, in part, by targeted degradation of CDF1, a repressor of *CO* transcription (Imaizumi *et al.*, 2005). Consistent with their proposed function in targeting substrates for degradation by the ubiquitin-proteasome system, ZTL and LKP2 have been shown to interact with known components of the SCF (Skp, Cullin, F-box) E3 ubiquitin ligase complex via their F-box motif (Han *et al.*, 2004; Yasuhara *et al.*, 2004). Furthermore, reduced levels of the ZTL-interacting SCF component AtRBX1 phenocopies the lengthened circadian period phenotype of *ztl* mutants, demonstrating the functional relevance of ZTL-SCF interactions in *Arabidopsis* (Han *et al.*, 2004).

The kelch-domain repeats found in ZTL, FKF1 and LKP2 form a β-propeller structure thought to be involved in mediating protein interactions. Protein interaction motifs situated C-terminal to the F-box are considered to determine target specificity for the F-box protein (Smalle and Vierstra, 2004). This also seems to be the case for members of the ZTL family as kelch repeats of FKF1 and LKP2 have been shown to be necessary for their interaction with CDF1 (Imaizumi *et al.*, 2005). Similarly, the C-terminal kelch repeats of LKP2 have been shown to be sufficient for mediating an interaction with CO (Fukamatsu *et al.*, 2005). Interestingly, ZTL has been reported to interact with the C-terminal regions of phyB and cry1 in yeast and *in vitro* (Jarillo *et al.*, 2001a). This interaction may therefore serve to confer a light-dependent regulation of ZTL activity in *Arabidopsis*. Given recent findings, however, it is seems very likely that such a regulation involves the N-terminal LOV domain present in all members of the ZTL/ADO family.

ZTL, FKF1 and LKP2 represent unique F-box proteins in that they possess a phototropin-like LOV domain at their N-terminus. Their LOV domains contain all the eleven conserved residues necessary for flavin binding, including the essential cysteine required for photochemical reactivity (Crosson *et al.*, 2003). Noticeably, the LOV domains of ZTL, FKF1 and LKP2, unlike the phototropin LOV domains, contain an additional amino acid insert within the α'A-αC loop. The functional significance of these extra nine amino acids is currently unknown. Recently, the LOV domains of these three family members have been shown to function as light sensors and exhibit similar photochemical and properties (Imaizumi *et al.*, 2003; Nakasako *et al.*, 2005). The LOV domain of FKF1 and it homologues binds FMN and undergoes a blue light-activated photochemical reaction analogous to that observed for the phototropin LOV domains. Curiously, all three LOV domains fail to revert back to their dark state, in stark contrast to the phototropin LOV domains. It is not yet known how this inability to recover to the dark state is related to the physiological functions of the ZTL/ADO family, but demonstration of photochemical reactivity leading to the formation of a flavin-cysteinyl adduct provides strong evidence that these family members function as blue light receptors in *Arabidopsis*. In addition to its light-sensing role, the LOV domains of ZTL/ADO family members may also serve to recruit specific target proteins since this region has been shown to be necessary for the interaction with particular clock-associated proteins in a yeast two-hybrid assay (Yasuhara *et al.*, 2004; Fukamatsu *et al.*, 2005).

It is worth noting that a search of the *Arabidopsis* genome has uncovered a unique LOV-containing protein that is unrelated to the phototropins or the ZTL/ADO family. This protein referred to as twin LOV protein 1 (TLP1) or PAS/LOV (Crosson *et al.*, 2003) contains a conventional PAS domain followed by a phototropin-like LOV domain (Figure 3.1). To date, nothing is known about the activity or function of this protein or whether its LOV domain can bind a flavin chromophore. It will now be of interest to establish whether PAS/LOV represents a further and as yet uncharacterized blue light receptor in *Arabidopsis*.

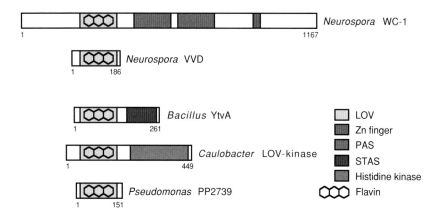

Figure 3.4 Protein structures of LOV-containing proteins found in fungi and bacteria. Two LOV-containing proteins found in the filamentous fungus *Neurospora* are shown: white collar-1 (WC-1) and VIVID (VVD). Protein structures of three bacterial LOV-containing proteins are also shown: *Bacillus* YtvA, *Caulobacter* LOV-kinase and the product of *Pseudomonas* gene PP2739. Each of these proteins contains a single LOV domain that functions as the binding site for a blue light absorbing flavin cofactor. Additional domain structures within these proteins are indicated.

3.6.2 LOV-containing proteins in fungi

Blue light acts to regulate a number of developmental and physiological processes in the filamentous fungus *Neurospora crassa*. Such processes include phototropism of the perithecial tips, carotenoid biogenesis, sexual fruiting body formation and circadian clock entrainment (Liu *et al.*, 2003). All known blue light responses are absent in *Neurospora* carrying mutations in the *white collar-1* (*wc-1*) and *wc-2* genes demonstrating the central importance of their encoded proteins in blue light sensing (Dunlap and Loros, 2005). These genes encode PAS-domain-containing transcription factors with GATA type zinc-finger DNA-binding domains (Ballario *et al.*, 1996; Linden and Macino, 1997). WC-1 is an obvious candidate for a blue light receptor as it contains a LOV domain in addition to two conventional PAS domains (Figure 3.4). WC-2, on the other hand, contains a single PAS domain. WC-1 and WC-2 proteins have been shown to dimerize in the nucleus via their PAS domains to form a White collar complex (WCC) that in turn acts to control the transcription of light-regulated genes (Ballario *et al.*, 1998; Talora *et al.*, 1999; Schwerdtfeger and Linden, 2000; Denault *et al.*, 2001; Cheng *et al.*, 2002).

The LOV domain of WC-1 contains all the conserved residues necessary for flavin binding and like the LOV domain of *Arabidopsis* ZTL and it homologues, contains an amino acid extension within the α'A-αC loop (Crosson *et al.*, 2003). In contrast to the LOV domains of the ZTL/ADO family, WC-1 has been shown to bind FAD as a chromophore not FMN. Froehlich *et al.* (2002) reported that WC-1 in the presence of FAD binds WC-2 to form an active WCC *in vitro*. Concurrently, He *et al.* (2002) showed that WCC purified from *Neurospora* binds FAD in stoichiometric amounts. Similarly, He and Liu (2005) have found that WCC binds FAD when expressed in insect cells. Taken together, these findings provide convincing evidence that WC-1 binds FAD as a chromophore and functions as a blue light receptor in *Neurospora*. Yet, direct measurements of the photochemical properties of the WC-1 LOV domain remain to be determined. Nevertheless, mutational analysis has clearly shown that the presence and activity of the LOV domain is essential for WC-1 activity (Ballario *et al.*, 1996; He *et al.*, 2002; Cheng *et al.*, 2003). In addition, He and Liu (2005) have shown that WC-1 at least *in vitro* is unable to recover to the dark state by measuring the ability of the WCC to respond to a second pulse of light. Thus, it is tempting to speculate that the photochemical properties of the WC-1 LOV domain may be analogous to those observed for ZTL and its homologues, which do not exhibit active photocycles *in vitro* (Imaizumi *et al.*, 2003).

Even though WC-1 is required for all known blue light responses (Dunlap and Loros, 2005), a second LOV-domain containing photoreceptor known as VIVID (VVD) has been identified in *Neurospora* (Heinzten *et al.*, 2001). VVD is a small cytosolic protein consisting mostly of a LOV domain (Figure 3.4) and plays an important role in mediating photoadaptive responses in *Neurospora* (Heinzten *et al.*, 2001; Schwerdtfeger and Linden, 2000; 2003; Shrode *et al.*, 2001). More recently, VVD has been shown to play a major role in facilitating circadian clock entrainment (Elvin *et al.*, 2005). When expressed and purified from *E. coli*, VVD binds a flavin chromophore that forms a flavin-cysteinyl adduct when irradiated with blue light (Schwerdtfeger and Linden, 2003). Similar to the phototropin LOV domains, the conserved active site cysteine is essential for the photochemical reactivity and function of VVD (Cheng *et al.*, 2003; Schwerdtfeger and Linden, 2003). Intriguingly, VVD is able to bind both FAD and FMN when expressed in *E. coli*. The significance of this is not clear at present, but raises the question as to the identity of the chromophore bound by VVD *in vivo*. Nonetheless, Cheng *et al.* (2003) have demonstrated that the LOV domain of VVD can partially replace the function of the WC-1 LOV domain, suggesting that these domains are, at least in part, functionally interchangeable. Indeed, the LOV domain of VVD is very similar to that of WC-1 in that it contains an 11 amino acid insert within the α'A-αC loop (Crosson *et al.*, 2003). Recombinant VVD differs from phototropin LOV domains in that it exhibits an extremely slow photocycle *in vitro* ($\sim$5 h). How this slow photocycle relates to VVD function in *vivo* is not known. The long recovery rate of VVD may be due to the fact that its photocycle was recorded at 4°C (Schwerdtfeger and Linden, 2003). Another possibility is that the extension in the α'A-αC loop may account for this photochemical behavior, given that ZTL and its homologues share a similar extension and show no appreciable dark recovery *in vitro* (Imaizumi *et al.*, 2003).

Homologues of WC-1 have been identified in ascomycetes other than *Neurospora*, including the soil fungus *Trichoderma atroviride* (Casas-Flores *et al.*, 2004) and the truffle-forming ascomycete *Tuber borchii* in which blue light inhibits hyphal growth (Ambra *et al.*, 2004). A protein similar to VVD named Envoy has been identified in the ascomycete *Hypocrea jecorina* where it functions to mediate the effects of light on cellulase gene expression (Schmoll *et al.*, 2005). In addition, homologues of WC-1 have also been identified in basidiomycetes, including the human pathogenic fungus *Cryptococcus neoformans* (Idnurm and Heitman, 2005; Lu *et al.*,

2005). Blue light inhibits sexual filamentation in this organism, which is mediated by the *Cryptococcus* equivalent of WC-1. Mutation of a gene closely related to *Neurospora wc-2* also results in a light-insensitive mating phenotype in *Cryptococcus* (Idnurm and Heitman, 2005). Similarly, overexpression of *Cryptococcus wc-1* and *wc-2* causes a dramatic inhibition of sexual filamentation upon light exposure (Lu *et al.*, 2005). WC-1 and WC-2 from *Cryptococcus* interact strongly in yeast, suggesting that they likely function in a manner similar to their *Neurospora* counterparts (Idnurm and Heitman, 2005). However, WC-1 from *Cryptococcus* lacks the DNA binding domain that is present in *Neurospora* WC-1. The absence of a DNA binding domain appears to be a structural feature common to WC-1-like proteins from basidiomycetes (Idnurm and Heitman, 2005). For example, Dst1 is required for fruiting body photomorphogenesis in the basidiomycete *Coprinus cenereus* and exhibits homology to WC-1 but contains no DNA-binding motif (Terashima *et al.*, 2005). Thus, WC-2 counterparts presumably mediate DNA binding of white-collar components in basidiomycetes.

3.6.3   LOV-containing proteins in bacteria

Genome sequencing projects have now revealed that LOV-containing proteins are also present in bacteria (Crosson *et al.*, 2003). Losi (2004) recently reported the identification of 29 LOV-containing proteins from the genomes of 24 bacterial species. These proteins are highly diverse and typically contain a single LOV domain coupled to a specific output domain, such as kinases, phosphodiesterases, response regulators, DNA-binding motifs, and regulators of stress sigma factors (Losi, 2004). YtvA is a small protein that acts as positive regulator of the general stress transcription factor σ^B in *Bacillus subtilis* (Akbar *et al.*, 2001). The N-terminal of YtvA contains a canonical LOV domain followed by a STAS domain (Figure 3.4). STAS domains are generally found in bacterial sulfate transporters and antisigma factor antagonists (Aravind and Koonin, 2000) and have been suggested to possess nucleoside triphosphate binding activity that is presumably important for domain function (Losi, 2004). Losi *et al.* (2002) demonstrated that YtvA expressed and purified from *E. coli* binds FMN and undergoes a blue light-activated photocycle analogous to that of the phototropin LOV domains. In contrast to the phototropin LOV domains, the LOV domain of YtvA exhibits a relatively slow photocyle (~1 h) (Losi *et al.*, 2003). In a related study, Losi (2004) reported similar photochemical properties for a LOV-kinase protein from *Caulobacter crescentus* (Figure 3.4). More recently, a small protein consisting mostly of a LOV domain encoded by the *Pseudomonas putida* gene PP2739 (Figure 3.4) has been shown to bind a flavin cofactor and exhibit photochemical properties similar to those of YtvA and the phototropin LOV domains (Krauss *et al.*, 2005). It therefore seems likely that many, if not all, of these LOV-containing proteins identified in bacteria bind flavin and show the same photochemical reactivity.

An important question to address now relates to the biological functions of these proteins in their bacterial hosts. In the case of YtvA, blue/UV-A wavelengths may be required to trigger stress-induced responses in *Bacillus*. Yet, the effects of blue light on the life cycle of *Bacillus*, a non-photosynthetic soil organism, are not known. Nevertheless, the presence of LOV domain-containing proteins throughout various kingdoms of life, including bacteria, clearly demonstrate that this functional light sensor is not only restricted to plants but has been conserved throughout evolution. Whatever these bacterial proteins do, it is tempting to speculate that their mechanism of photoactivation may be similar in terms of signal transmission from the LOV-core to the output domain in question. FTIR spectroscopy studies indicate that formation of the flavin-cysteinyl adduct within the LOV domain of YtvA gives rise to a conformational change in the C-terminal STAS domain (Bednarz *et al.*, 2004). Further spectroscopic analysis monitoring tryptophan fluorescence also provides evidence for an interaction between the LOV and STAS domain of YtvA (Losi *et al.*, 2004, 2005). Moreover, both *Bacillus* YtvA (Losi *et al.*, 2005) and *Pseudomonas* gene product PP2739 (Krauss *et al.*, 2005) are reported to contain an α-helix C-terminal to the LOV domain similar to the Jα-helix identified by Harper *et al.* (2003) in plant phototropins. Bacterial LOV proteins may therefore represent convenient paradigms for further elucidating the mechanisms by which light-sensitive LOV domains act to regulate protein activity.

3.7 Conclusions and future perspectives

Since the discovery of the first phototropin gene less than 10 years ago, the increase in knowledge of these blue light receptors has been exponential. A great deal of information has already been obtained with respect to the photochemical and bio-chemical properties of the phototropins, in addition to their physiological roles. Yet, important questions still remain to be addressed regarding their mode of action and the nature of the signaling events that couple phototropin activation to specific pho-toresponses. A major challenge for future research will be to unravel the processes associated with phototropin signaling and how these relate to components that have already been identified, including increases in cytosolic calcium, 14-3-3 proteins and members of the NRL family. Moreover, the identification of the LOV-sensing motif in proteins other than the phototropins greatly expands the possible avenues for future research. Why do the LOV domains of ZTL/ADO family members ex-hibit a truncated photocycle and how does this relate to their physiological function? What is the function of the novel PAS/LOV protein in *Arabidopsis*? What are the biological roles of the diverse range of LOV-containing proteins identified from the genomes of various bacterial species? And do these bacterial LOV domains exert their effects on specific output domains in a similar manner? Clearly, much work re-mains to be done and the coming decade should undoubtedly yield exciting advances in our knowledge of phototropin receptors and other LOV-containing proteins.

Acknowledgements

I am very grateful to both Winslow R. Briggs and Trevor E. Swartz for their careful review of the manuscript and their helpful comments.

References

Ahmad, M., Jarillo, J.A., Smirnova O. and Cashmore, A. (1998) *Nature* **392**, 720–723.
Akbar, S., Gaidenko, T.A., Kang, C.M., O'Reilly M., Devine K.M. and Price C.W. (2001) *J. Bacteriol.* **183**, 1329–1338.
Ambra R., Grimaldi B., Zamboni S., Filetici P., Macino G. and Ballario P. (2004) *Fungal Genet. Biol.* **41**, 688–697.
Aravind L. and Koonin E.V. (2000) *Curr. Biol.* **10**, R53–55.
Ataka K., Hegemann P. and Heberle J. (2003) *Biophys. J.* **84**, 466–474.
Babourina O., Godfrey L. and Voltchanskii K. (2004) *Ann. Bot. (Lond.)* **94**, 187–194.
Babourina O., Newman I. and Shabala S. (2002) *Proc. Natl. Acad. Sci. USA* **99**, 2433–2438.
Ballario P., Talora C., Galli D., Linden H. and Macino G. (1998) *Mol. Microbiol.* **29**, 719–729.
Ballario P., Vittorioso P., Magrelli A., Talora C., Cabibbo A. and Macino G. (1996) *EMBO J.* **15**, 1650–1657.
Baum G., Long J.C., Jenkins G.I. and Trewavas A.J. (1999) *Proc. Natl. Acad. Sci. USA* **96**, 13554–13559.
Bednarz T., Losi A., Gärtner W., Hegemann P. and Heberle J. (2004) *Photochem. Photobiol. Sci.* **3**, 575–579.
Blakeslee J.J., Bandyopadhyay A., Peer W.A., Makam S.N. and Murphy A.S. (2004) *Plant Physiol.* **134**, 28–31.
Bögre L., Okresz L., Henriques R. and Anthony R.G. (2003) *Trends Plant Sci.* **8**, 424–431.
Briggs W.R., Beck C.F., Cashmore A.R., Christie J.M., Hughes J., Jarillo J., Kagawa T., Kanegae H., Liscum E., Nagatani A., Okada K., Salomon M., Rüdiger W., Sakai T., Takano M., Wada M. and Watson J.C. (2001a) *Plant Cell* **13**, 993–997.
Briggs W.R. and Christie J.M. (2002) *Trends Plant Sci.* **7**, 204–210.
Briggs W.R., Christie J.M. and Salomon M. (2001b) *Antiox. Redox Signal.* **3**, 775–788.
Briggs W.R., Christie J.M. and Swartz T.E. (2005) Phototropins. In: *Photomorphogenesis in Plants and Bacteria: Function and Signal Transduction Mechanisms* (eds E. Schäfer and F. Nagy), pp. 223–252. Kluwer, Dordrecht.
Casas-Flores S., Rios-Momberg M., Bibbins M., Ponce-Noyola P. and Herrera-Estrella A. (2004) *Microbiology* **150**, 3561–3569.
Celaya R.B. and Liscum E. (2005) *Photochem. Photobiol.* **81**, 73–80.
Cheng P., He Q., Yang Y., Wang L. and Liu Y. (2003) *Proc. Natl. Acad. Sci. USA* **100**, 5938–5943.
Cheng P., Yang Y., Gardner K.H. and Liu Y. (2002) *Mol. Cell Biol.* **22**, 517–524.
Christie J.M. and Briggs W.R. (2001) *J. Biol. Chem.* **276**, 11457–11460.
Christie J.M. and Briggs W.R. (2005) Blue light sensing and signaling by the phototropins. In: *Handbook of Photosensory Receptors* (eds W.R. Briggs and J.L. Spudich), Wiley-VCH, Weinheim, pp. 277–304.
Christie J.M., Reymond P., Powell G., Bernasconi P., Reibekas A.A., Liscum E. and Briggs W.R. (1998) *Science* **282**, 1698–1701.
Christie J.M., Salomon M., Nozue K., Wada M. and Briggs W.R. (1999) *Proc. Natl. Acad. Sci. USA* **96**, 8779–8783.
Christie J.M., Swartz T.E., Bogomolni R. and Briggs W.R. (2002) *Plant J.* **32**, 205–219.
Corchnoy S.B., Swartz T.E., Lewis J.W., Szundi I., Briggs W.R. and Bogomolni R.A. (2003) *J. Biol. Chem.* **278**, 724–731.
Crosson S. (2005) LOV domain structure, dynamics, and diversity. In: *Handbook of Photosensory Receptors* (eds W.R. Briggs and J.L. Spudich), Wiley-VCH, Weinheim, pp. 323–336.
Crosson S. and Moffat. K. (2001) *Proc. Natl. Acad. Sci. USA* **98**, 2995–3000.
Crosson S. and Moffat K. (2002) *Plant Cell* **14**, 1067–1075.
Crosson S., Rajagopal S. and Moffat K. (2003) *Biochemistry* **42**, 2–10.
Denault D.L., Loros J.J. and Dunlap J.C. (2001) *EMBO J.* **20**, 109–117.
Dietrich P., Sanders D. and Hedrich R. (2001) *J. Exp. Bot.* **52**, 1959–1967.

Dunlap J.C. (2005) Blue light receptors – beyond phototropins and cryptochromes. In: *Photomorphogenesis in Plants and Bacteria: Function and Signal Transduction Mechanisms* (eds E. Schäfer and F. Nagy), pp. 253–277. Kluwer, Dordrecht.

Dunlap J.C. and Loros, J.J. (2005) Neurospora photoreceptors. In: *Handbook of Photosensory Receptors* (eds W.R. Briggs and J.L. Spudich), Wiley-VCH, Weinheim, pp. 371–389.

Durr H., Salomon M. and Rüdiger W. (2005) *Biochemistry* **44**, 3050–3055.

Eisinger W., Bogomolni R.A. and Taiz L. (2003) *Am. J. Bot.* **90**, 1560–1566.

Eitoku T., Nakasone Y., Matsuoka D., Tokutomi S. and Terazima M. (2005) *J. Am. Chem. Soc.* **127**, 13238–13244.

Elliot R.C., Platten D., Watson J.C. and Reid J.B. (2004) *J. Plant Physiol.* **161**, 265–270.

Elvin M., Loros J.J., Dunlap J.C. and Heintzen C. (2005) *Genes Dev.* **19**, 2593–2605.

Emi T., Kinoshita T., Sakamoto K., Mineyuki Y. and Shimazaki K. (2005) *Plant Physiol.* **138**, 1615–1626.

Esmon C.A., Tinsley, A.G., Ljung K., Sandberg G., Hearne L.B. and Liscum E. (2006) *Proc. Natl. Acad. Sci. USA* **103**, 236–241.

Federov R., Schlichting I., Hartmann E., Domratcheva T., Fuhrmann M. and Hegemann P. (2003) *Biophys. J.* **84**, 2474–2482.

Ferl R.J. (2004) *Physiol. Plant.* **120**, 173–178.

Folta K.M. and Kaufman L.S. (2003) *Plant Mol Biol.* **51**, 609–618.

Folta K.M., Lieg E.J., Durham T. and Spalding E.P. (2003) *Plant Physiol.* **133**, 1464–1470.

Folta K.M. and Spalding E.P. (2001) *Plant J.* **26**, 471–478.

Friml J., Wisniewska J., Benkova E., Mendgen K. and Palme K. (2002) *Nature* **415**, 806–809.

Froehlich A., Liu Y., Loros J.J. and Dunlap J.C. (2002) *Science* **297**, 815–819.

Fukamatsu Y., Mitsui S., Yasuhara M., Tokioka Y., Ihara N., Fujita S. and Kiyosue T. (2005) *Plant Cell Physiol.* **46**, 1340–1349.

Gallagher S., Short T.W., Ray P.M., Pratt L.H. and Briggs W.R. (1988) *Proc. Natl. Acad. Sci. USA* **85**, 8003–8007.

Geisler M. and Murphy A.S. (2006) *FEBS Lett.* **580**, 1094–1102.

Gong W., Hao B., Mansy S.S., Gonzalez G., Gilles-Gonzalez M.A. and Chan M.K. (1998) *Proc. Natl. Acad. Sci. USA* **95**, 15177–15182.

Haga K., Takano M., Neumann R. and Iino M. (2005) *Plant Cell* **17**, 103–115.

Hager A. and Brich M. (1993) *Planta* **189**, 567–576.

Hager A., Brich M. and Balzen I. (1993) *Planta* **190**, 120–126.

Han L., Mason M., Risseeuw E.P., Crosby W.L. and Somers D.E. (2004) *Plant J.* **40**, 291–301.

Harada A., Sakai T. and Okada K. (2003) *Proc. Natl. Acad. Sci. USA* **100**, 8583–8588.

Harper S.M., Christie J.M. and Gardner K.H. (2004) *Biochemistry* **43**, 16184–16192.

Harper S.M., Neil L.C. and Gardner K.H. (2003) *Science* 301, 1541–1544.

Harper R.M., Stowe-Evans E.L., Luesse D.R., Muto H., Tatematsu K., Watahiki M.K., Yamamoto K. and Liscum E. (2000) *Plant Cell* **12**, 757–770.

He Q., Cheng P., Yang Y., Wang L., Gardner K.H. and Liu Y. (2002) *Science* **297**, 840–843.

He Q. and Liu Y. (2005) *Genes Dev.* **19**, 2888–2899.

Heintzen C., Loros J.J. and Dunlap J.C. (2001) *Cell* **104**, 453–464.

Hoff W.D., Xie A., Van Stokkum I.H., Tang X.J., Gural J., Kroon A.R. and Hellingwerf K.J. (1999) *Biochemistry* **38**, 1009–1017.

Holzer W., Penzkofer A., Fuhrmann M. and Hegemann P. (2002) *Photochem. Photobiol.* **75**, 479–487.

Huala E., Oeller P.W. Liscum E., Han I.-S., Larsen E. and Briggs W.R. (1997) *Science* **278**, 2121–2123.

Huang K. and Beck C.F. (2003) *Proc. Natl. Acad. Sci. USA* **100**, 6269–6274.

Huang K., Merkle T. and Beck C.F. (2002) *Physiol. Plant.* **114**, 613–622.

Idnurm A. and Heitman J. (2005) *PLoS Biol.* **3**, e95.

Iino M. (2001) Phototropism in higher plants In: *Photomovement* (eds D.-P. Häder and M. Lebert), pp. 659–811.

Iino M. (2006) *Curr. Opin. Plant Biol.* **9**, 89–93.

Imaizumi T., Schultz T.F., Harmon F.G., Ho L.A. and Kay S.A. (2005) *Science* **309**, 293–297.

Imaizumi T., Tran H.G., Swartz T.E., Briggs W.R. and Kay S.A. (2003) *Nature* **426**, 302–306.
Inada S., Ohgishi M., Mayama T., Okada K. and Sakai T. (2004) *Plant Cell* **16**, 887–896.
Iwata T., Nozaki D., Tokutomi S., Kagawa T., Wada M. and Kandori H. (2003) *Biochemistry* **42**, 8183–8191.
Iwata T., Tokutomi S. and Kandori H. (2005) *J. Am. Chem. Soc.* **124**, 11840–11841.
Janoudi A.K., Gordon W.R., Wagner D., Quail P. and Poff K.L. (1997) *Plant Physiol.* **113**, 975–979.
Jarillo J.A., Capel J., Tang R.-H., Yang H.-Q., Alonso J.M., Ecker J.R. and Cashmore A.R. (2001a) *Nature* **410**, 487–490.
Jarillo J.A., Gabrys H., Capel J., Alonso J.M., Ecker J.R. and Cashmore A.R. (2001b) *Nature* **410**, 592–594.
Kadota A., Sato, Y. and Wada M. (2000) *Planta* **210**, 932–937.
Kagawa T. (2003) *J. Plant Res.* **116**, 77–82.
Kagawa T., Kasahara M., Abe T., Yoshida S. and Wada M. (2004) *Plant Cell Physiol.* **45**, 416–426.
Kagawa T., Sakai T., Suetsugu N., Oikawa K., Ishiguro S., Kato T., Tabata S., Okada K. and Wada M. (2001) *Science* **291**, 2138–2141.
Kagawa T. and Wada M. (2000) *Plant Cell Physiol.* **41**, 84–93.
Kanegae H., Tahir M., Savazzini F., Yamamoto K., Yano M., Sasaki T., Kanegae T., Wada M. and Takano M. (2000) *Plant Cell Physiol.* **4**, 415–423.
Kasahara M., Kagawa T., Oikawa K., Suetsugu N., Miyao M. and Wada M. (2002a) *Nature* **420**, 829–832.
Kasahara M., Swartz T.E., Olney M.O., Onodera A., Mochizuki N., Fukuzawa H., Asamizu E., Tabata S., Kanegae H., Takano M., Christie J.M., Nagatani A. and Briggs W.R. (2002b) *Plant Physiol.* **129**, 762–773.
Kasahara M., Kagawa T., Sato Y., Kiyosue T. and Wada M. (2004) *Plant Physiol.* **135**, 1388–1397.
Kawai H., Kanegae T., Christensen S., Kiyosue T., Sato Y., Imaizumi T., Kadota A. and Wada M. (2003) *Nature* **421**, 287–290.
Kay C.W., Schleicher E., Kuppig A., Hofner H., Rüdiger W., Schleicher M., Fischer M., Bacher A., Weber S. and Richter G. (2003) *J. Biol. Chem.* **278**, 10973–10982.
Kennis J.T., Crosson S., Gauden M., van Stokkum I.H., Moffat K. and van Grondelle R. (2003) *Biochemistry* **42**, 3385–3392.
Kennis J.T., van Stokkum I.H., Crosson S., Gauden M., Moffat K. and van Grondelle R. (2004) *J. Am. Chem. Soc.* **126**, 4512–4513.
Kinoshita T., Doi M., Suetsugu N., Kagawa T., Wada M. and Shimazaki K. (2001) *Nature* **414**, 656–660.
Kinoshita T., Emi T., Tominaga M., Sakamoto K., Shigenaga A., Doi M. and Shimazaki K. (2003) *Plant Physiol.* **133**, 1453–1463.
Kinoshita T. and Shimazaki K. (1999) *EMBO J.* **18**, 5548–5558.
Kinoshita T. and Shimazaki K. (2001) *Plant Cell Physiol.* **42**, 424–432.
Kinoshita T. and Shimazaki K. (2002) *Plant Cell Physiol.* **43**, 1359–1365.
Kiyosue T. and Wada M. (2000) *Plant J.* **23**, 807–815.
Knieb E., Salomon M. and Rüdiger W. (2004) *Planta* **218**, 843–851.
Knieb E., Salomon M. and Rüdiger W. (2005) *Photochem. Photobiol.* **81**, 177–182.
Kong S.-G., Suzuki T., Tamura K., Mochizuki N., Hara-Nishimura I. and Nagatani A. (2006) *Plant J.*, in press.
Kottke T., Heberle J., Hehn D., Dick B. and Hegemann P. (2003) *Biophys. J.* **84**, 1192–2001.
Krauss U., Losi A., Gärtner W., Jaeger K.E. and Eggert T. (2005) *Phys. Chem. Chem. Phys.* **7**, 2804–2811.
Lariguet P. and Fankhauser C. (2004) *Plant J.* **40**, 826–834.
Lascève G., Leymarie J., Olney M.A., Liscum E., Christie J.M., Vavasseur A. and Briggs W.R. (1999) *Plant Physiol.* **120**, 605–614.
Linden H. and Macino G. (1997) *EMBO J.* **16**, 98–109.
Liscum E. and Briggs W.R. (1995) *Plant Cell* **7**, 473–485.
Liscum E. and Briggs W.R. (1996) *Plant Physiol.* **112**, 291–296.

Liscum E. and Stowe-Evans E.L. (2000) *Photochem. Photobiol.* **72**, 273–282.

Liu Y., He Q. and Cheng P. (2003) *Cell Mol. Life Sci.* **60**, 2131–2138.

Losi A. (2004) *Photochem. Photobiol. Sci.* **3**, 566–574.

Losi A., Ghiraldelli E., Jansen S. and Gärtner W. (2005) *Photochem. Photobiol.* **81**, 1145–1152.

Losi A., Polverini E., Quest B. and Gärtner W. (2002) *Biophys. J.* **82**, 2627–2634.

Losi A., Quest B. and Gärtner W. (2003) *Photochem. Photobiol. Sci.* **2**, 759–766.

Losi A., Ternelli E. and Gärtner W. (2004) *Photochem. Photobiol.* **80**, 150–153.

Lu Y.K., Sun K.H. and Shen W.C. (2005) *Mol. Microbiol.* **56**, 480–491.

Mao J., Zhang Y.C., Sang Y., Li Q.H. and Yang H.Q. (2005) *Proc. Natl. Acad. Sci. USA* **102**, 12270–12275.

Más P., Kim W.-I., Somers D.E. and Kay S.A. (2003) *Nature* **426**, 567–570.

Matsuoka D. and Tokutomi S. (2005) *Proc. Natl. Acad. Sci. USA* **102**, 13337–13342.

Motchoulski A. and Liscum E. (1999) *Science* **286**, 961–964.

Nakasako M., Matsuoka D., Zikihara K. and Tokutomi S. (2005) *FEBS Lett.* **579**, 1067–1071.

Nelson D.C., Lasswell J., Rogg L.E., Cohen M.A. and Bartel B. (2000) *Cell* **101**, 331–340.

Noh B., Bandyopadhyay A., Peer W.A., Spalding E.P. and Murphy A.S. (2003) *Nature* **424**, 999–1002.

Nozue K., Kanegae T., Imaizumi T., Fukada S., Okamoto H., Yeh K.C., Lagarias J.C. and Wada M. (1998) *Proc. Natl. Acad. Sci. USA* **95**, 15826–15830.

Nozue K., Christie J.M., Kiyosue T., Briggs W.R. and Wada M. (2000) *Plant Physiol.* **122**, 1457.

Nozaki D., Iwata T., Ishikawa T., Todo T., Tokutomi S. and Kandori H. (2004) *Biochemistry* **43**, 8373–8379.

Oikawa K., Kasahara M., Kiyosue T., Kagawa T., Suetsugu N., Takahashi F., Kanegae T., Niwa Y., Kadota A. and Wada M. (2003) *Plant Cell* **15**, 2805–2815.

Ohgishi M., Saji K., Okada K. and Sakai T. (2004) *Proc. Natl. Acad. Sci. USA* **101**, 2223–2228.

Onodera A., Kong S.G., Doi M., Shimazaki K., Christie J., Mochizuki N. and Nagatani A. (2005) *Plant Cell Physiol.* **46**, 367–374.

Palmer J.M., Short T.W., Gallagher S. and Briggs W.R. (1993) *Plant Physiol.* **102**, 1211–1218.

Reymond P., Short T.W. and Briggs W.R. (1992a) *Plant Physiol.* **100**, 655–661.

Reymond P., Short T.W., Briggs W.R. and Poff K.L. (1992b) *Proc. Natl. Acad. Sci. USA* **89**, 4718–4721.

Richter G., Weber S., Romisch W., Bacher A., Fischer M. and Eisenreich W. (2005) *J. Am. Chem. Soc.* **127**, 17245–17252.

Rutter J., Michnoff C.H., Harper S.M., Gardner K.H. and McKnight S.L. (2001) *Proc. Natl. Acad. Sci. USA* **98**, 8991–8996.

Sakai T., Kagawa T., Kasahara M., Swartz T.E., Christie J.M., Briggs W.R., Wada M. and Okada K. (2001) *Proc. Natl Acad. Sci. USA* **98**, 6969–6974.

Sakai T., Wada T., Ishiguro S. and Okada K. (2000) *Plant Cell* **12**, 225–236.

Sakamoto K. and Briggs W.R. (2002) *Plant Cell* **14**, 1723–1735.

Salomon M., Christie J.M., Knieb E., Lempert U. and Briggs W.R. (2000) *Biochemistry* **39**, 9401–9410.

Salomon M., Eisenreich W., Dürr H., Schleicher E., Knieb E., Massey V., Rüdiger W., Müller F., Bacher A. and Richter G. (2001) *Proc. Natl. Acad. Sci. USA* **98**, 12357–12361.

Salomon M., Knieb E., von Zeppelin T. and Rüdiger W. (2003) *Biochemistry* **42**, 4217–4225.

Salomon M., Zacherl M., Luff L. and Rüdiger W. (1997a) *Plant Physiol.* **115**, 493–500.

Salomon M., Zacherl M. and Rüdiger W. (1997b) *Bot. Acta* **110**, 214–216.

Salomon M., Zacherl M. and Rüdiger W. (1997c) *Plant Physiol.* **115**, 485–491.

Salomon M., Zacherl M. and Rüdiger W. (1996) *Planta* **199**, 336–342.

Salomon M., Lempert U. and Rüdiger W. (2004) *FEBS Lett.* **572**, 8–10.

Schleicher E., Kowalczyk R.M., Kay C.W., Hegemann P., Bacher A., Fischer M., Bittl R., Richter G. and Weber S. (2004) *J. Am. Chem. Soc.* **126**, 11067–11076.

Schmoll M., Franchi L. and Kubicek C.P. (2005) *Eukaryot. Cell* **4**, 1998–2007.

Schultz T.F., Kiyosue T. Yanofsky M., Wada M., and Kay S.A. (2001) *Plant Cell* **13**, 2659–2670.

Schultz T.F. (2005) The ZEITLUPE family of putative photoreceptors. In: *Handbook of Photosensory Receptors* (eds W.R. Briggs and J.L. Spudich), Wiley-VCH, Weinheim, pp. 337–347.

Schwerdtfeger C. and Linden H. (2000) *Eur. J. Biochem.* **267**, 414–422.

Schwerdtfeger C. and Linden H. (2003) *EMBO J.* **22**, 4846–4855.

Short T.W. and Briggs W.R. (1990) *Plant Physiol.* **92**, 179–185.

Short T.W., Porst M., Palmer J.M., Fernbach E. and Briggs W.R. (1994) *Plant Physiol.* **104**, 1317–1324.

Short T.W., Reymond P. and Briggs W.R. (1993) *Plant Physiol.* **101**, 647–655.

Shrode L.B., Lewis Z.A., White L.D., Bell-Pedersen D. and Ebbole D.J. (2001) *Fungal Genet. Biol.* **32**, 169–181.

Smalle J. and Vierstra R.D. (2004) *Annu. Rev. Plant. Biol.* **55**, 555–590.

Somers D.E., Schultz T.F., Milnamow M., and Kay S. (2000) *Cell* **101**, 319–329.

Somers D.E., Kim W.Y. and Geng R. (2004) *Plant Cell* **16**, 769–782.

Stoelzle S., Kagawa T., Wada M., Hedrich R. and Dietrich P. (2003) *Proc. Natl. Acad. Sci. USA* **100**, 1456–1461.

Stowe-Evans E.L., Harper R.M., Motchoulski A.V. and Liscum E. (1998) *Plant Physiol.* **118**, 1265–1275.

Stowe-Evans E.L., Luesse D.R. and Liscum E. (2001) *Plant Physiol.* **126**, 826–834.

Suetsugu N., Mittmann F., Wagner G., Hughes J. and Wada M. (2005) *Proc. Natl. Acad. Sci. USA* **102**, 13705–13709.

Suetsugu N. and Wada M. (2005) Photoreceptor gene families in lower plants. In: *Handbook of Photosensory Receptors* (eds W.R. Briggs and J.L. Spudich),Wiley-VCH, Weinheim, pp. 349–369.

Swartz T.E., Corchnoy S.B., Christie J.M., Lewis J.W., Szundi I., Briggs W.R. and Bogomolni R.A. (2001) *J. Biol. Chem.* **276**, 36493–36500.

Swartz T.E., Wenzel P.J., Corchnoy S.B., Briggs W.R. and Bogomolni R.A. (2002) *Biochemistry* **41**, 7182–7189.

Swartz T.E. and Bogomolni R.A. (2005) LOV-domain photochemistry. In: *Handbook of Photosensory Receptors* (eds W.R. Briggs and J.L. Spudich), Wiley-VCH, Weinheim, pp. 305–322.

Takemiya A., Inoue S., Doi M., Kinoshita T. and Shimazaki K. (2005) *Plant Cell* **17**, 1120–1127.

Talbott L.D., Shmayevich I.J., Chung Y., Hammad J.W. and Zeiger E. (2003) *Plant Physiol.* **133**, 1522–1529.

Talora C., Franchi L., Linden H., Ballario P. and Macino G. (1999) *EMBO J.* **18**, 4961–4968.

Tatematsu K., Kumagai S., Muto H., Sato A., Watahiki M.K., Harper R.M., Liscum E. and Yamamoto K.T. (2004) *Plant Cell* **16**, 379–393.

Taylor B.L. and Zhulin I.B. (1999) *Microbiol. Mol. Biol. Rev.* **63**, 479–506.

Tepperman J.M., Zhu T., Chang H.S., Wang X. and Quail P.H. (2001) *Proc. Natl. Acad. Sci. USA* **98**, 9437–9442.

Terashima K., Yuki K., Muraguchi H., Akiyama M. and Kamada T. (2005) *Genetics* **171**, 101–108.

Ueno K., Kinoshita T., Inoue S., Emi T. and Shimazaki K. (2005) *Plant Cell Physiol.* **46**, 955–963.

Wada M., Kagawa T. and Sato Y. (2003) *Annu. Rev. Plant Biol.* **54**, 455–468.

Watson J.C. (2000) *Adv. Bot. Res.* **32**, 149–184.

Whippo C.W. and Hangarter R.P. (2003) *Plant Physiol.* **132**, 1499–1507.

Yasuhara M., Mitsui S., Hirano H., Takanabe R., Tokioka Y., Ihara N., Komatsu A., Seki M., Shinozaki K. and Kiyosue T. (2004) *J. Exp. Bot.* **55**, 2015–2027.

Part II Photoreceptor signal transduction

4 Phytochrome-interacting factors

Peter H. Quail

4.1 Introduction

Signal transfer from photoactivated phytochrome (phy) to downstream cellular components logically requires direct interaction of the photoreceptor molecule with one or more primary signalling partners (Quail, 2000; Moller *et al.*, 2002; Schäfer and Bowler, 2002; Gyula *et al.*, 2003; Chen *et al.*, 2004; Quail, 2006a,b). One approach to identifying such phytochrome-interacting factors (PIFs) is to screen or assay for proteins that physically bind to the photoreceptor molecule, using various biochemical or molecular interaction assays. However, because physical interaction alone does not establish functional biological relevance, there is a need to assess the necessity of such PIFs to phy signalling in the cell, using genetic or reverse genetic disruption of the interactor's activity, coupled with the monitoring of visible and/or molecular phenotypes for any discernable perturbations of photoresponsiveness. Over 20 proteins have been reported in the literature to interact with one or more members of the phy family, primarily phyA and phyB. The complexity of the collective pattern of interactions presented by these studies is summarized schematically in the molecular interaction map in Figure 4.1. This chapter examines the methodology and data documenting these interactions, and evaluates the extent to which evidence has been provided that these proteins function as phy signalling intermediates in the living cell.

4.2 Methodology

4.2.1 Initial identification of PIFs

The yeast two-hybrid (Y2H) system (Phizicky and Fields, 1995; Brent and Finley Jr, 1997) has been used both in non-targeted screens of cDNA expression libraries, and in targeted interaction tests with pre-selected proteins to identify phy-interacting proteins. The former permits open-ended identification and cloning of candidate signalling partners from among potentially the full spectrum of cellular proteins expressed in the library, whereas the latter restricts the selection to cloned proteins chosen on the basis of a pre-formulated hypothesis. An alternative approach, not yet reported for the phy system, but likely to increase in prominence, is the proteomic strategy of affinity purification of the phy molecule from cellular extracts, followed by identification of associated proteins by mass spectrometry (Aebersold and Mann, 2003; Kirkpatrick *et al.*, 2005).

Molecular Interaction Map

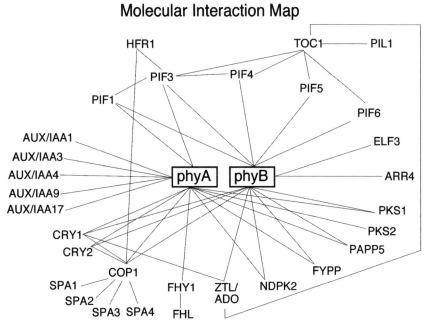

Figure 4.1 Molecular interaction map. Connecting lines depict physical interactions that have been reported between the phy photoreceptor molecules and various putative signalling components.

4.2.2   *Subsequent assay and characterization of the interaction*

The Y2H system has also been used in both plate and liquid-assay configurations to verify, quantify and dissect the molecular interaction between the photoreceptor and candidate interactors. The power and sensitivity of this system for detecting weak interactions is well known, but it is also notorious for generating false positives (Phizicky and Fields, 1995; Brent and Finley, 1997; Serebriiskii *et al.*, 2000). Considerable care is needed to rigorously exclude interactions of questionable biological relevance. A powerful advantage of the phy system in this regard is that it is reasonable (albeit not necessary) to expect that functionally relevant interacting proteins may bind differentially to the Pr and Pfr conformers of the photoreceptor molecule. On the basis of an earlier demonstration that fully photoactive phy can be reconstituted in yeast cells expressing the phy polypeptide by supplying the chromophore exogenously (Li and Lagarias, 1994), a Y2H system has been developed that allows us to test the capacity of candidate interactors to bind the photoreceptor molecule in a red/far-red (R/FR) reversible fashion (Shimizu-Sato *et al.*, 2002).

In vitro molecular and biochemical methods have also been used to assess phy-PIF physical interactions. One commonly used class of these methods involves affinity-matrix-based 'pull-down' or co-precipitation assays. In these assays, the bait protein (e.g. phy) is immobilized on an insoluble matrix (bead) via an antibody or affinity ligand (e.g. nickel) directed at the native, or epitope-tagged, bait protein, and the loaded matrix is mixed with the prey proteins and is pelleted (Ni *et al.*,

1999; Khanna *et al.*, 2004). The presence, absence and quantity of prey molecules physically bound to the bait are then determined by methods that usually include either Western blots using antibodies to the native or epitope-tagged prey protein or measurement of pre-incorporated radioactive label. In each case, the capacity for differential binding of the Pr and Pfr forms of the phy molecule to the candidate PIF is tested. These assays have been used to detect interactions between pairs of recombinant proteins synthesized in cell-free systems or *Escherechia coli* and proteins synthesized in the plant, by co-precipitation from crude tissue extracts. Although interactions detected by co-precipitation from crude plant extracts are frequently interpreted as demonstrating '*in vivo*' association (i.e. an interaction that exists before the cell is ruptured and retained in the extract), it should be noted that the procedure does not exclude the possibility of an artifactual, post-homogenization association. Such post-homogenization binding of phy molecules to other cellular components has a long history in the phy field (Quail, 1975).

A second class of *in vitro* biochemical interaction assay that has been reported involves the measurement of an enzymatic activity associated with the phy-interactor pair, monitored either as a phosphotransfer reaction (Fankhauser *et al.*, 1999) or an alteration of an intrinsic catalytic activity in the interactor protein (Choi *et al.*, 1999). A third class of assay that has been employed involves the use of cytochemical or biophysical methods to monitor colocalization and/or physical interaction within the plant cell. This approach involves the coexpression of phy and interactor proteins in transgenic plants, each fused to a different green fluorescent protein (GFP)-variant that emits a different wavelength of fluoresced light. Fluorescence microscopy is then used to assess the relative localization of the two molecules in the living cell. Co-emission of both fluoresced wavelengths from the same apparent physical position in the cell is interpreted as evidence of *in vivo* colocalization (Bauer *et al.*, 2004). However, this procedure has insufficient resolution to evaluate physical interaction between the two proteins. For this purpose, fluorescence resonance energy transfer procedure is employed (Mas *et al.*, 2000). By this method, closely associated molecules are detected by transfer of excitation energy absorbed by one of the fluorescent protein pair (shorter wavelength emission) to the other (longer wavelength emission) one, causing the latter to fluoresce in response to indirect photoexcitation.

4.2.3 Reverse genetic assessment of functional relevance to phy signalling

The availability of strategies and community resources providing targeted disruption of the expression of any gene of interest has increased dramatically over the last few years, especially in *Arabidopsis*. These include T-DNA and transposon insertion, antisense and RNAi expression, and Tilling or Delete-a-Gene technology (Henikoff and Comai, 2003). This increased access to mutant collections with near genome-wide coverage, especially that of the SALK collection from the Ecker laboratory (Alonso *et al.*, 2003), is reflected in the increasingly routine phenotypic analysis of such mutants for aberrant photoresponsiveness in order to examine the necessity of putative phy signalling partners to the regulatory activity of the photoreceptor in the cell. A large number of studies from multiple laboratories have used the seedling-de-etiolation process in *Arabidopsis* as a model system for this purpose. Consequently, much of the information we currently have about the molecular and cellular basis of the signalling process has come from such studies. Because light-induced de-etiolation in wild-type seedlings involves concomitant, reciprocal responses in hypocotyl cells (inhibition of longitudinal expansion) and cotyledon cells (stimulation of expansion), this behaviour provides a clear diagnostic, visible phenotypic marker of the normal photomorphogenic process (Quail, 2002a). Disruption of early steps in the normal photosensory perception or signalling pathways can be anticipated to perturb the photoinduced expansion of these two cell types reciprocally. Thus, mutations causing a perturbation specific to light signalling events (Halliday *et al.*, 1999; Huq *et al.*, 2000) can be readily distinguished from others that more globally or non-specifically affect cell expansion responses per se (Okamoto *et al.*, 2001; Ullah *et al.*, 2001). For example, mutationally induced, global inhibition of cell expansion will produce light-grown seedlings with shorter hypocotyls and smaller cotyledons than wild type, as distinct from the shorter hypocotyls and larger cotyledons than wild type, expected of light-signalling specific mutations. Awareness of this distinction is critical, because without further analysis (Ullah *et al.*, 2001; Jones *et al.*, 2003) a short-hypocotyl phenotype can be erroneously interpreted as indicating direct involvement of the mutated component in the normal phy signalling process.

4.3   phỳ-interactors

4.3.1   PIF3

PIF3 was originally identified in a Y2H screen, using the non-chromophoric C-terminal domain of phyB as bait (Ni *et al.*, 1998). However, subsequent *in vitro* pull-down experiments with recombinant proteins showed that (a) PIF3 binds with much higher affinity to the full-length and to the isolated, chromophoric, N-terminal domain of the chromophore-conjugated, photoactive photoreceptor than to the C-terminal domain, (b) this binding is photoreversibly specific to the Pfr conformer and (c) PIF3 binds to the Pfr form of both phyA and phyB, but with higher apparent affinity for phyB (Ni *et al.*, 1999; Zhu *et al.*, 2000; Huq and Quail, 2002). PIF3 was identified (Ni *et al.*, 1998) as a member of the 162-member basic helix-loop-helix (bHLH) transcription factor family of *Arabidopsis* (Bailey *et al.*, 2003; Toledo-Ortiz *et al.*, 2003). It was shown to be constitutively nuclear and to bind in sequence-specific fashion to a G-box DNA core motif (CACGTG) present in numerous light-responsive gene promoters, and phyB was shown to bind to the DNA-bound PIF3 molecule specifically and reversibly upon photoconversion to the active Pfr conformer (Martínez-García *et al.*, 2000). Other work demonstrated that phy molecules are induced to translocate from the cytoplasm into the nucleus upon Pfr formation (Nagy and Schafer, 2002; Nagatani, 2004), and that this translocation is necessary for phy regulatory activity *in vivo* (Huq *et al.*, 2003). Collectively, these data were interpreted to indicate the potential existence of a direct signalling pathway from the photoreceptor to target genes, whereby light-induced Pfr formation leads to rapid translocation into the nucleus, where it binds to promoter-bound PIF3 and alters the transcription of target genes (Martínez-García *et al.*, 2000; Tepperman *et al.*, 2001; Quail, 2002a,b).

In the absence of known insertional knockout mutants of PIF3 at the time, the evidence that PIF3 was functionally necessary for phy signalling was derived from the aberrant photoresponsive visible phenotypic behaviour and gene expression pattern of antisense-*PIF3*-expressing transgenic *Arabidopsis* lines, primarily a line designated A22. This line exhibited long hypocotyls (Ni *et al.*, 1998) and reduced induction of a subset of rapidly photoresponsive genes (in particular *CCA1* and *LHY*) in response to light signals, and PIF3 was shown to bind to the G-box element present in the promoters of these genes (Martínez-García *et al.*, 2000). The robust nature of the visible hyposensitive phenotype was interpreted to indicate that PIF3 functions positively and pleiotropically in transducing light signals to the genes that drive seedling de-etiolation and the circadian clock (Martínez-García *et al.*, 2000; Quail, 2002a,b). Moreover, it was speculated that the molecular mechanism by which this signalling might occur could involve the phy molecule functioning as an integral light-switchable component of transcriptional regulatory complexes directly at the promoters of light-responsive genes (Quail, 2002a,b).

Recent studies with bona fide knockout *pif3* mutants have substantially altered important aspects of this model and have provided exciting new insights into the primary mechanism of phy signalling. Three laboratories have reported that *pif3* mutants exhibit shorter hypocotyls than wild type when grown in prolonged continuous red (Rc) (Halliday *et al.*, 1999; Kim *et al.*, 2003a; Bauer *et al.*, 2004; Monte *et al.*, 2004), in direct contrast to the long-hypocotyl phenotype of the original A22 *PIF3*-antisense line (Ni *et al.*, 1998). This phenotype of the A22 line now appears to be due to an inadvertent mutation at a locus other than *PIF3* (E. Monte and P. Quail, unpublished). These data have been interpreted as indicating that PIF3 acts negatively in regulating this visible phenotype (Kim *et al.*, 2003; Bauer *et al.*, 2004; Duek and Fankhauser, 2005) rather than being a positive regulator necessary for phy-induced de-etiolation, as initially concluded (Ni *et al.*, 1998; Halliday *et al.*, 1999; Quail, 2002a,b). This reassessment also includes a reinterpretation of the phenotype of the *poc1* mutant, identified previously in a forward genetic screen as carrying a T-DNA insertion in the promoter region of the *PIF3* gene (Halliday *et al.*, 1999). This mutant exhibited a short-hypocotyl phenotype in prolonged Rc, but this was initially interpreted as being caused by mutagenically induced overexpression of PIF3, rather than disruption of expression, as now appears to be the case (Bauer *et al.*, 2004; Monte *et al.*, 2004). It is noteworthy that, although all authors report a short hypocotyl in these *pif3* mutants, concomitant enhancement of cotyledon expansion has not been consistently observed in response to prolonged Rc (Halliday *et al.*, 1999; Kim *et al.*, 2003; Bauer *et al.*, 2004; Monte *et al.*, 2004), raising the concern that the evidence for PIF3 function as a mediator of phy action in this longer term phenotype is not robustly reproducible. Nevertheless, it is clear that the original conclusion that PIF3 functions pleiotropically as a centrally positioned mediator of the global, phy-induced, seedling de-etiolation process was in error.

Instead, PIF3 appears to have a more specialized role in mediating phy-induced regulation of rapidly photoresponsive genes encoding chloroplast-targeted protein products. This conclusion is based on a genome-wide, microarray-based, expression-profiling study of the *pif3* mutant, showing that the majority of a subset of genes that are induced within 1 h of exposure of dark-grown seedlings to Rc, and are dependent on PIF3 for this induction, encode such plastid-destined polypeptides (Monte *et al.*, 2004). These include *SIGE*, a regulatory subunit of the chloroplast RNA polymerase, which could have a central function in phy-regulated plastid-genome transcription (Monte *et al.*, 2004). These data thus provide evidence that PIF3 has a critical positive function in early phy-induced chloroplast biogenesis at the initiation of the de-etiolation process upon first exposure of dark-grown seedlings to light. On the other hand, PIF3 does not appear to participate as a pivotal mediator of the phy-regulated expression of a diverse, master set of transcription-factor genes defined in microarray experiments as previously hypothesized (Tepperman *et al.*, 2001).

Of more profound importance to the ultimate understanding of the molecular mechanism of phy action was the discovery by Bauer *et al.* (2004) that light induces rapid degradation of the nuclear-localized PIF3 protein in a manner redundantly dependent on phyA, phyB and phyD, upon initial irradiation of dark-grown seedlings. This basic observation has been subsequently confirmed by others (Figure 4.2) and the degradation shown to be inhibited by MG132, indicating that degradation is likely mediated via the 26S-proteosome system (Monte *et al.*, 2004; Park *et al.*, 2004). Preliminary evidence suggesting light-induced ubiquitination of PIF3 has also been presented (Park *et al.*, 2004). However, because of the omission of a critical unirradiated control from these experiments, the evidence is not rigorous and must therefore be confirmed. Using phyB-CFP (cyan fluorescent protein) and PIF3-YFP (yellow fluorescent protein) that coexpressed in transgenic *Arabidopsis*, Bauer *et al.* (2004) demonstrated that these two molecules undergo rapid, light-induced colocalization into subnuclear foci, referred to as speckles, suggesting that they may interact directly in the nucleus. Recent data support this suggestion and provide evidence that photoactivation of the phy molecule induces rapid, intranuclear phosphorylation of PIF3, in a manner dependent on this direct interaction, and that this modification tags the transcription factor for degradation via the ubiquitin-proteosome system (UPS), possibly localized in the nuclear speckles (Al-Sady *et al.*, 2006). The twin questions of whether transphosphorylation of PIF3 (and possibly other target proteins) is the primary biochemical mechanism of signal transfer from the activated photoreceptor to its signalling partners and whether the phy molecule itself is, or is a subunit of, the protein kinase responsible are currently the focus of intense research interest. The signalling-initiated, UPS-mediated degradation of primary transduction components is emerging as a widely utilized general mechanism across many of the major plant signalling systems (Sullivan *et al.*, 2003; Moon *et al.*, 2004; Dharmasiri *et al.*, 2005; Hoecker, 2005; Huq, 2006).

The rapid degradation of PIF3 could indicate that this factor functions only transiently at the initial dark-to-light transition experienced by etiolated seedlings, with a primary role in regulating genes necessary for chloroplast biogenesis. In principle, it is possible that PIF3 functions constitutively in dark-grown seedlings, either as a

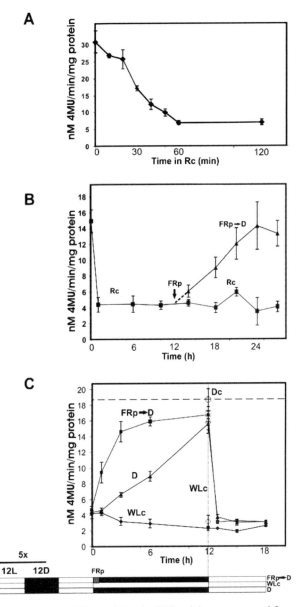

Figure 4.2 Light-regulation of PIF3 protein levels. GUS activity was measured fluorometrically in extracts of transgenic *Arabidopsis* seedlings expressing GUS-PIF3 fusion protein driven by the constitutive 35S CaMV promoter. (A) Rapid Rc-induced degradation of GUS-PIF3 in 4-day, dark-grown seedlings transferred to Rc (10 μmol m^{-2} s^{-1}) for 2 h. (B) Rc-induced GUS-PIF3 degradation and reaccumulation in darkness after a far-red pulse (FRp → D) in seedlings grown for 4 days in the dark before transfer to Rc. (C) phy-regulated PIF3 protein levels in green seedlings grown for 5 days in white-light/dark (L/D) diurnal cycles (12:12) before transfer to 12-h darkness with (FRp → D) or without (D) a preceding FRp, or continued maintenance in continuous white light (WLc) and subsequent exposure to WLc again at 12 h for all treatments. Open symbols at the 12-h time-point are from an identical parallel experiment, which included control seedlings maintained in continuous darkness throughout (Dc; open square) (Monte *et al.*, 2004).

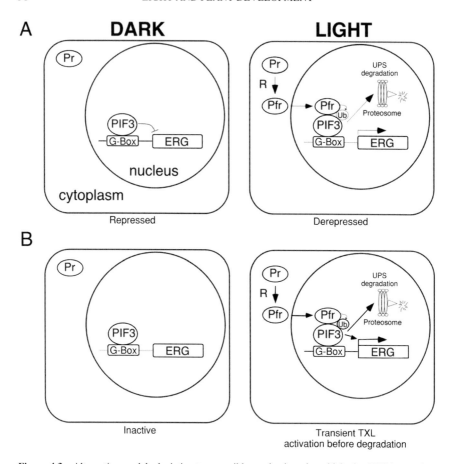

Figure 4.3 Alternative models depicting two possible mechanisms by which phy-PIF3 interactions might regulate target gene expression in response to light signals. (A) PIF3 binds to G-box motifs in the promoters of target genes and represses expression in darkness. Light triggers Pfr formation and translocation into the nucleus where the photoreceptor binds to PIF3, inducing its ubiquitination and degradation via the UPS system, thereby derepressing expression of the target gene. (B) PIF3 bound to target gene promoters is inactive in regulating expression in darkness. Light-triggered Pfr formation, nuclear translocation and binding to PIF3 induces ubiquitination of the bHLH protein, which both activates the transcription factor and flags it for proteosomal degradation, thereby inducing transient transcriptional activation of the target gene. ERG = early-response gene; Ub = ubiquitin.

positive regulator of genes necessary for skotomorphogenesis or as a negative regulator of genes necessary for photomorphogenesis, and that light-induced degradation reverses this activity (Figure 4.3A). However, the absence of a visible phenotype or significant perturbation of gene expression profiles in dark-grown, *pif3*-null mutants (Monte *et al.*, 2004) argues against this possibility, unless there is functional redundancy for this activity in darkness. An interesting alternative possibility is that the phy-induced PIF3 phosphorylation and/or ubiquitination transiently activate the transcriptional activity of the bHLH molecule prior to degradation (Figure 4.3B). There is emerging evidence for such mechanisms of transient activation of

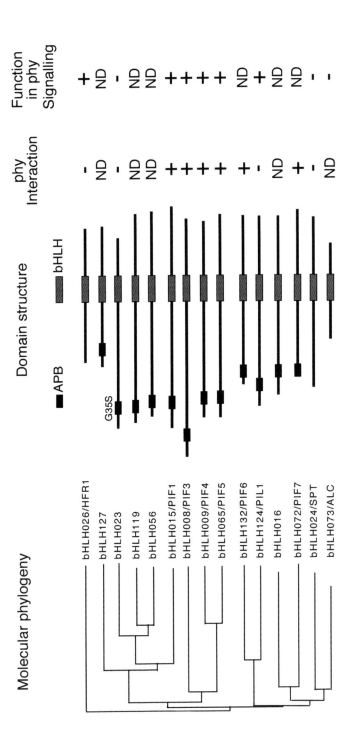

Figure 4.4 Several *Arabidopsis* Subfamily-15 bHLH proteins are phy-interactors and/or are involved in phy signalling. Molecular phylogeny and domain structure of PIF3-related bHLH proteins (Khanna *et al.*, 2004). The proteins are designated according to their bHLH numbers (Bailey *et al.*, 2004) and synonyms where applicable. Schematic representations of the polypeptides, aligned at their conserved bHLH domains, are depicted, and the presence of the APB (active phyB binding) motif, conserved in multiple members, is indicated, including the G35S substitution of the invariant G in bHLH023. Pfr-specific interaction with one or more phys is indicated as (+), and lack of interaction as (−). Evidence of functional involvement in phy signalling *in vivo* is indicated as (+), and lack of such evidence as (−). ND = not determined.

transcription factors en route to UPS-mediated degradation in a variety of eukaryotic systems (Lipford and Deshaies, 2003; Lipford *et al.*, 2005). This mechanism predicates sustained transcriptional activity on repeated or constant signal input and transcription factor replenishment, thereby permitting rapid adjustment to fluctuating or altered signal input. There is evidence that the light-induced degradation of PIF3 results rapidly in a new, lower steady-state level of the protein in sustained light, such as that experienced during a normal day–night cycle, and that degradation ceases rapidly upon Pfr removal and return of plants to darkness, with consequent reaccumulation to high levels over a 12-h night period (Monte *et al.*, 2004) (Figures 4.2B and 4.2C). This rapidly reversible, phy-induced, dynamic regulation of PIF3 levels suggests that, rather than acting only briefly and transiently during the initial phases of seedling de-etiolation, PIF3 remains potentially functionally important in fully green seedlings. This activity may account for the phenotype of *pif3*-null mutants observed under prolonged Rc irradiations (Halliday *et al.*, 1999; Kim *et al.*, 2003; Bauer *et al.*, 2004; Monte *et al.*, 2004).

4.3.2   Other bHLH transcription factors and the active phyB binding domain

Following the identification of PIF3 as a potential phy signalling partner, a comprehensive bioinformatics analysis of the emerging *Arabidopsis* genome sequence revealed the presence of 162 predicted bHLH genes, the second largest transcription factor family in the genome (Bailey *et al.*, 2003; Toledo-Ortiz *et al.*, 2003). Phylogenetic analysis showed that 14 of these predicted protein sequences cluster with PIF3 in a subclade, designated Subfamily 15 (Toledo-Ortiz *et al.*, 2003) (Figure 4.4). This sequence-relatedness to PIF3 has prompted examination of the remaining members of Subfamily 15 for involvement in phy signalling. Two of these factors, SPT (spatula) and ALC, identified independently in genetic screens as functioning in gynoecium development (Heisler *et al.*, 2001; Rajani and Sundaresan, 2001), appear to have no direct involvement in photomorphogenesis (Khanna *et al.*, 2004) and are not discussed further here. Of the remainder, although the extent of the evidence varies, there are indications that eight members of the subfamily (designated PIF1, PIF3, PIF4, PIF5, PIF6, PIF7, HFR1 and PIL1) appear to have some degree of activity in photomorphogenesis, including the six designated as PIFs, which have been shown to interact physically with one or more phys, whereas the other five either have no apparent activity in photomorphogenesis (bHLH023, which does not bind to the photoreceptor) or are still being investigated (bHLH127, bHLH119, bHLH056 and bHLH016).

Interestingly, several of these factors, PIF4, PIF1, PIL1 (phytochrome-interacting factor 3-like 1) and HFR1 (long hypocotyl in far-red), were identified through more than one line of investigation. PIF4 was identified separately in a forward genetic screen for mutants hypersensitive to Rc in the de-etiolation process and in a Y2H screen for factors that interact with PIF3 (Huq and Quail, 2002). The *pif4*-null mutant (originally designated *srl2*) exhibits shorter hypocotyls and larger cotyledons than wild type in prolonged Rc, suggesting that this factor acts negatively in phy-induced de-etiolation (Huq and Quail, 2002). Like PIF3, the PIF4 protein binds selectively to the Pfr form of phyB, but with lower affinity than PIF3, and exhibits little detectable binding to phyA. PIF4 is constitutively nuclear, heterodimerizes with PIF3 and binds as a homodimer and heterodimer to the G-box DNA motif (Toledo-Ortiz *et al.*, 2003). This observation, together with the heterodimerization of other bHLHs with PIF3 (see below) raises the possibility of combinatorial amplification of the number of configurations in which bHLH family members might participate in regulating photomorphogenesis (Quail, 2000; Toledo-Ortiz *et al.*, 2003).

PIF1 was isolated in a Y2H screen for PIF3 interactors, in addition to being targeted for reverse genetic analysis because of its close homology to PIF3 (Huq *et al.*, 2004; Oh *et al.*, 2004). Like PIF3 and PIF4, PIF1 is constitutively nuclear and binds selectively and photoreversibly to the Pfr form of phyB. However, in addition, and in contrast to PIF3 and PIF4, PIF1 also binds robustly to the Pfr form of phyA (Huq *et al.*, 2004). This strong binding to the active form of both phyA and phyB makes PIF1 unique among the bHLHs thus far examined. PIF1 appears to function in dark-grown seedlings to suppress accumulation of excess levels of protochlorophyllide, which become potentially lethal upon first exposure of the seedlings to light, thereby suggesting a critical role of this factor in seedling survival and early competitiveness upon emergence from subterranean darkness following natural soil germination of seeds (Huq *et al.*, 2004). PIF1 (also called PIL5) also acts in darkness to suppress seed germination and this activity is reversed by light, mediated by one or more phys, leading to germination (Oh *et al.*, 2004). Together, these data suggest that PIF1 may function as a repressor of certain aspects of photomorphogenesis in darkness and that phyA and phyB repress this activity upon photoactivation. Consistent with this notion, recent evidence indicates that, like PIF3, the phy system induces rapid, UPS-mediated degradation of the PIF1 protein in response to exposure of dark-grown seedlings to light (Shen *et al.*, 2005). This observation raises the possibility that the phys may target multiple bHLH family members for light-induced proteolysis via a mechanism similar to that for PIF3 (Huq, 2006).

Sequence alignments of the *Arabidopsis* bHLH proteins revealed the presence of a conserved motif in the N-terminal region of 12 of the 15 Subfamily-15 members and absence from all other members of the superfamily (Khanna *et al.*, 2004) (Figure 4.4). *In vitro* protein-interaction assays showed that six of the members containing this motif (PIF1, PIF3, PIF4, PIF5, PIF6 and PIF7) bind selectively and reversibly to the Pfr form of phyB, whereas the three other members thus far tested that do not bind to phyB either do not contain this motif (SPT) or naturally lack one or more of the otherwise invariant residues in the motif (HFR1 and bHLH023) (Fairchild *et al.*, 2000; Khanna *et al.*, 2004; P. Leivar, E. Monte and P. Quail, unpublished). Targeted substitution mutagenesis of the invariant amino acids in the active motif eliminated phyB binding to the full-length bHLH protein, and interaction assays with the isolated motif segment showed that the photoactivated receptor can bind in conformer-specific fashion to this peptide sequence alone (Khanna *et al.*, 2004). These data establish that this motif, designated APB (active phyB binding) motif, is both necessary and sufficient for Pfr-specific binding of phyB to a subset of the bHLH PIFs. Conversely, the absence or non-conserved variants of the motif in SPT, HFR1 and bHLH023 appear to account for the lack of phy-binding to these factors. The functional relevance of the APB domain to PIF4 activity *in vivo* has been established by the failure of mutant PIF4 protein, carrying targeted, non-phyB-binding, APB substitutions, to rescue the *pif4* mutant when expressed transgenically (Khanna *et al.*, 2004). PIL1, on the other hand, is the single deviant from this general pattern in that it contains an apparently conserved APB domain but does not bind robustly to either phyA or phyB (Khanna *et al.*, 2004). The reason for this observation is as yet unknown. The potential phy-binding activities of the remaining members of Subfamily 15 are yet to be reported.

Independently of its sequence-relatedness to PIF3, PIL1 was identified as a factor of interest in phy signalling based on the striking phy-regulated expression of the *PIL1* gene. Early microarray-based expression profile studies identified *PIL1* as displaying rapid and robust repression of expression in response to initial exposure of dark-grown seedlings to Rc or FRc (continuous far-red) light (Tepperman *et al.*, 2001, 2004). Conversely, in a separate microarray study, *PIL1* expression rapidly increased in light-grown plants upon exposure to shade-avoidance conditions (Salter *et al.*, 2003). Together these data suggest that the Pfr form of the photoreceptor acts to repress *PIL1* gene expression from initially high levels in etiolated seedlings, but that rapid derepression occurs upon Pfr removal or reduction in level imposed by the FR-rich irradiation of vegetative shade. This regulation appears to be highly dynamic and reversible in response to shade conditions and to interact with the circadian clock in light-grown plants (Salter *et al.*, 2003; Yamashino *et al.*, 2003). This behaviour of the *PIL1* gene contrasts with that of other bHLH gene-family members, including *PIF3*, which appears to be constitutively expressed at the transcriptional level, *PIF4*, which is rapidly induced by both Rc and FRc (Huq and Quail, 2002), and *HFR1*, which is induced by FRc, but repressed by Rc (Fairchild *et al.*, 2000; Sessa *et al.*, 2005). This diversity is indicative of a complex, multilevel regulatory network involved in the control of the levels of the Subfamily-15 bHLH proteins by the phy family.

HFR1, the most divergent member of the Subfamily-15 bHLHs, was identified in three independent forward genetic screens as acting positively and specifically in phyA signalling under FRc (Fairchild *et al.*, 2000; Fankhauser and Chory, 2000; Soh *et al.*, 2000). This factor is constitutively nuclear, but lacks the normal basic region of the bHLH domain, suggesting that it may not be able to bind to DNA, or may recognize a DNA motif divergent from the G-box recognized by other Subfamily 15 members (Fairchild *et al.*, 2000; Toledo-Ortiz *et al.*, 2003). Because HFR1 can heterodimerize with PIF3, it may function to inhibit or change the DNA-binding-site specificity of other bHLHs (Fairchild *et al.*, 2000). In addition to photoregulation of *HFR1* at the transcriptional level, as mentioned above, HFR1 protein levels are regulated at the protein level through controlled degradation. However, in contrast to PIF3 and PIF1, HFR1 is maintained at low levels in the dark through UPS-mediated degradation, and induced to accumulate to high levels in the light by abrogation of this process (Duek *et al.*, 2004). The data indicate that HFR1 is constitutively phosphorylated in darkness, that the constitutively photomorphogenic 1 (COP1) E3-ligase binds to and ubiquitinates HFR1, targeting it for proteosomal degradation, and that COP1 preferentially recognizes the phosphorylated form of HFR1 in a manner that requires the N-terminal 45 amino acids of HFR1 (Duek *et al.*, 2004; Jang *et al.*, 2005; Yang *et al.*, 2005). The mechanism by which the phys abrogate the extant, COP1-mediated degradation of HFR1 is as yet unknown. However, the HFR1 protein was found not to bind to phyA or phyB (Fairchild *et al.*, 2000), suggesting a mechanism not requiring direct interaction with the photoreceptor may be involved.

4.3.3   Nucleoside diphosphate kinase 2

The enzyme nucleoside diphosphate kinase 2 (NDPK2) was initially isolated in a Y2H screen using the non-chromophoric C-terminal domain of *Arabidopsis* phyA as bait (Choi *et al.*, 1999). This protein, which appears to localize to both cytoplasm and nucleus, was subsequently shown, by *in vitro* cross-linking and pull-down experiments, to interact with the Pfr form of biochemically purifed oat phyA protein at a higher level than the Pr form (Choi *et al.*, 1999; Im *et al.*, 2004; Kim *et al.*, 2004). In addition, the Pfr form of the oat phy in these preparations enhanced the intrinsic enzymatic activity of the NDPK2 protein 1.7-fold when co-incubated *in vitro*, whereas the Pr form had no detectable effect on this activity (Choi *et al.*, 1999; Kim *et al.*, 2004). More recent data show that dCDP strongly enhances the selective binding to the Pfr form, and that NDPK2 can bind to phyB as well as phyA (Shen *et al.*, 2005). These data indicate that the *Arabidopsis* NDPK2 protein is capable of physical interaction with both phyA and phyB in a conformer-selective fashion and that Pfr induces a relatively small but significant enhancement of the intrinsic gamma-phosphate-exchanging enzymatic activity of the protein. A recent analysis has examined the biochemical basis of this phenomenon in some detail, showing that Pfr binding alters the pKa for a critical His residue in the catalytic site of the enzyme (Shen *et al.*, 2005). Collectively, the data are consistent, therefore, with the interaction being molecularly selective. A recent report presents evidence that artificial, *in vitro* phosphorylation of S598 of oat phyA by exogenously added protein kinase A reduces the interaction of this photoreceptor with NDPK2 (as well as PIF3) (Kim *et al.*, 2005). However, the relevance of this observation to mechanisms of phy signalling is unclear, as this Ser residue is not conserved in most other phyA (or other phy) proteins thus far sequenced (including *Arabidopsis* phyA).

An *ndpk2* T-DNA insertional mutant displayed reduced sensitivity to both Rc and FRc as regards hook opening and cotyledon separation, but little or no perturbation of hypocotyl responsiveness (Choi *et al.*, 1999). These data suggest a possible positive functional role in phyA and phyB signalling in a subset of de-etiolation response parameters. However, little additional information has been presented since this initial study, and the molecular function of NDPK2 in phy signalling remains unclear.

4.3.4   Phytochrome kinase substrate 1

Phytochrome kinase substrate 1 (PKS1) is a novel, constitutively cytoplasmic protein that was isolated in an early Y2H screen using the 160 amino acids at the C-terminus of *Arabidopsis* phyA as bait (Fankhauser *et al.*, 1999). Subsequent *in vitro* binding studies showed that PKS1 could bind to full-length phyA and phyB proteins, but with no difference in apparent affinity for either Pr or Pfr conformers, nor for the apoprotein. By contrast, biochemically purified preparations of recombinant oat phyA catalyzed *in vitro* phosphorylation of the N-terminal half (215 residues) of PKS1 on Ser and Thr residues at a level 2.14-fold higher with the phyA present as Pfr than as Pr (Fankhauser *et al.*, 1999). In addition, evidence was presented for the presence of a phosphorylated species of PKS1 in *Arabidopsis* seedlings grown in prolonged Rc that was absent from dark-control seedlings. These data are consistent with the attractive proposal that the phy molecule is itself an autonomous, light-activated protein kinase that phosphorylates PKS1 preferentially in the Pfr form, and that this transphosphorylation of substrates, such as PKS1, may represent the primary biochemical mechanism of signal transfer from the photoreceptor to its targets. This proposal is strengthened by the clear molecular phylogenetic evidence in recent years that the eukaryotic plant phys have evolved from prokaryotic progenitors that are canonical two-component His-kinases (Bhoo *et al.*, 2001; Montgomery and Lagarias, 2002). However, as has been discussed (Quail, 2000, 2002a,b, 2006a,b), rigorous molecular genetic evidence in support of this general proposal is still lacking for the plant phys, there is contrary evidence indicating that the putative kinase domain of the phy molecule is dispensable for seedling de-etiolation (Krall and Reed, 2000; Matsushita *et al.*, 2003), and the potential functional role of PKS1 in phy signalling is yet to be directly assessed.

   Recent evidence from a study with a *pks1* mutant suggests that PKS1 functions in conjunction with a related protein, PKS2, in a phyA-mediated very low fluence mode to provide homeostasis to phyA signalling (Lariguet *et al.*, 2003). The amino acid sequence of the PKS1 protein does not appear to provide insight into its molecular function. However, the constitutively cytoplasmic localization of the protein, coupled with earlier evidence that overexpressed PKS1 appeared to act negatively in phyB signalling, has led to the suggestion that PKS1 may function to anchor the phy molecule in the cytoplasm in the Pr form, with Pfr-induced phosphorylation leading to release of the photoreceptor for translocation into the nucleus (Fankhauser *et al.*, 1999; Fankhauser, 2000).

4.3.5   Type 5 protein phosphatase

An *Arabidopsis* Type 5 serine/threonine protein phosphatase, designated type 5 protein phosphatase (PAPP5), was recently isolated in a Y2H screen using the full-length *Arabidopsis* PHYA apoprotein as bait, and subsequently shown by a variety of *in vitro* interaction assays to interact *in vitro* with the *Arabidopsis* PHYA and PHYB apoproteins, and in partially Pfr-selective fashion with both biochemically purified oat phyA and a transgenically expressed phyB-GFP fusion protein in cell extracts (Ryu *et al.*, 2005). In addition, transient coexpression of PAPP5-CFP with transgenically expressed phyB-YFP provided evidence that the subcellular localization of PAPP5 followed that of phyB, being initially cytoplasmic in darkness, but in the nucleus colocalized with phyB in nuclear speckles after transfer to white light for 15 h. Data were also presented that PAPP5 was able to dephosphorylate pre-phosphorylated oat phyA *in vitro*. A series of further experiments involving *in vivo* spectrophotometric measurements of *Arabidopsis* phy levels, and *in vitro* binding of NDPK2 to oat phyA, were interpreted to suggest that PAPP5 controls the flux of light information to downstream photoresponses through regulation of phy stability and binding affinity towards NDPK2 (Ryu *et al.*, 2005). However, the spectrophotometric data cited in support of these conclusions are less than robust, and parallel Western blot analysis of phy protein levels were not presented. In addition, because, as mentioned above, the phosphorylatable serine (S598) shown to be critical to NDPK2 binding affinity towards oat phyA is lacking in all *Arabidopsis* phys, the relevance of the PAPP5 identified here to any phy signalling through NDPK2 in *Arabidopsis* remains to be established. Thus, although evidence is presented that *papp5* mutants of *Arabidopsis* exhibited reduced photoresponsiveness in various phy-regulated processes (Ryu *et al.*, 2005), the mechanism by which PAPP5 may participate in these responses remains to be clarified.

4.3.6   *Protein phosphatase 2A*

A protein phosphatase 2A, designated FyPP, was isolated in a Y2H screen of a pea cDNA library using the C-terminal domain of *Arabidopsis* phyA as bait (Kim *et al.*, 2002). Subsequent *in vitro* interaction assays revealed no compelling differential affinity of FyPP for the Pr and Pfr forms of phyA and phyB (<1.3-fold). However, pre-phosphorylated oat phyA was dephosphorylated more rapidly in the Pfr than the Pr form by the recombinant pea FyPP, suggesting differential recognition of the two phy conformers by the enzyme. An *Arabidopsis* mutant null for AtFyPP3 was shown to flower early in long days (Kim *et al.*, 2002), but a direct link between the *in vitro* measured enzymatic activity of the pea FyPP towards the oat phyA molecule and phy-regulated flowering in *Arabidopsis* remains to be demonstrated.

4.3.7   *Early flowering 3*

Mutants at the *ELF3* locus were initially identified in a forward genetic screen for early flowering mutants in *Arabidopsis* and later shown to have reduced photoresponsiveness to Rc and FRc during seedling de-etiolation, and to be involved in regulation of the circadian clock (Reed *et al.*, 2000; Liu *et al.*, 2001). The early flowering 3 (ELF3) protein has a novel sequence, and cell fractionation data suggest that it is nuclear localized (Liu *et al.*, 2001). A Y2H screen with ELF3 as bait yielded the C-terminal domain of PHYB as an interactor, and *in vitro* pull-down assays showed that the ELF3 protein is capable of binding to full-length recombinant phyB produced in yeast (Liu *et al.*, 2001). However, no difference in binding to the Pr and Pfr forms of the photoreceptor were detected, and the functional significance of this physical interaction to phyB activity in regulating flowering and seedling photomorphogenesis remains to be established.

4.3.8   Far-red elongated hypocotyl 1

Far-red elongated hypocotyl 1(FHY1) is unique among the phy-interacting proteins characterized thus far. The *FHY1* locus was one of the first identified as specifically involved in phyA signalling under FRc in a forward genetic screen for long-hypocotyl mutants (Whitelam *et al*., 1993). Subsequent cloning of the locus revealed FHY1 to be a novel, plant-specific protein, found, by FHY1-GFP fusion protein expression, to be localized in cytoplasm and nucleus (Desnos *et al*., 2001; Zeidler *et al*., 2004), and an NLS (nuclear localization signal) in the protein was shown to be necessary for FHY1 function in phyA signalling (Zeidler *et al*., 2004). Recently, evidence from analysis of the subcellular localization of a phyA-GFP fusion protein expressed in a *fhy1* mutant background has established that FHY1 is specifically required for the light-induced accumulation of phyA, but not phyB, in the nucleus (Hiltbrunner *et al*., 2005). In addition, the FHY1 protein interacts with phyA selectively in the Pfr form in *in vitro* pull-down experiments, and in Y2H assays with photoactive phyA. Moreover, the two proteins were shown to be induced by light to colocalize in nuclear bodies when phyA-CFP and FHY1-YFP were transiently co-expressed in mustard seedlings. Collectively, the data suggest that FHY1 is involved in the light-induced translocation of phyA into the nucleus by virtue of its capacity to bind selectively to the Pfr form (Hiltbrunner *et al*., 2005). As the phyA protein itself does not appear to have a conventional NLS sequence, FHY1 may function to 'piggy-back' the photoreceptor into the nucleus after light-induced binding in the cytoplasm. Alternatively, FHY1 may function to retain phyA in the nucleus after translocation via another mechanism (Hiltbrunner *et al*., 2005). Recently, FHL (FHY1-like), the only close homolog of FHY1 in *Arabidopsis*, was also shown to be involved in phyA-mediated responsiveness to FRc (Zhou *et al*., 2005). The data indicate that FHY1 and FHL act at least partially redundantly to facilitate full phyA activity, thereby raising the possibility that FHL may also function in phyA nuclear transport.

4.4   Pre-selected interaction targets

In addition to the components identified in the open-ended library screens for phy-interacting proteins described above, an array of other proteins, pre-selected as potential direct phy targets based on a variety of rationales, have been reported to interact with the photoreceptor molecule in interaction assays.

4.4.1   Arabidopsis *response regulator 4*

Given the apparent evolution of the eukaryotic plant phys from the His-kinase domain containing prokaryotic phys, Sweere *et al*. (2001) reasoned that the plant phys may have retained the capacity to molecularly recognize plant response-regulator-related proteins. Using recombinant phyA and phyB produced in yeast and ligated to chromophore *in vitro*, these authors found that both Pr and Pfr forms of phyB bound equally well to *Arabidopsis* response regulator 4 (ARR4) in pull-down experiments, whereas phyA showed no binding. Subsequent Y2H assays indicated that the extreme N-terminal 173 residues of phyB were responsible for this binding. Despite the lack of binding selectivity between the Pr and Pfr conformers, evidence was presented that ARR4 inhibited Pfr-to-Pr dark reversion, both in yeast cells and when overexpressed in *Arabidopsis* seedlings, thereby stabilizing the phyB molecule in its active Pfr form (Sweere *et al.*, 2001). These ARR4-overexpressing seedlings exhibited hypersensitive inhibition of hypocotyl elongation in Rc, leading to the conclusion that this was due to the maintenance of higher levels of the biologically active conformer of phyB by interaction with the ARR4 protein. However, although a speculative model proposing crosstalk regulation of phyB signalling through cytokinin regulation of an ARR4-containing, two-component system has been suggested (Lohrmann and Harter, 2002; Grefen and Harter, 2004), compelling evidence for the function of the endogenous ARR4 molecule in phy signalling remains to be presented.

4.4.2 Zeitlupe

Zeitlupe (ZTL) was originally identified in genetic screens for components affecting the circadian clock, and subsequently identified as an F-box protein that targets the central oscillator protein TOC1 for regulated degradation via the UPS system (Somers *et al.*, 2000; Mas *et al.*, 2003). On the basis of the evidence that the phys control the circadian clock, ZTL (also called ADO1 (ADAGIO1)) was tested for physical interaction with phyB (and the blue-light photoreceptor cry1) using Y2H and *in vitro* pull-down assays. Initial studies showed that ZTL could interact with the non-chromophoric C-terminal domain of PHYB (and CRY1) in both assays (Jarillo *et al.*, 2001). However, a recent more in-depth examination of the phyB interaction in a Y2H assay, while reproducing the original observations, failed to detect any interaction of ZTL with the full-length, photoactive phyB molecule, irrespective of its presence as Pr or Pfr, and observed no effect of an array of *ztl* mutations on the interaction with the PHYB C-terminal domain (Kevei *et al.*, 2006). Consequently, the functional significance of the latter interaction is yet to be established.

4.4.3 Cryptochrome 1 and 2

On the basis of genetic and photobiological evidence of physiological crosstalk between the phy and cry signalling pathways, Ahmad *et al.* (1998) sought evidence of direct molecular interactions between the two photoreceptor families. Using the same recombinant oat phyA produced in yeast as was used to test PKS1 phosphorylation (Fankhauser *et al.*, 1999), Ahmad *et al.* (1998) provided evidence that these phyA preparations could also phosphorylate recombinant cryptochrome 1 and 2 (CRY1 and CRY2) proteins on serine residues *in vitro*. However, in contrast to

PKS1, no difference in the degree of CRY phosphorylation was observed between the Pr and Pfr forms of phyA. Data from a Y2H assay showed that the C-terminal domain of *Arabidopsis* PHYA is capable of direct interaction with the C-terminal domain of CRY1 (Ahmad *et al.*, 1998). *In vivo* radioactive phosphate labelling studies with dark-adapted *Arabidopsis* plants provided evidence that transgenically expressed, His-tagged CRY1, affinity-purified from these plants, was phosphorylated in plants given a 1-min R pulse followed by immediate extraction, but not in plants given either no light or FR irradiation simultaneously with the R pulse. These data were interpreted as indicating that phy induces very rapid CRY1 phosphorylation *in vivo*, potentially enhancing the activity of the blue-light photoreceptor. Given that both the phy and cry proteins are or become nuclear localized, the opportunity for light-induced, direct interaction in the nucleus exists. In support of this possibility, a study using fluorescence resonance energy transfer microscopy has provided evidence that phyB and cry2 interact physically *in vivo* in nuclear speckles that are induced in a light-dependent manner (Mas *et al.*, 2000). This study also showed that overexpressed cry2 co-immunoprecipitated with phyB from extracts of *Arabidopsis* plants, but no test of whether this interaction is light dependent was performed. Collectively, the data are consistent with the capacity of the two photoreceptors to interact in the nucleus, but direct evidence that this interaction is involved in regulatory crosstalk is yet to be presented. Similarly, evidence of the biological relevance of the reported phy-stimulated phosphorylation of cry1 and cry2 is lacking.

4.4.4   AUX/IAA proteins

On the basis of the premise that the phys may regulate cell expansion rates via the auxin system, several members of the AUX/IAA family were examined for direct interaction with the photoreceptor molecule *in vitro*, using recombinant AUX/IAA proteins produced in *E. coli* and oat phyA produced in yeast (Colon-Carmona *et al.*, 2000). Pull-down experiments provided evidence of *in vitro* binding of *Arabidopsis* IAA17 and pea IAA4 to oat phyA, but the chromophoric state of the photoreceptor was not reported. Using the same yeast-produced, photoactive, oat phyA preparations shown to phosphorylate PKS1, CRY1 and CRY2 (Ahmad *et al.*, 1998; Fankhauser *et al.*, 1999), evidence was also presented that these preparations could also phosphorylate recombinant *Arabidopsis* IAA3, IAA17, IAA1, IAA9 and pea IAA4 *in vitro* (Colon-Carmona *et al.*, 2000). However, no significant difference was reported in the level of phosphorylation of these proteins when the phyA molecule was present as Pr or Pfr. The relevance of these *in vitro* interactions to phyA signalling *in vivo* remains to be established.

4.4.5   COP1

Soon after the first physical detection of phy by difference spectroscopy in living plant tissue, it was discovered that the levels of this spectroscopically measurable molecule (now known to be essentially exclusively phyA), although high in dark-grown tissue, dropped dramatically upon exposure to light, as a result of the Pfr form of the photoreceptor being rapidly labile in the cell (Hendricks *et al.*, 1962).

Subsequent studies over the ensuing years established that this observation was due to selective proteolysis of the Pfr form of phyA at a rate that was about 100-fold greater than for the Pr form. However, the mechanism underlying this degradation remained unknown until the pioneering work of Vierstra and colleagues who discovered that the Pfr conformer of phyA is rapidly ubiquitinated following its light-induced formation, as a prelude to proteolysis (Shanklin *et al.*, 1987; Jabben *et al.*, 1989).

COP1 was originally identified in a forward genetic screen for mutants exhibiting constitutive photomorphogenesis in darkness (Deng *et al.*, 1992), and eventually identified as a ubiquitin E3 ligase that functions to suppress photomorphogenesis in darkness by targeting activators of de-etiolation, such as the transcription factor HY5, for degradation via the 26S proteosome (Osterlund *et al.*, 2000; Saijo *et al.*, 2003; Seo *et al.*, 2003). Initial tantalizing evidence of a possible direct connection between COP1 and the phys came from a report that COP1 interacted with the non-chromophoric C-terminal domain of PHYB (as well as with CRY1) in a targeted Y2H assay (Yang *et al.*, 2001). However, the C-terminal domain of phyA was reported not to bind to COP1 in this study, leaving the significance of the observed interaction to be elucidated. More recently, Seo *et al.* (2004) have shown that *cop1* mutants exhibit a strongly reduced rate of light-induced degradation of phyA *in vivo* and that this degradation is likely proteosome-mediated. In addition, evidence is provided that recombinant COP1 can polyubiquitinate either recombinant *Arabidopsis* PHYA apoprotein or both Pr and Pfr conformers of biochemically purified pea phyA, about equally, in *in vitro* assays, that COP1 can bind to these photoreceptor molecules in *in vitro* pull-down assays, and that COP1 colocalizes with phyA in nuclear bodies in transfected onion cells. The data are interpreted to indicate that COP1 functions as an E3 ligase targeting phyA for degradation via the UPS pathway (Seo *et al.*, 2004). While the mutant molecular phenotype is compelling, the absence of evidence of conformer-specific phyA-COP1 interaction in the binding and ubiquitination assays in this study needs to be addressed experimentally in order to explain the light-induced nature of the degradation process *in vivo*. The authors suggest that one possibility is that the Pfr-dependent step in this process is phyA translocation into the nucleus, providing physical access to COP1, rather than conformer-specific intermolecular recognition. Alternatively, accessory proteins, such as members of the SPA1 quartet (SPA1 through SPA4), which are known to bind to and work in concert with COP1 (Hoecker, 2005), might modulate the specificity of the COP1 E3 ligase towards different substrates such as phyA (Seo *et al.*, 2004).

4.5 Perspective

Numerous studies of the phy system over the years have identified a variety of different facets of the photoreceptor's molecular properties and behaviour in the cell potentially relevant to its photosensory function. Apart from dimerization, chromophore ligation and light-induced conformer switching, which are autonomous properties intrinsic to the phy molecule itself, there is evidence that the photoreceptor is subject to, or can engage in, at least five other definable activities, all

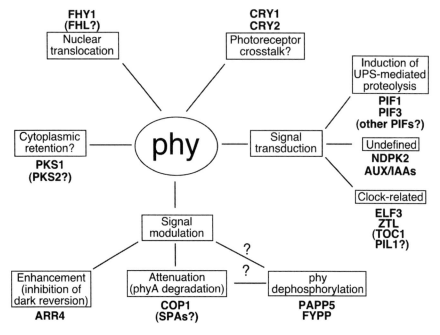

Figure 4.5 Schematic summary of putative or established functions of various phy-interacting proteins in multiple facets of phy cellular and molecular activities. Question marks indicate uncertainty in the postulated function.

implying the necessity of intermolecular interactions: cytoplasmic retention, nuclear translocation, photoreceptor crosstalk, signal transduction and signal modulation. The research discussed in this chapter has begun to provide insight into at least some of the molecular components in the cell that are, or may be, engaged in these activities via direct physical interaction with the photoreceptor molecule. Figure 4.5 summarizes these findings schematically. Examination of the data indicates that significant progress has been made in recent times in elucidating the underlying molecular mechanisms in some areas, such as nuclear translocation, signal trans-duction and signal attenuation, whereas in others definitive data providing evidence of the biological relevance and mechanistic basis of these phenomena are sparse. Nevertheless, the continued application of the combined power of the molecular-genetic, biochemical, proteomic and cytological tools that are available can be anticipated to yield additional exciting advances in these areas in the near future.

Acknowledgements

I thank many colleagues who have contributed to the work from this laboratory cited here and Jim Tepperman for figure preparation. This research was supported by NIH grant GM47475, DOE-BES grant DE-FG03-87ER13742 and USDA CRIS 5335-21000-017-00D.

References

Aebersold, R. and Mann, M. (2003) Mass spectrometry-based proteomics. *Nature* **422**, 198–207.

Ahmad, M., Jarillo, J.A., Smirnova, O. and Cashmore, A.R. (1998) The CRY1 blue light photoreceptor of *Arabidopsis* interacts with phytochrome A in vitro. *Mol. Cell* **1**, 939–948.

Al-Sady, B., Ni, W., Kircher, S., Schäfer, E. and Quail, P.H. (2006) Photoactivated phytochrome induces rapid PIF3 phosphorylation as a prelude to proteasome-mediated degradation. *Mol. Cell* **23**, 439–446.

Alonso, J.M., Stepanova, A.N., Leisse, T.J., *et al.* (2003) Genome-wide insertional mutagenesis of *Arabidopsis thaliana*. *Science* **301**, 653–657.

Bailey, P.C., Martin, C., Toledo-Ortiz, G., *et al.* (2003) Update on the basic helix-loop-helix transcription factor gene family in *Arabidopsis thaliana*. *Plant Cell* **15**, 2497–2501.

Bauer, D., Viczian, A., Kircher, S., *et al.* (2004) Constitutive photomorphogenesis 1 and multiple photoreceptors control degradation of phytochrome interacting factor 3, a transcription factor required for light signalling in *Arabidopsis*. *Plant Cell* **16**, 1433–1445.

Bhoo, S.H., Davis, S.J., Walker, J., Karniol, B. and Vierstra, R.D. (2001) Bacteriophytochromes are photochromic histidine kinases using a biliverdin chromophore. *Nature* **414**, 776–779.

Brent, R. and Finley, R.L., Jr. (1997) Understanding gene and allele function with two-hybrid methods. *Annu. Rev. Genet.* **31**, 663–704.

Chen, M., Chory, J. and Fankhauser, C. (2004) Light signal transduction in higher plants. *Annu. Rev. Genet.* **38**, 87–117.

Choi, G., Yi, H., Lee, J., Kwon, Y.-K., Soh, M.S., Shin, B., Luka, Z., Hahn, T.-R. and Song, P.-S. (1999) Phytochrome signalling is mediated through nucleoside diphosphate kinase 2. *Nature* **401**, 610–613.

Colon-Carmona, A., Chen, D.L., Yeh, K.C. and Abel, S. (2000) Aux/IAA proteins are phosphorylated by phytochrome in vitro. *Plant Physiol.* **124**, 1728–1738.

Deng, X.-W., Matsui, M., Wei, N., *et al.* (1992) *COP1*, an *Arabidopsis* photomorphogenic regulatory gene, encodes a protein with both a Zn-binding motif and a Gß homologous domain. *Cell* **71**, 791–801.

Desnos, T., Puente, P., Whitelam, G.C. and Harberd, N.P. (2001) FHY1: a phytochrome A-specific signal transducer. *Genes Dev.* **15**, 2980–2990.

Dharmasiri, N., Dharmasiri, S., Weijers, D., *et al.* (2005) Plant development is regulated by a family of auxin receptor F box proteins. *Dev. Cell* **9**, 109–119.

Duek, P.D., Elmer, M.V., Van Oosten, V.R. and Fankhauser, C. (2004) The degradation of HFR1, a putative bHLH class transcription factor involved in light signalling, is regulated by phosphorylation and requires COP1. *Curr. Biol.* **14**, 2296–2301.

Duek, P.D. and Fankhauser, C. (2005) bHLH class transcription factors take centre stage in phytochrome signalling. *Trends Plant Sci.* **10**, 51–54.

Fairchild, C.D., Schumaker, M.A. and Quail, P.H. (2000) *HFR1* encodes an atypical bHLH protein that acts in phytochrome A signal transduction. *Genes Dev.* **14**, 2377–2391.

Fankhauser, C. (2000) Phytochromes as light-modulated protein kinases. *Semin. Cell Dev. Biol.* **11**, 467–473.

Fankhauser, C. and Chory, J. (2000) RSF1, an *Arabidopsis* locus implicated in phytochrome A signalling. *Plant Physiol.* **124**, 39–45.

Fankhauser, C., Yeh, K.C., Lagarias, J.C., Zhang, H., Elich, T.D. and Chory, J. (1999) PKS1, a substrate phosphorylated by phytochrome that modulates light signalling in *Arabidopsis*. *Science* **284**, 1539–1541.

Grefen, C. and Harter, K. (2004) Plant two-component systems: principles, functions, complexity and cross talk. *Planta* **219**, 733–742.

Gyula, P., Schaefer, E. and Nagy, F. (2003) Light perception and signalling in higher plants. *Curr. Opin. Plant Biol.* **6**, 446–452.

Halliday, K.J., Hudson, M., Ni, M., Qin, M. and Quail, P.H. (1999) *poc1*: an *Arabidopsis* mutant perturbed in phytochrome signalling due to a T-DNA insertion in the promoter of *PIF3*, a gene encoding a phytochrome-interacting, bHLH protein. *Proc. Natl. Acad. Sci. USA* **96**, 5832–5837.

Heisler, M.G.B., Atkinson, A., Bylstra, Y.H., Walsh, R. and Smyth, D.R. (2001) SPATULA, a gene that controls development of carpel margin tissues in *Arabidopsis*, encodes a bHLH protein. *Development* **128**, 1089–1098.

Hendricks, S.B., Butler, W.L. and Siegelmann, H.W. (1962) A reversible photoreaction regulating plant growth. *J. Phys. Chem.* **66**, 2550–2555.

Henikoff, S. and Comai, L. (2003) Single-nucleotide mutations for plant functional genomics. *Annu. Rev . Plant Biol.* **54**, 375–401.

Hiltbrunner, A., Viczian, A., Bury, E., *et al.* (2005) Nuclear accumulation of the phytochrome A photoreceptor requires FHY1. *Curr. Biol.* **15**, 2125–2130.

Hoecker, U. (2005) Regulated proteolysis in light signalling. *Curr. Opin. Plant Biol.* **8**, 469–476.

Huq, E. (2006) Degradation of negative regulators: a common theme in hormone and light signalling networks? *Trends Plant Sci.* **11**, 4–7.

Huq, E., Al-Sady, B., Hudson, M., Kim, C., Apel, K. and Quail, P.H. (2004) Phytochrome-interacting factor 1, a basic helix-loop-helix transcription factor, is a critical regulator of the chlorophyll biosynthetic pathway. *Science* **305**, 1937–1941.

Huq, E., Al-Sady, B. and Quail, P.H. (2003) Nuclear translocation of the photoreceptor phytochrome B is necessary for its biological function in seedling photomorphogenesis. *Plant J.* **35**, 660–664.

Huq, E., Kang, Y., Qin, M. and Quail, P.H. (2000) *SRL1*: a new locus specific to the phyB signaling pathway in *Arabidopsis*. *Plant J.* **23**, 1–11.

Huq, E. and Quail, P.H. (2002) PIF4, a phytochrome-interacting bHLH factor, functions as a negative regulator of phytochrome B signalling in *Arabidopsis*. *EMBO J.* **21**, 2441–2450.

Im, Y.J., Kim, J.-I., Shen, Y., *et al.* (2004) Structural analysis of *Arabidopsis thaliana* nucleoside diphosphate kinase-2 for phytochrome-mediated light signalling. *J. Mol. Biol.* **343**, 659–670.

Jabben, M., Shanklin, J. and Vierstra, R.D. (1989) Ubiquitin-phytochrome conjugates – pool dynamics during *in vivo* phytochrome degradation. *J. Biol. Chem.* **264**, 4998–5005.

Jang, I.-C., Yang, J.-Y., Seo, H.S. and Chua, N.-H. (2005) HFR1 is targeted by COP1 E3 ligase for post-translational proteolysis during phytochrome A signalling. *Gen. Dev.* **19**, 593–602.

Jarillo, J.A., Capel, J., Tang, R.-H., *et al.* (2001) An *Arabidopsis* circadian clock component interacts with both CRY1 and phyB. *Nature* **410**, 487–490.

Jones, A.M., Ecker, J.R. and Chen, J.-G. (2003) A reevaluation of the role of the heterotrimeric G protein in coupling light responses in *Arabidopsis*. *Plant Physiol* **131**, 1623–1627.

Kevei, E., Gyula, P., Hall, A., *et al.* (2006) Forward genetic analysis of the circadian clock separates the multiple functions of ZEITLUPE. *Plant Physiol.* **140**, 933–945.

Khanna, R., Huq, E., Kikis, E.A., Al-Sady, B., Lanzatella, C. and Quail, P.H. (2004) A novel molecular recognition motif necessary for targeting photoactivated phytochrome signalling to specific basic helix-loop-helix transcription factors. *Plant Cell* **16**, 3033–3044.

Kim, D.-H., Kang, J.-G., Yang, S.-S., Chung, K.-S., Song, P.-S. and Park, C.-M. (2002) A phytochrome-associated protein phosphatase 2A modulates light signals in flowering time control in *Arabidopsis*. *Plant Cell* **14**, 3043–3056.

Kim, J., Bhinge, A.A., Morgan, X.C. and Iyer, V.R. (2005) Mapping DNA-protein interactions in large genomes by sequence tag analysis of genomic enrichment. *Nat. Methods* **2**, 47–53.

Kim, J., Yi, H., Choi, G., Shin, B., Song, P.-S. and Choi, G. (2003a) Functional characterization of phytochrome interacting factor 3 in phytochrome-mediated light signal transduction. *Plant Cell* **15**, 2399–2407.

Kim, J.-I., Shen, Y., Han, Y.-J., *et al.* (2004) Phytochrome phosphorylation modulates light signalling by influencing the protein-protein interaction. *Plant Cell* **16**, 2629–2640.

Kim, J.H., Choi, D. and Kende, H. (2003b) The AtGRF family of putative transcription factors is involved in leaf and cotyledon growth in *Arabidopsis*. *Plant J.* **36**, 94–104.

Kirkpatrick, D.S., Gerber, S.A. and Gygi, S.P. (2005) The absolute quantification strategy: a general procedure for the quantification of proteins and post-translational modifications. *Methods* **35**, 265–273.

Krall, L. and Reed, J.W. (2000) The histidine kinase-related domain participates in phytochrome B function but is dispensable. *Proc. Natl. Acad. Sci. USA* **97**, 8169–8174.

Lariguet, P., Boccalandro, H.E., Alonso, *et al.* (2003) A growth regulatory loop that provides homeostasis to phytochrome A signalling. *Plant Cell* **15**, 2966–2978.

Li, L. and Lagarias, J.C. (1994) Phytochrome assembly in living cells of the yeast *Saccaromyces cerevisiae. Proc. Natl. Acad. Sci. USA* **91**, 12535–12539.

Lipford, J.R. and Deshaies, R.J. (2003) Diverse roles for ubiquitin-dependent proteolysis in transcriptional activation. *Nat. Cell Biol.* **5**, 845–850.

Lipford, J.R., Smith, G.T., Chi, Y. and Deshaies, R.J. (2005) A putative stimulatory role for activator turnover in gene expression. *Nature* **438**, 113–116.

Liu, X.L., Covington, M.F., Fankhauser, C., Chory, J. and Wagner, D.R.Y. (2001) ELF3 encodes a circadian clock-regulated nuclear protein that functions in an *Arabidopsis* phyB signal transduction pathway. *Plant Cell* **13**, 1293–1304.

Lohrmann, J. and Harter, K. (2002) Plant two-component signalling systems and the role of response regulators. *Plant Physiol.* **128**, 363–369.

Martínez-García, J.F., Huq, E. and Quail, P.H. (2000) Direct targeting of light signals to a promoter element-bound transcription factor. *Science* **288**, 859–863.

Mas, P., Devlin, P.F., Panda, S. and Kay, S.A. (2000) Functional interaction of phytochrome B and cryptochrome 2. *Nature* **408**, 207–211.

Mas, P., Kim, W.Y., Somers, D.E. and Kay, S.A. (2003) Targeted degradation of TOC1 by ZTL modulates circadian function in *Arabidopsis thaliana. Nature* **426**, 567–570.

Matsushita, T., Mochizuki, N. and Nagatani, A. (2003) Dimers of the N-terminal domain of phytochrome B are functional in the nucleus. *Nature* **424**, 571–574.

Moller, S.G., Ingles, P.J. and Whitelam, G.C. (2002) The cell biology of phytochrome signalling. *New Phytol.* **154**, 553–590.

Monte, E., Tepperman, J.M., Al-Sady, B., *et al.* (2004) The phytochrome-interacting transcription factor, PIF3, acts early, selectively, and positively and light-induced chloroplast development. *Proc. Natl. Acad. Sci. USA* **101**, 16091–16098.

Montgomery, B.L. and Lagarias, J.C. (2002) Phytochrome ancestry: sensors of bilins and light. *Trends Plant Sci.* **7**, 357–366.

Moon, J., Parry, G. and Estelle, M. (2004) The ubiquitin-proteosome pathway and plant development. *Plant Cell* **16**, 3181–3195.

Nagatani, A. (2004) Light-regulated nuclear localization of phytochromes. *Curr. Opin. Plant Biol.* **7**, 708–711.

Nagy, F. and Schafer, E. (2002) Phytochromes control photomorphogenesis by differentially regulated, interacting signalling pathways in higher plants. *Annu. Rev. Plant Biol.* **53**, 329–355.

Ni, M., Tepperman, J.M. and Quail, P.H. (1998) PIF3, a phytochrome-interacting factor necessary for normal photoinduced signal transduction, is a novel basic helix-loop-helix protein. *Cell* **95**, 657–667.

Ni, M., Tepperman, J.M. and Quail, P.H. (1999) Binding of phytochrome B to its nuclear signalling partner PIF3 is reversibly induced by light. *Nature* **400**, 781–784.

Oh, E., Kim, J., Park, E., Kim, J.-I., Kang, C. and Choi, G. (2004) PIL5, a phytochrome-interacting basic helix-loop-helix protein, is a key negative regulator of seed germination in *Arabidopsis thaliana. Plant Cell* **16**, 3045–3058.

Okamoto, H., Qu, L. and Deng, X.-W. (2001) Does EID1 aid the fine-tuning of phytochrome A signal transduction in *Arabidopsis? Plant Cell* **13**, 1983–1986.

Osterlund, M.T., Hardtke, C.S., Wei, N. and Deng, X.W. (2000) Targeted destabilization of HY5 during light-regulated development of *Arabidopsis. Nature* **405**, 462–466.

Park, E., Kim, J., Lee, Y., *et al.* (2004) Degradation of phytochrome interacting factor 3 in phytochrome-mediated light signalling. *Plant Cell Physiol.* **45**, 968–975.

Phizicky, E.M. and Fields, S. (1995) Protein-protein interactions: methods for detection and analysis. *Microbiol. Rev.* **59**, 94–123.

Quail, P.H. (1975) Particle-bound phytochrome: association with a ribonucleoprotein fraction from Cucurbita L. *Planta* **123**, 223–234.

Quail, P.H. (2000) Phytochrome interacting factors. *Semin. Cell Dev . Biol.* **11**, 457–466.

Quail, P.H. (2002a) Phytochrome photosensory signalling networks. *Nat. Rev. Mol. Cell Biol.* **3**, 85–93.

Quail, P.H. (2002b) Photosensory perception and signalling in plant cells: new paradigms? *Curr. Opin. Cell Biol.* **14**, 180–188.

Quail, P.H. (2006a) General introduction. In: *Photomorphogenesis in Plants and Bacteria*, 3rd edn (eds Schäfer, E. and Nagy, F.). Springer, Dordrecht, The Netherlands.

Quail, P.H. (2006b) Phytochrome signal transduction network. In: *Photomorphogenesis in Plants and Bacteria*, 3rd edn (eds Schäfer, E. and Nagy, F.). Springer, Dordrecht, The Netherlands.

Rajani, S. and Sundaresan, V. (2001) The *Arabidopsis* myc/bHLH gene ALCATRAZ enables cell separation in fruit dehiscence. *Curr. Biol.* **11**, 1914–1922.

Reed, J.W., Nagpal, P., Bastow, R.M., *et al.* (2000) Independent action of ELF3 and phyB to control hypocotyl elongation and flowering time. *Plant Physiol.* **122**, 1149–1160.

Ryu, J.S., Kim, J.-I., Kunkel, T., Kim, B.C., Cho, D.S. and Hong, S.H.E.A. (2005) Phytochrome-specific type 5 phosphatase controls light signal flux by enhancing phytochrome stability and affinity for a signal transducer. *Cell* **120**, 395–406.

Saijo, Y., Sullivan, J.A., Wang, H.Y., *et al.* (2003) The COP1-SPA1 interaction defines a critical step in phytochrome A-mediated regulation of HY5 activity. *Genes Dev.* **17**, 2642–2647.

Salter, M.G., Franklin, Keara A. and Whitelam, Garry C. (2003) Gating of the rapid shade-avoidance response by the circadian clock in plants. *Nature* **426**, 680–683.

Schäfer, E. and Bowler, C. (2002) Phytochrome-mediated photoperception and signal transduction in higher plants. *EMBO Rep.* **3**, 1042–1048.

Seo, H.S., Watanabe, E., Tokutomi, S., Nagatani, A. and Chua, N.-H. (2004) Photoreceptor ubiquiti-nation by COP1 E3 ligase desensitizes phytochrome A signalling. *Genes Dev.* **18**, 617–622.

Seo, H.S., Yang, J.-Y., Ishikawa, M., Bolle, C., Ballesteros, M.L. and Chua, N.-H. (2003) LAF1 ubiquitination by COP1 controls photomorphogenesis and is stimulated by SPA1. *Nature* **423**, 995–999.

Serebriiskii, I., Estojak, J., Berman, M. and Golemis, E.A. (2000) Approaches to detecting false positives in yeast two-hybrid systems. *Biotechniques* **28**, 328–336.

Sessa, G., Carabelli, M., Sassi, M., *et al.* (2005) A dynamic balance between gene activation and repression regulates the shade avoidance response in *Arabidopsis*. *Genes Dev.* **19**, 2811–2815.

Shanklin, J., Jabben, M. and Vierstra, R.D. (1987) Red light-induced formation of ubiquitin-phytochrome conjugates: identification of possible intermediates of phytochrome degradation. *Proc. Natl. Acad. Sci. USA* **84**, 359–363.

Shen, H., Moon, J. and Huq, E. (2005) PIF1 is regulated by light-mediated degradation through the ubiquitin-26S proteasome pathway to optimize photomorphogenesis of seedlings in *Arabidopsis*. *Plant J.* **44**, 1023–1035.

Shimizu-Sato, S., Huq, E., Tepperman, J.M. and Quail, P.H. (2002) A light-switchable gene promoter system. *Nat. Biotechnol.* **20**, 1041–1044.

Soh, M.S., Kim, Y.M., Han, S.J. and Song, P.S. (2000) REP1, a basic helix-loop-helix protein, is required for a branch pathway of phytochrome A signalling in *Arabidopsis*. *Plant Cell* **12**, 2061–2073.

Somers, D.E., Schultz, T.F., Milnamow, M. and Kay, S.A. (2000) ZEITLUPE encodes a novel clock-associated PAS protein from *Arabidopsis*. *Cell* **101**, 319–329.

Sullivan, J.A., Shirasu, K. and Deng, X.W. (2003) The diverse roles of ubiquitin and the 26S proteasome in the life of plants. *Nat. Rev. Genet.* **4**, 948–958.

Sweere, U., Eichenberg, K., Lohrmann, J., *et al.* (2001) Interaction of the response regulator ARR4 with the photoreceptor phytochrome B in modulating red light signalling. *Science* **294**, 1108–1111.

Tepperman, J.M., Hudson, M.E., Khanna, R., *et al.* (2004) Expression profiling of phyB mutant demonstrates substantial contribution of other phytochromes to red-light-regulated gene expression during seedling de-etiolation. *Plant J.* **38**, 725–739.

Tepperman, J.M., Zhu, T., Chang, H.S., Wang, X. and Quail, P.H. (2001) Multiple transcription-factor genes are early targets of phytochrome A signalling. *Proc. Natl. Acad. Sci. USA* **98**, 9437–9442.

Toledo-Ortiz, G., Huq, E. and Quail, P.H. (2003) The *Arabidopsis* basic helix-loop-helix transcription factor family. *Plant Cell* **15**, 1749–1770.

Ullah, H., Chen, J.G., Young, J.C., Im, K.H., Sussman, M.R. and Jones, A.M. (2001) Modulation of cell proliferation by heterotrimeric G protein in *Arabidopsis*. *Science* **292**, 2066–2069.

Whitelam, G.C., Johnson, E., Peng, J., *et al.* (1993) Phytochrome A null mutants of Arabidopsis display a wild-type phenotype in white light. *Plant Cell* **5**, 757–768.

Yamashino, T., Matsushika, A., Fujimori, T., *et al.* (2003) A link between circadian-controlled bHLH factors and the APRR1/TOC1 quintet in *Arabidopsis thaliana*. *Plant Cell Physiol.* **44**, 619–629.

Yang, H.-Q., Tang, R.-H. and Cashmore, A.R. (2001) The signalling mechanism of *Arabidopsis* CRY1 involves direct interaction with COP1. *Plant Cell* **13**, 2573–2587.

Yang, J., Lin, R., Sullivan, J., *et al.* (2005) Light regulates COP1-mediated degradation of HFR1, a transcription factor essential for light signalling in *Arabidopsis*. *Plant Cell* **17**, 804–821.

Zeidler, M., Zhou, Q., Sarda, X., Yau, C.-P. and Chua, N.-H. (2004) The nuclear localization signal and the C-terminal region of FHY1 are required for transmission of phytochrome A signals. *Plant J.* **40**, 355–365.

Zhou, Q., Hare, P.D., Yang, S.W., Zeidler, M., Huang, L.-F. and Chua, N.-H. (2005) FHL is required for full phytochrome A signalling and shares overlapping functions with FHY1. *Plant J.* **43**, 356–370.

Zhu, Y., Tepperman, J.M., Fairchild, C.D. and Quail, P. (2000) Phytochrome B binds with greater apparent affinity than phytochrome A to the basic helix-loop-helix factor PIF3 in a reaction requiring the PAS domain of PIF3. *Proc. Natl. Acad. Sci. USA* **97**, 13419–13424.

5 Phosphorylation/dephosphorylation in photoreceptor signalling

Cathrine Lillo, Trudie Allen and Simon Geir Møller

5.1 Introduction

The phosphorylation and dephosphorylation of proteins represents a universal means of regulating protein activity in prokaryotic and eukaryotic cells and is recognised as essential in numerous signal transduction chains.

The eukaryotic protein kinase superfamily comprises one of the largest super-families of proteins where they transfer the γ-phosphate of a purine nucleotide triphosphate to the hydroxyl groups of their protein substrates. Although the eukaryotic protein kinases can be divided into serine/threonine (Ser/Thr) kinases and tyrosine kinases, both subgroups share a conserved catalytic core (Hanks and Hunter, 1995). There are also a number of conserved regions in the catalytic domain of eukaryotic kinases: (i) a glycine-rich stretch close to a lysine residue important for ATP binding and (ii) a region located in the central part of the catalytic domain, harbouring a conserved aspartic acid residue, important for catalytic activity (Hanks and Hunter, 1995). From the three main groups of plant photoreceptors, phytochromes, cryptochromes (crys) and phototropins, only phototropins harbour these protein kinase signatures and therefore belong to the eukaryotic protein kinase superfamily.

In prokaryotes, protein kinases different from the eukaryotic type were long known, and these kinases phosphorylate a nitrogen atom of a histidine residue (receptor domain) and an acyl group of an aspartate residue (response domain) (Klumpp and Krieglstein, 2002). These two activities are found within the same protein and are referred to as the two-component system. Following the triggering signal, the histidine residue is phosphorylated by the receptor domain (autophosphorylation), and the response domain thereafter catalyses transfer of the phosphoryl group to the conserved aspartate residue. Until 1993 it was thought that such kinases only existed in bacteria; however, the yeast osmosensor (Ota and Varshavsky, 1993) and the ethylene receptor in *Arabidopsis* were then identified as two-component kinases (Chang *et al.*, 1993).

Some prokaryotic and all eukaryotic histidine kinases have an additional receiver domain that senses the signal from the histidine kinase domain (HKD). Many histidine kinases also have phosphatase activity. Phytochromes are reminiscent of such two-component kinases. However, the phytochromes do not have the five (H, N, G1, F, G2) conserved signature motifs found in typical histidine kinases. *Arabidopsis* phytochrome C possesses the conserved His (H motif), which is necessary for receiving the phosphoryl group. The other signature motifs are important for binding of ATP, but they are not conserved in the phytochromes (Hwang *et al.*, 2002).

The crys exhibit kinase activity, but do not resemble the eukaryotic superfamily, nor do they resemble the prokaryotic type two-component kinases.

5.1.1   The photoreceptors autophosphorylate, but the classical activation loop is not involved

Phosphorylation is a common way of regulating kinases. In the eukaryotic kinases activation often requires phosphorylation of a special segment: an activation loop in the centre of the kinase domain (Nolen *et al.*, 2004). Phosphorylation of this loop can be carried out by an upstream kinase or by autophosphorylation (Lochhead *et al.*, 2005). Phytochromes, crys and phototropins all autophosphorylate, but only phototropins resemble common kinases and are therefore the only candidates for this general activation mechanism. In phototropin 1 (phot1) of *Avena sativa*, the phosphorylation sites were mapped, and eight different sites identified (Figure 5.1).

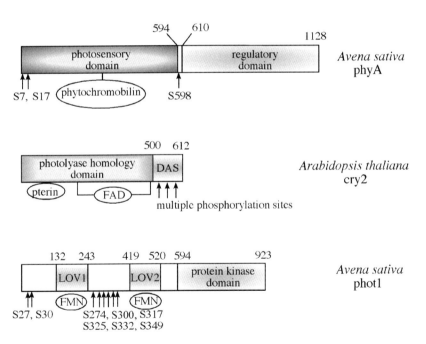

Figure 5.1   Phosphorylation sites in the three main classes of photoreceptors. In *Avena sativa* phyA three phosphorylation sites have been identified, two in the very amino terminal end S7, S17, and another site, S598, in the hinge between the N-terminal photosensory domain and the C-terminal regulatory/histidine kinase domain (Lapko *et al.*, 1999). The exact phosphorylation sites in crys have not been determined, but it is known that *Arabidopsis* cry2 is phosphorylated on multiple sites in the C-terminal (DAS) region (Shalitin *et al.*, 2002). The phosphorylation sites have been mapped in *Avena sativa* phot1 revealing two phosphorylation sites in the N-terminal end at S27 and S30, and six phosphorylation sites in the hinge between the LOV1 and LOV2 domain, S274, S300, S317, S325, S332, and S349 (Salomon *et al.*, 2003).

These sites were, however, all located in the N-terminal chromophore binding part of the protein, and not in the C-terminal kinase part of the enzyme (Salomon *et al.*, 2003). The phot1 activation mechanism must differ from the common mechanism described for many eukaryotic kinases because phosphorylation does not take place in the kinase domain. In the classical two-component system autophosphorylation activates signalling, but since plant phytochromes generally lack the conserved His that is autophosphorylated, their activation mechanism must be different. Although the classical signatures involved in eukaryotic and bacterial two-component kinases are not conserved in the photoreceptors, autophosphorylation, as a means of establishing and amplifying a signal, is conserved in a wide range of signalling pathways including photoreceptor signal transduction.

5.1.2   *Phosphatases in photoreceptor signalling*

For a signal to function in a physiologically meaningful way there must not only be a mechanism that triggers the signal, but also one that turns it off. The effects of protein kinases are generally counteracted by protein phosphatases, although sometimes deactivation is not achieved by dephosphorylation, but rather by protein degradation. Indeed, this type of control can be an efficient way of removing the phosphorylated protein, and negating the signal. Indeed, there are many examples that phosphorylation induces rapid degradation of proteins (del Pozo and Estelle, 2000; Lieu *et al.*, 2000; Hoecker, 2005). The eukaryotic protein phosphatases can be divided into three distinct gene families, when referring to their catalytic subunit. Two of these, the PPP and PPM families, dephosphorylate phosphoserine and phosphothreonine residues, whereas the third group, PTP, dephosphorylates phosphotyrosine residues (Barford *et al.*, 1998). The PPP family is further divided into PP1, PP2A, PP2B, and some novel protein phosphatases such as PP5 and PP7A also belong to this family (see Section 5.2.5). The Ser/Thr phosphatases are all metalloenzymes with Zn^{2+}, Fe^{2+} (Fe^{3+}) or Mn^{2+} at the active centre, and they dephosphorylate their substrates in a single step, using a metal-activated nucleophilic water molecule. In contrast, the PTP enzymes catalyse dephosphorylation by use of a cysteinyl-phosphate intermediate. The different groups of phosphatases are characterised by their requirements for different ions, and by being deactivated by certain inhibitors. PP2B requires Ca^{2+} for activity, whereas PP2A and PP1 do not have special requirements for ion cofactors. PP1 and PP2A are both inhibited by okadaic acid. Most protein phosphates are made up of several protein subunits, and these subunits can belong to different families. The PP2A phosphatases comprise three subunits: a catalytic subunit, a structural subunit and a regulatory subunit. The catalytic and structural subunits are strongly conserved throughout eukaryotes, whereas the regulatory subunits belong to different families with total lack of sequence similarity (Haynes *et al.*, 1999).

The phosphatases found to be involved in regulating the phosphorylation state of phytochromes, flower-specific phytochrome-associated protein phosphatase (FYPP) and phytochrome-associated protein phosphatase 5 (PAPP5), belong to or are closely related to the PP2A family (see Section 5.2.5). Protein phosphatase 7 (PP7), which probably acts downstream of cryptochrome, is a novel phosphatase (del Pozo and Estelle, 2000; Møller et al., 2003). Other phosphatases are likely to be involved in photoreceptor signal transduction, for example a phosphatase that act on phototropin. However, the identity of these proteins has yet to be established. Indeed, a fuller characterisation of phosphatases involved in light signalling will enhance our understanding of this environmentally controlled network.

5.2 Phytochromes

5.2.1 Phosphorylation of phytochrome

Phytochromes are photoreceptors responsible for red/far-red (R/FR) reversible plant responses and there is evidence that phytochrome can, itself, be phosphorylated. Wong et al. (1986) demonstrated that A. sativa phytochrome, in either the Pr or Pfr form, could be phosphorylated by mammalian cAMP-dependent protein kinase (kinase A). Additionally, extracts of the Pr form acted as a substrate for protein kinase G, protein kinase C and a polycation-dependent protein kinase isolated in association with purified phytochrome. The sites for phosphorylation in Pr and Pfr appeared to be spatially distinct (Wong et al., 1986). Utilising synthesised peptides, McMichael and Lagarias (1990) identified two candidate phosphorylation sites in the A. sativa phytochrome. They went on to demonstrate that within these specifically the serine residues Ser17 and Ser598 were phosphorylated in vitro; these sites were phosphorylated preferentially in the Pr and Pfr form respectively. Analysis of oat phyA (phytochrome A), using fast atom bombardment mass spectrometry, suggested that in vivo the main site of phosphorylation is Ser7 (Lapko et al., 1997). Ser7 was found to be phosphorylated in the Pr and Pfr forms of phytochrome; however Ser598, previously identified in vitro, was found in vivo to be phosphorylated in seedlings exposed to R light but not in those grown in the dark (Lapko et al., 1999) (Figure 5.1). This suggests that phosphorylation at Ser598 may be a light-dependent event.

Phytochrome autophosphorylation, demonstrated using purified oat phyA extracts, probably provided the first evidence that phytochrome may act as a kinase. Experiments performed by Yeh and Lagarias (1998) provided some support for such a role. Incubation of recombinant oat and green algal phytochromes (expressed in yeast) with $[\gamma\text{-}^{32}P]$ ATP exhibited greater labelling in the Pfr versus the Pr form. These experiments suggested that Pfr was preferentially phosphorylated. The acid stability and base lability of these phosphorylations were characteristic of phosphoserine or phosphothreonine (Yeh and Lagarias, 1998). The observed regulation of phosphorylation by light led the authors to conclude that at least oat phyA phytochrome is a Ser/Thr kinase.

The Ser/Thr kinase activity of phytochrome was not predicted given that it had no sequence homology to Ser/Thr kinases. Instead, sequence analysis of eukaryotic phytochrome revealed that the C-terminal domain shared sequence similarity with bacterial sensor proteins (Schneider-Poetsch et al., 1991). Alignment of plant phytochrome sequence with that of Cph1 (Cyanobacterial phytochrome), which was reported to have histidine–kinase activity, revealed two divergent domains with homology to histidine kinases (Yeh and Lagarias, 1998). The first of these contains two PAS (PER/ARNT/SIM) motifs and is referred to as the PAS-related domain (PRD); the second of these is the histidine kinase-related domain (HKRD) (Yeh and Lagarias, 1998). The sequences of the PRD and HKRD from phytochrome are more similar to the HKD of Cph1 than each other, hence the proposal that they arose by duplication of an ancestral HKD with similarity to Cph1 (Yeh and Lagarias, 1998). Cph1 operates as the light sensor of a two-component sensory system (Yeh *et al.*, 1997). Although experimental evidence provides the possibility that plant phytochrome acts as a kinase, we still lack the unequivocal evidence that demonstrates this function *in vivo*. There are, however, other reasons to doubt that phytochrome is a genuine protein kinase. First, as mentioned above, the HKRD of phytochrome does not contain the consensus sequences normally associated with such activity. Secondly, there are concerns that the kinase activity observed in the studies mentioned could be due to a protein with kinase activity that is closely associated with phytochrome. Thirdly, truncated phyB, lacking the HKRD, could still act to control inhibition of hypocotyl elongation and flowering, though point mutations within the HKRD did affect these responses (Krall and Reed, 2000).

If phytochromes do behave as kinases, evidence for their physiological role could be suggested by the enhanced very low fluence response observed in mutants lacking the serine-rich domain and reduced high irradiance response for hypocotyl elongation (Casal *et al.*, 2002). The Lm-2 accession of *Arabidopsis* has a single phyA amino acid substitution (Met548Thr), and Lm-2 seedlings are impaired in FR responses. Similarly, plants with the same substitution in phyB have altered physiological responses to R light. This seemingly important amino acid residue has, however, not been identified as important for phosphorylation, yet autophosphorylation of *Avena* Lm-2 was reduced. Additionally, light-induced degradation of phyA did not occur in Lm-2 (Maloof *et al.*, 2001). Collectively these data underline the complexity of phytochrome phosphorylation/dephosphorylation in light signalling.

5.2.2   *Phytochrome kinase substrate 1*

Phytochrome kinase substrate 1 (PKS1) was identified by yeast two-hybrid screening using the 160 amino acid C-terminal of phyA as bait (Fankhauser *et al.*, 1999). PKS1, in fact, interacts with both phyA and phyB, in either the Pr or Pfr form. The phosphorylation of PKS1 was examined using a fusion of PKS1 with glutathione S-transferase (GST). The GST–PKS1 fusion was phosphorylated by recombinant oat phyA; both the phosphorylation and autophosphorylation of phyA were increased by R light, i.e. with phytochrome conversion to Pfr (Fankhauser *et al.*, 1999). These results suggest an *in vivo* kinase activity of phyA. Overexpression of PKS1 led to less inhibition of hypocotyl elongation in W (white) and R light, suggesting a role for PKS1 in phyB responses. Consistently, the W light phenotype of PKS1 overexpression was much reduced in a *phyB* mutant background, and was enhanced in a *phyA* mutant when compared with wild type. A role for PKS1, and for PKS2, the closest homologue, in phyA signalling has been demonstrated using *pks1* and *pks2* mutants (Lariguet *et al.*, 2003). These mutants are affected in the very low fluence response, and they display altered phenotypic responses for cotyledon opening, inhibition of hypocotyl elongation and the FR block of greening under pulses of FR, but not continuous FR (FRc). The *pks1 pks2* double mutant lacked the phenotypes associated with the single mutants, suggesting that PKS1 and PKS2 act in an antagonistic manner. Expression of *PKS1* is induced rapidly, and transiently, by W light; protein levels mimic this pattern with a 2 h delay for W and R light. Under FRc *PKS1* mRNA followed a similar pattern to that of W and R light; however, PKS1 protein levels increased presumably because of increased protein stability. So it appears that phytochrome signalling through PKS1 (and PKS2) is regulated at many levels, with phosphorylation being perhaps just one of these.

5.2.3 Nucleoside diphosphate kinase 2

Nucleoside diphosphate kinases (NDPKs) are enzymes found in prokaryotic and eukaryotic organisms; they catalyse the transfer of γ-phosphate to NDP from ATP. NDPKs have been characterised from a number of plant species, including rice, where they control coleoptile elongation (Pan *et al.*, 2000), oat (Sommer and Song, 1994), pea (Finan *et al.*, 1994), tomato (Harris *et al.*, 1994) and spinach (Nomura *et al.*, 1991). In *Arabidopsis* several NDPK isoforms exist, and the regulation of NDPK by light has been demonstrated; *NDPK Ia* transcript is induced by UV-B, this enzyme may play a role in histidine biosynthesis (Zimmermann *et al.*, 1999). However, only NDPK2 has implicated in physiological responses associated with phytochrome signalling.

NDPK2 was identified in the same manner as PKS1, using a yeast two-hybrid screen with the phyA C-terminal domain as bait. However, unlike PKS1, NDPK2 binds preferentially to the Pfr from of phyA (Choi *et al.*, 1999) and phyB (Shen *et al.*, 2005). Neither NDPK1 nor NDPK3 was observed to have any interaction with phytochrome. The NDPK isoforms share 72%–75% homology, with the C-terminal extension displaying the most variability; domain swap experiments supported the specificity for interaction with phytochrome being conferred by this region (Im *et al.*, 2004). At a structural level, the C-terminal extensions provide each NDPK isoform with unique side chain extensions. Evidence for interaction of phyA with NDPK2, occurring through the C-terminal PRD of phyA, came from binding assays and immunoprecipitation with phyA deletion constructs, and from reduced yeast two-hybrid interaction of phyA with missense mutations within this region (Choi *et al.*, 1999; Im *et al.*, 2004). These results were corroborated by Shen *et al.* (2005) who, using phyA C-terminal mutants, identified PAS domain A of phyA as the site of interaction with NDPK2. By spectrophotometrically measuring the decrease in NADH in different concentrations of phytochrome, Choi *et al.* (1999) demonstrated that NDPK2 activity was increased by phyA in the Pfr form. This change in activity was achieved by R-light-dependent reduction of the K_m value of NDPK2. In hyperactive *ndpk2* mutants the reduction of ATP and GDP K_m values was less than in wild type, substantiating a link between greater activity of NDPK2 and increased nucleotide affinity (Shen *et al.*, 2005). The ability of NDPK2 to bind

GDP was not increased by Pfr, so increase in NDPK2 activity does not arise from increased binding potential. Detailed studies by Shen *et al.* (2005) showed that Pfr increases the activity of NDPK2 by lowering the pK_a values of His197, which is found in the nucleotide-binding pocket of NDPK. The lower pK_a leads to increased activity by enabling phosphorylation or dephosphorylation to occur more easily. Observations that autophosphorylation of NDPK2 was increased by addition of Pfr, and that NDPK2 phosphotransferring ability to myelin basic protein (MBP) was conferred by Pfr, provided evidence for both increased phosphorylation and dephosphorylation by lowered His197 pK_a. Additionally, the presence of the substrate MBP increased autophosphorylation of NDPK2.

The physiological effects of altered NDPK2 levels are illustrated by the phenotype of *ndpk2* mutants. Under FRc or Rc *ndpk2* seedlings displayed reduced inhibition of hypocotyl elongation, despite having shorter hypocotyls than wild type in the dark (Choi *et al.*, 2005). Hook opening and cotyledon expansion was also reduced (Choi *et al.*, 1999); together these results suggest a role for NDPK2 in both phyA- and phyB-mediated responses. A mechanism for altered physiological responses was suggested by Choi *et al.* (2005), who observed that induction of the auxin responsive genes IAA4 and IAA17 by auxin was reduced in *ndpk2* mutants. These workers proposed that NDPK2 participates in auxin-regulated responses, at least partly by regulating auxin transport.

Moon *et al.* (2003) demonstrated that NDPK2 is also involved in responses to oxidative stress. From plants overexpressing NDPK2, proteins were autophosphorylated at higher levels. AtMPK3 and AtMPK6, *Arabidopsis* mitogen activated protein kinases (MAPKs), were shown to be targets of NDPK2 phosphorylation. NDPK2 enhances the ability of AtMPK3 to phosphorylate its substrate MBP. Transgenic plants that were overexpressing NDPK2 had greater resistance to cold, salt and reactive oxygen species stress. Together these data demonstrate that NDPK2 is not only involved in light signalling but also in other important developmental responses.

5.2.4   FYPP

Kim and coworkers (2002) identified a PP2A-related catalytic subunit, designated FYPP, that predominately expressed in floral organs and influenced flowering time. FYPP loss-of-function mutants and antisense plants exhibited an accelerated flowering phenotype in long days (Kim *et al.*, 2002) whereas sense plants flowered slightly later.

The interaction with phytochrome (C-terminus) was shown by coexpression and protein interaction in yeast and by *in vitro* immunoprecipitation assays using a recombinant FYPP–GST fusion protein. FYPP associated with both phyA and phyB but this interaction was 1.4 times stronger with phyB than phyA. FYPP bound preferentially to the Pfr form of phytochrome, although there was only a 30% difference in binding when compared to the Pr form.

The specificity towards the Pfr form and the observed red light (see Section 5.3) effects on activity is relatively small. It is, however, to be expected that FYPP also contains a structural subunit, although these have not yet been identified, and true *in vivo* effects may therefore be much more prominent. Therefore, the regulatory properties and physiological mechanisms for FYPP and phytochrome interactions are far from solved and the various effects found may vary or even be different in a putative FYPP complex.

5.2.5 PAPP5

Ryu and coworkers (2005) identified a type 5 phosphatase PAPP5 that interacted with both phyA and phyB. The type 5 phosphatases (PP5) are closely related to PP2A phosphatases and are inhibited by okadaic acid. In contrast to PP2As, the PP5 phosphatases consist of only a single peptide chain that includes both regulatory and catalytic functions (Chinkers, 2001). These phosphatases are found in all eukaryotes examined so far, from yeast to humans. They are characterised by an N-terminal TPR (tetratricopeptide repeat) domain and a catalytic domain similar to PP2A/PP1 catalytic domains. The PAPP5 was identified by yeast two-hybrid screening, using phyA as bait. Both phyA and phyB were shown to bind to PAPP5 by *in vitro* immunoprecipitation assay. The TPR domain serves both regulatory functions and mediates protein–protein interactions. For interaction with phytochrome the TPR region of PAPP5 was shown to be necessary and sufficient. As expected for a protein of functional importance, binding of PAPP5 to phytochrome strongly depends on the isoform of phytochrome, showing much higher affinity towards the Pfr form. Furthermore, assays using oat phyA suggested that phytochrome was a substrate for PAPP5. Studies of PAPP5 loss-of-function mutants and overexpression lines confirmed that PAPP5 is involved in both phyA- and phyB-mediated processes. Generally, overexpressing plants were hypersensitive to R and FR light whereas loss-of-function mutants were hyporesponsive. In a similar fashion to *phyB* null mutants, the *papp5* loss-of-function mutants flowered earlier than the wild type when grown in long days (Ryu *et al.*, 2005). Thus, *papp5* and *fypp* (see Section 5.2.4) loss-of-function mutant phenotypes may result from reduced phytochrome activity brought about by enhanced phytochrome phosphorylation status.

PAPP5 has also been shown to enhance phyA-NDPK2 binding. When phosphorylated oat phyA Pfr was incubated with PAPP5, its affinity for NDPK2 increased more than sixfold. This result is in accordance with the finding that PAPP5 could dephosphorylate oat phytochrome Ser598, because phosphorylation of this amino acid was previously shown to hinder binding of NDPK2 (Kim *et al.*, 2004). PAPP5 can also dephosphorylate the other two phosphorylation sites Ser7 and Ser17 on phytochrome.

A working model consistent with the various experimental findings can now be constructed (Ryu *et al.*, 2005), depicting phytochromes in three different stages: Pr, Pfr phosphorylated and Pfr non-phosphorelated/dephosphorylated. The flux of light information and signalling would increase with photoconversion from Pr to Pfr phosphorylated, and further increased with Pfr dephosphorylation.

5.3 Cryptochromes

5.3.1 Cryptochrome phosphorylation

Cryptochromes are photolyase-like blue light receptors and have been shown to become phosphorylated both *in vivo* and *in vitro* (see Chapters 2 and 3). *In vivo* *Arabidopsis* cry1 and cry2 are phosphorylated in response to blue light as shown by feeding plants with $^{32}PO_4{}^{3-}$ (Shalitin *et al.*, 2002, 2003). For cry2, maximum labelling was seen after 10–15 min of blue light exposure, and for cry1 about 40 min of illumination gave maximum phosphorylation. Interestingly, after further exposure to blue light the concentration of phosphorylated cry2 decreased without any increase in non-phosphorylated cry2, showing that cry2 was not dephosphory-lated but rather degraded. Phosphorylation is therefore most likely a trigger for cry2 degradation. Phosphorylated cry1, on the other hand, appears to be more stable in the light, although interpretations of the results can be ambiguous since synthesis versus degradation has not been thoroughly investigated. Phytochrome has previously been shown to interact with and phosphorylate crys (Ahmad *et al.*, 1998). However, examination of different phytochrome mutants, including double and triple mutants, did not provide support for this proposed phytochrome function. However, the possibility that phytochrome phosphorylates cryptochrome cannot be excluded because a null mutant lacking all five phytochromes in Arabidopsis was not available(Shalitin *et al.*, 2002).

Attempts to set up an *in vitro* phosphorylation assay for cry1 revealed that heterologously expressed cry1 readily autophosphorylated, and the phosphorylation was strongly enhanced by blue light (Shalitin *et al.*, 2003). It was further shown that *in vitro* autophosphorylation occurred on serine residues; it required the flavin adenine dinucleotide (FAD) cofactor, and blue light dependency was confirmed (Bouly *et al.*, 2003).

Links between cryptochrome phosphorylation and cryptochrome function have recently been demonstrated (Shalitin *et al.*, 2003). Shalitin and coworkers isolated nine *cry1* missense mutants showing cryptochrome-deficient characteristics. Each of the mutants expressed the full-length CRY1 apoprotein but these mutated CRY1 proteins failed to phosphorylate *in vivo*. Hence cry1 phosphorylation is closely associated with cry1 function. Functional phosphorylation of cry1 was further confirmed by the work of Zeugner and coworkers (2005). In their work they mutated two of the three tryptophans conserved between photolyases and crys and required for flavin-reducing electron transfer chain in *Escherichia coli* photolyase. These tryptophans were clearly important for the intrinsic electron transfer also in *Arabidopsis* cry1 (Giovani *et al.*, 2003; Zeugner *et al.*, 2005) as the mutations led to a phenotype reminiscent of cry1-deficient plants. This work also established a link between the photoinduced electron transfer reaction and autophosphorylation. When tested *in vitro*, the tryptophan mutated proteins retained basal (light-independent) autophosphorylation, although reduced compared to the wild type. The tryptophan mutated protein therefore retained the capacity for undergoing autophosphorylation, but the stimulation of phosphorylation by blue light was completely suppressed, indicating that intraprotein electron transfer is necessary for stimulation of autophosphorylation of cryptochrome.

Intriguingly cry1 has no obvious kinase domain, but still autophosphorylates. The precise phosphorylation sites have, however, not been determined (Figure 5.1). Bouly and coworkers (2003) showed that cry1 as well as cry2 indeed does bind ATP as confirmed by several experimental approaches. Crystal structure determination of the *Arabidopsis* photolyase-like domain of cry1 revealed binding of an ATP analogue near the FAD cofactor (Brautigam *et al.*, 2004). Three crystal structures of the crytpochrome/photolyase superfamily have been studied: *Arabidopsis* cry1, *E. coli* CDP photolyase and *Synechococcus* CRY-DASH (Brautigam *et al.*, 2004; Lin and Todo, 2005; Chapter 2). Whilst there are clear similarities between these three-dimensional structures, differences were also apparent. For instance cry1 does not have the positively charged groove along the surface where DNA binds in proteins with photolyase activity, and the FAD-access cavity is larger and deeper in *Arabidopsis* cry1 compared with *E. coli* photolyase (Lin and Todo, 2005). Apparently, this cavity in photolyases, which binds the pyridine dimer in need of repair, has evolved into a cavity that binds ATP in *Arabidopsis* cryptochrome. It has further been suggested that the C-terminal part of cry1 can bend onto this ATP-binding domain, resulting in phosphorylation of the C-terminal end (see Chapter 2 for detailed information). Crystallisation of the holocryptochrome will shed light on this suggested mechanism.

5.3.2 Phosphorylation of the C-terminal end is necessary for signal transduction

The C-terminal end of either *Arabidopsis* CRY1 or CRY2 was shown to mediate a constitutive light response when fused to GUS and transformed into wild type *Arabidopsis* (Yang *et al.*, 2000), a phenotype similar to the constitutive light response exhibited by many (*constitutive photomorphogenic*) *cop* mutants. It was also shown that this C-terminal CRY end was constitutively phosphorylated *in vivo* in both blue light and darkness (Shalitin *et al.,* 2002), confirming the assumption that phosphorylation is part of the cry signalling mechanism. The N-terminal end is, however, also important for phosphorylation of cryptochrome because mutations in the N-terminal region abolish phosphorylation. Furthermore, dimerisation of cryptochrome based on the N-terminal domain studies was shown to be necessary for phosphorylation of the cryptochrome. Experiments showed that in seedlings expressing the GUS–C-terminal CRY fusion protein, a multimer was formed and GUS was able to confer a change in the C-terminal CRY, which resulted in constitutive phosphorylation (Sang *et al.*, 2005).

Recently it was also confirmed that the ATP-binding domain (photolyase homology region) of *Arabidopsis* cry1 interacts with the C-terminal region (Partch *et al.*, 2005). The C-terminal domain was found to lack secondary structures like α-helices and β-sheets, showing intrinsic disorder. Partch and coworkers pointed out that such disordered regions are more common in signal transduction and regulatory proteins than in metabolic and biosynthetic components as they readily interact with multiple other proteins in a thermodynamically efficient way. The ordered tertiary structure of the C-terminal domain was increased when interacting with the ATP-binding domain, but in the response to blue light and concomitant phosphorylation the C-terminal domain underwent a conformational change and was apparently released from the ATP-binding domain. The STAESS region in the DAS motif (DQXVP-acidic-STAESS) was shown to be the site of conformational rearrangement (Partch *et al.*, 2005).

A picture of cryptochrome function is emerging based on autophosphorylation being an important step in the transmission of the blue light signal. cry is inactive and stable in the dark, but in response to blue light exposure phosphorylation is triggered, leading to conformational changes in the C-terminal end, resulting in active cryptochrome and possibly simultaneously also unstable cryptochrome. cry1 and cry2 may, however, be different with respect to stability in light and darkness (Lin and Shalitin, 2003; Partch *et al.*, 2005). The phosphorylated tail probably interacts with COP1 and hinders COP1 E3 ubiquitin ligase activity. As a result COP1 is less effective in degrading nuclear transcription factors like HY5 (long hypocotyl 5). A phosphatase may be required for reversion of phosphorylated to non-phosphorylated cryptochrome; however, this may not be essential for cry2 signalling as the cry2 protein is rapidly degraded in the light.

5.4   Phototropins

Phosphorylation of a 120-kDa membrane-bound protein was long recognised as an early sign of phototropism (Briggs and Huala, 1999). Subsequently this protein was shown to be the blue light receptor kinase phototropin (Huala *et al.*, 1997; Chapter 3). Indeed, the phototropin C-terminal part was shown to contain the sequence motifs typical of eukaryotic protein kinases, and is closely related to the PvPK (*Phaseolus vulgaris* protein kinase) group of serine–threonine kinases (Huala *et al.*, 1997; Hardie, 1999).

Phosphorylation of phototropin has been studied *in vitro* and *in vivo*, and the concentration of phosphorylated phototropin rapidly reaches a maximum level in response to blue light irradiation (Short *et al.*, 1994). In etiolated *Pisum sativum* plasma membrane preparations, blue light induced maximum labelling of the 120-kDa protein from $[\gamma$-^{32}P] ATP in 2–5 min, followed by gradual loss of phosphorylation during the next 15 min (Short *et al.*, 1994). Similarly, incubation of microsomal membranes from *Arabidopsis* showed a high incorporation of $[\gamma$-^{32}P] ATP into phototropin within 2 min (Liscum and Briggs, 1995). *In vivo* assays using *Vicia faba* guard cell protoplasts showed that labelling of phototropin with ^{32}P-orthophosphate peaked after about 1 min in response to blue light, and then decreased during the next 20 min of the experiment (Kinoshita *et al.*, 2003). The disappearance of phosphorylated phototropin would require either involvement of a phosphatase or rapid degradation of the phosphorylated phototropin. Testing of a general phosphatase inhibitor (NaF) or searching for degradation products did not reveal how the phosphorylated product disappeared (Short *et al.*, 1994). From recent work by Salomon and coworkers (2003) on *A. sativa*, it has been deduced that a phosphatase(s) is probably involved, because different phosphorylation sites within phot1 had different turnover rates *in vivo*. The properties, classification and identification of a phosphatase(s) involved require further investigation.

Arabidopsis phot1 expressed in insect cells has been shown to retain autophosphorylation and kinase activity in response to blue light (Briggs and Christie, 2002; Chapter 3). The N-terminal domain of phototropin contains two light oxygen voltage (LOV) domains and these were assumed to undergo a conformational change in response to the formation of a cysteinyl adduct between a conserved cysteine (corresponding to residue 39 within each LOV domain) and the chromophore flavin mononucleotide (FMN). This conformational change would then activate the kinase in the C-terminal part of the protein (Kasahara *et al.*, 2002). Mutation analysis of this cysteine in *Arabidopsis* phototropins showed that it was only the cysteine in LOV2 that was necessary for increasing phot1 as well as phot2 kinase activity and hypocotyl curvature (Christie *et al.*, 2002). Studies of isolated LOV domains confirmed that the adduct between the cysteine residue and a carbon of the isoalloxazine ring of FMN was formed in response to blue light irradiation. A model for the conformational change was later suggested on the basis of mutation analysis and NMR spectroscopy, indicating that displacement of a special α-helix (Jα) is the critical event in regulation of the kinase activity (Harper, 2004). In this model the Jα helix interacts with the kinase domain to lock it into an inactive conformation, and upon illumination Jα becomes displaced and the interaction between LOV2, Jα and the kinase domain changes (Harper, 2004). Some other possible interpretations regarding changes in domain interactions in terms of kinase activation were also suggested (Harper, 2004). The model where the LOV2 domain is a dark state inhibitor, and light activation displaces the LOV2 domain away from the kinase domain is further supported by the results of Matsuoka and Tokutomi (2005), who clearly demonstrated inhibition of the kinase domain by LOV2 in darkness.

Early work on maize and pea indicated phosphorylation of phototropin at multiple sites (Palmer *et al.*, 1993; Short *et al.*, 1994) but the actual domains phosphorylated were first identified using extracts of etiolated oat coleoptiles. The extracts were incubated with [γ-^{32}P] ATP, exposed to tryptic digestion and the products were analysed by two-dimensional thin-layer electrophoresis (Salomon *et al.*, 2003). In these studies the tissue was also pre-treated with blue light to induce *in vivo* phosphorylation. By comparing several differently exposed samples, it was concluded that certain sites were phosphorylated more quickly than others. The response to light intensity was also different for different sites, and the rate of dephosphorylation in darkness depended on the site in question. *In vitro* it was found that a PKA kinase resulted in the same pattern of phosphorylation and was therefore used to specifically identify the phosphorylation sites in cloned (N-terminal) phot1 fragments which were mutated by site-directed mutagenesis. Two phosphorylation sites were identified in the N-terminal domain upstream of LOV1 (S27 and S30), and six phosphorylation sites were identified in the hinge between LOV1 and LOV2 (Figure 5.1). No other domains were found to be phosphorylated (Salomon *et al.*, 2003).

Although phototropin autophosphorylation activity has been long recognised, recent findings have shown that a general kinase substrate, i.e. casein, could be phosphorylated by phototropin (Matsuoka and Tokutomi 2005). By expressing different fragments of *Arabidopsis* phot2 in *E. coli*, it was shown that the LOV2 domain inhibited the casein kinase activity and that this inhibition was abolished by blue light irradiation.

Phototropins mediate stomatal opening, and 14-3-3 proteins bind to phosphorylated phototropin from *V. faba* guard cells (Kinoshita *et al.*, 2003). Furthermore, phosphorylation of a serine in the hinge between LOV1 and LOV2 was essential for binding of 14-3-3. It is still unknown how phosphorylation is involved in signal transduction from phototropin. Since it has now been shown that phototropin can phosphorylate different substrates (Matsuoka and Tokutomi, 2005), it will be exciting to discover the real *in vivo* protein substrates for phototropins. Furthermore, the function of 14-3-3 proteins in the phototropin signalling needs to be clarified.

5.5   Is phosphorylation/dephosphorylation important for downstream events?

Many additional components associated with phytochrome signalling have been identified, examples include FHY1 (Whitelam *et al.*, 1993), FAR1 (Hudson *et al.*, 1999), RED1 (Wagner *et al.*, 1997), HFR1 (Fairchild *et al.*, 2000), PIF3 (phytochrome interacting factor 3) (Ni *et al.*, 1998), SPA1 (Hoecker *et al.*, 1998), PAT1 (Bolle *et al.*, 2000), EID1 (Empfindlicher Im Dunkelroten Licht) (Dieterle, 2001), whilst in terms of cryptochrome and phototropin signalling data is sparse. The mechanism for action of these signalling intermediates vary from transcriptional regulation to protein degradation; however, it is clear that phosphorylation and dephosphorylation also play key roles in regulating downstream events.

5.5.1   HY5

The HY5 protein is a bZIP (basic leucine zipper) transcription factor that has been shown to promote photomorphogenesis (Oyama *et al.*, 1997) and negatively regulate auxin signalling (Cluis *et al.*, 2004). The activity of HY5 is regulated by light through interaction with COP1. In the dark HY5 is targeted for degradation, this process is mediated by COP1 E3 ligase (Hardtke *et al.*, 2000; Osterlund *et al.*, 2000). The light regulation of this interaction/degradation arises in part from the nucleocytoplasmic partitioning of COP1. In the dark COP1 is nuclear localised, and following transfer to light conditions nuclear levels gradually deplete (von Armin and Deng, 1994). Additionally, regulation of photomorphogenesis through HY5 occurs via phosphorylation. Hardtke *et al.* (2000) demonstrated that protein extracts of *Arabidopsis* phosphorylate GST–HY5 fusions *in vitro*, with more activity occurring with extracts form dark-grown seedlings than light-grown ones; this appeared to be mediated by phytochrome. Endogenous HY5 was also shown to be phosphorylated; however, dephosphorylated HY5 interacted with COP1 four times more strongly than the phosphorylated version *in vitro*. Similarly, unphosphorylated HY5 was reported to bind more strongly to the promoters of *CHS1* and *RBCS1a*, genes that were previously identified as targets of HY5. The phosphorylation of HY5 may be mediated by CKII (casein kinase II), for which there is a consensus sequence in the COP1-binding domain. The actual site of HY5 phosphorylation is Ser36. In accordance with the greater interaction of unphosphorylated HY5 with COP1, this isoform was observed to undergo degradation. The detection of less unphosphorylated HY5 in dark-grown seedlings is consistent with the greater degradation of this isoform.

5.5.2   *Long hypocotyl in far-red light*

Long hypocotyl in far-red light (HFR1), which is involved in both phyA and cry1 signalling, was identified in a number of laboratories by isolation of a mutant that was long in FR ((RSF1) Fankhauser and Chory, 2000; (HFR1) Fairchild *et al.*, 2000; (REP1) Soh *et al.*, 2000). The gene encodes a bHLH (basic helix-loop-helix) protein that was demonstrated to interact with PIF3, but not with phyA or phyB (Fairchild *et al.*, 2000). Evidence for a role of phosphorylation in HFR1 stability was reported by Duek *et al.* (2004). Western analysis of hemagglutinin (HA-)tagged HFR1 revealed a second isoform present in light but downregulated in dark. The second band was not observed if immunoprecipitated samples were subjected to treatment with shrimp alkaline phosphatase, suggesting that this second isoform was phosphorylated and unstable in the dark. In the presence of proteasome inhibitors, phosphorylated HFR1 from dark-grown seedlings was more stable, indicating that this isoform is normally degraded by the 26S proteasome. Experiments to determine the stability of HFR1 in a *cop1* mutant background indicated that degradation of phosphorylated HFR1 is COP1-dependent. The lower detectable levels of the phosphorylated form of HFR1 in the dark are due to the rapid degradation of this isoform. Physical interaction between COP1 and HFR1 was demonstrated using yeast two-hybrid assays, with the N-terminus of HFR1 being the likely site of interaction. As for HY5 (see Section 5.5.1) COP1 appears to target HFR1 degradation by the 26S proteasome; however, there is a fundamental difference. In the case of HY5 the unphosphorylated protein is the target for COP1, yet HFR1 is targeted for degradation in its phosphorylated isoform.

5.5.3   *Circadian clock-associated and late elongated hypocotyl*

Circadian clock-associated (CCA1) and late elongated hypocotyl (LHY) are proteins integral to the circadian oscillator in plants regulating developmental responses such as hypocotyl elongation and flowering time. CCA1 and LHY are myb transcription factors (Wang *et al.*, 1997; Schaffer *et al.*, 1998) whose transcription and protein abundance oscillate in a circadian manner, peaking around dawn (see Chapter 8 for details). These two proteins bind and together repress expression of TOC1 (timing of CAB1) through binding to its promoter. CCA1 and LHY also feed back to repress their own transcription. In addition to this, the degradation of these proteins during the day gradually releases repression of TOC1, enabling its transcript to increase to a maximum around dusk. TOC1 in turn promotes expression of CCA1 and LHY, completing the cycle (Alabadi *et al.*, 2002).

The phytochromes A, B, D and E in *Arabidopsis* have all been demonstrated to provide information concerning the light environment to the circadian clock, primarily through analysis of mutants (Somers *et al.*, 1998; Devlin and Kay, 2000) Similarly the cry1 and cry2 have also been shown to be involved in light input to the clock (Devlin and Kay, 2000; Chapter 8).

Yeast two-hybrid screening, using CCA1 as bait, identified interaction of this protein with the subsequently named CKB3, a protein with homology to the regulatory β-subunit of protein kinase CK2 (Sugano *et al.*, 1998). CK2 was demonstrated to phosphorylate CCA1 and LHY (Sugano *et al.*, 1999) *in vitro*, and *in vivo* phosphorylation of CCA1 was shown to occur using *Arabidopsis* whole cell extracts. Additionally, the DNA-binding activity of the CCA complex was shown to require phosphorylation by CK2. A role for CBK3 phosphorylation of CCA1 and LHY in the circadian clock was also reported by Sugano *et al.* (1999). Transgenic plants overexpressing CKB3 had altered expression of *CCA1* and *LHY* as well as similarly changed expression of *CAT2*, *CAT3*, *CCR2* and *Lhcb1*1*, genes known to be outputs of the circadian clock. CK2 phosphorylation of CCA1 was shown by Daniel and Tobin (2004) to be required for CCA1 regulation of circadian rhythmicity. This was illustrated using transgenic plants overexpressing wild type or mutated CCA1. Plants overexpressing wild-type CCA1, which is phosphorylated, are arrhythmic for CCA1 and expression of other genes (*CAT2*, *CAT3*, *CCR2* and *Lhcb1*1*) (Wang and Tobin, 1998); however, overexpression of mutated CCA1, which is not phosphorylated, does not lead to arrhythmia, again highlighting the necessity for phosphorylation in regulation of circadian rhythms (Daniel and Tobin, 2004).

5.5.4   EID1

Dieterle *et al.* (2001) reported that EID1 is an F-box protein, forming part of the SCF (Skp, Cdc53, F-box) E3 ubiquitin ligase complex that targets proteins for degradation. EID1 does not target phyA itself for degradation, but may instead be involved in proteolysis of downstream components. The authors speculate that phosphorylation may play a role in this activity as F-box proteins often interact only with phosphorylated proteins; in this situation modification of proteins in a Pfr-dependent manner may regulate the abundance of that protein through interaction with EID1.

5.5.5   Aux/IAA

Aux/IAA genes are one family of auxin response genes that contain motifs for DNA binding of auxin response factors in their promoters. Using the SHY2 proteins of *Arabidopsis* (Reed *et al.*, 1998), Colón-Carmona *et al.* (2000) demonstrated that oat phyA and IAA proteins can interact *in vitro*. The authors also reported that Pr and Pfr forms of oat phyA could phosphorylate the N-terminal domain of Aux/IAA proteins. In wild-type seedlings, phosphorylation of IAA3 was not detected, whereas detection of phosphorylated SHY2-2 mutant protein was observed *in vivo*. However, a role for phosphorylation of Aux/IAA by phytochrome in regulating light and auxin interaction has not yet been established.

5.5.6 PP7

Little is known of the downstream events following light perception by cryptochrome. However, a Ser/Thr PP7 showing similarity to the *Drosophila* retinal degeneration C protein phosphatase has been identified based on its role in blue light signalling (Møller *et al.*, 2003). Transgenic plants deficient for PP7 exhibit loss of hypocotyl growth inhibition and limited cotyledon expansion specifically in response to blue light irradiation. Strikingly, these phenotypes are as dramatic as observed in the *hy4* mutant deficient for cry1, indicating that PP7 is of paramount importance for cryptochrome signalling. Although it is known that PP7 is indeed a nuclear-localised calcium-dependent phosphatase, how PP7 exerts its function is unknown. One possibility would be that PP7 dephosphorylates, and thereby stabilises proteins necessary for photomorphogenesis. In the absence of PP7, these proteins would remain phosphorylated and could be rapidly degraded by a pathway involving COP1.

5.5.7 Downstream of phototropin

Phototropins control important processes such as phototropism, chloroplast orientation and stomatal opening. It is not clear how phototropins influence these processes, but phosphorylation steps downstream of the photoreceptor are almost certainly involved. For example it has been suggested that phot1 together with interacting proteins (NPH3 and RPT2) may form a large complex that associates with the plasmalemma. This association has been shown to depend on its phosphorylation status and the complex would then lead to changes in auxin transport that results in differentiated growth and bending of the plant. However, this needs to be demonstrated (Esmon *et al.*, 2005).

Another phototropin-influenced process is stomatal opening, which depends on an active H^+-ATPase. The H^+-ATPase shows activation by blue light; however, for the *phot1 phot2* double mutant no activation is seen. Activation of the ATPase requires phosphorylation of the ATPase C-terminus and binding of 14-3-3 proteins (Ueno *et al.*, 2005). Although phototropins have been shown to phosphorylate other substrates (casein), phototropin does not phosphorylate the ATPase because phosphorylation takes place in the *phot1 phot2* mutant in response to the ATPase activator fussicoccin (Ueno *et al.*, 2005). Strikingly, both phototropins and the H^+-ATPase bind 14-3-3 proteins; however, the signalling cascade linking phototropins and regulation of the H^+-ATPase is not at all clarified.

A component that may be important in several signalling cascades starting from phototropins is actin. Actin filaments are involved in stomatal opening, and ion channels are known to be linked to actin filaments. Furthermore, chloroplast movement is a phototropin-controlled process known to involve actin filaments (Staiger, 2000). Actin function is influenced by specific interacting proteins, and such proteins, though not yet studied in plants, are regulated by phosphorylation in other organisms (Staiger, 2000). Actin depolymerising factors in maize and *Arabidopsis* are also regulated by phosphorylation, but the influence of blue light and other signals still needs to be examined (Staiger, 2000).

5.6   Conclusions

On the basis of our knowledge of signal transduction pathways in other organisms, it is clear that plants have a unique complement of phosphorylation and dephosphorylation mechanisms involved in photoreceptor signalling. Despite this it is becoming increasingly clear that phosphorylation and dephosphorylation of both the photoreceptors and downstream components represent important regulatory events ensuring appropriate signal flux in response to light (Figure 5.2). Insight into photoreceptor autophosphorylation and how interacting proteins influence phosphorylation status has shed light on the involvement of both kinases and phosphatases following initial

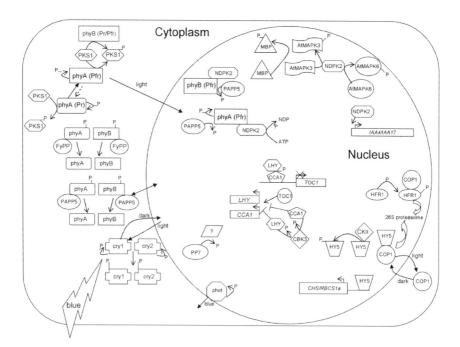

Figure 5.2   Phosphorylation events associated with light signalling. Phosphorylation and dephosphorylation of photoreceptors and their signalling intermediates play an important role in regulating plant growth responses to perceived light. The effect of a phosphorylation event on a protein differs; for example phosphorylation of HY5 prevents its degradation, whereas phosphorylated HFR1 is targeted to the 26S proteasome.

light perception. The fact that the phosphorylation status of downstream signalling intermediates can dramatically influence light responses, clearly indicates that phosphorylation and dephosphorylation are not limited to immediate early events.

Although recent progress in the field has provided a solid basis for future studies, it will now be important not only to identify additional kinases and phosphatases but also to elucidate the nature of the various phosphorylation/dephosphorylation events and integrate these into a coherent network.

References

Ahmad, M., Jarillo, J.A., Smirnova, O. and Cashmore, A.R. (1998) The CRY1 blue light photoreceptor of *Arabidopsis* interacts with phytochrome A in vitro. *Mol. Cell* **1**, 939–948.

Alabadi, D., Yanovsky, M.J., Mas, P., Harmer, S.L. and Kay, S.A. (2002) Critical role for CCA1 and LHY in maintaining circadian rhythmicity in *Arabidopsis. Curr. Biol.* **12**, 757–761.

Barford, D., Das, A.K. and Egloff, M.-P. (1998) The structure and mechanism of protein phosphatases: insight into catalysis and regulation. *Annu. Rev. Biophys. Biomol. Struct* **27**, 133–164.

Bolle, C., Koncz, C. and Chua, N.-H. (2000) PAT1, a new member of the GRAS family is involved in phytochrome A signal transduction. *Genes and Dev.* **14**, 1269–1278.

Bouly, J.-P., Giovani, B., Djamei, A., *et al.* (2003) Novel ATP-binding and autophosphorylation activity associated with *Arabidopsis* and human cryptochrome-1. *Eur. J. Biochem.* **270**, 2921–2928.

Brautigam, C.A., Smith, B.S., Ma, Z., Palnitkar, M., Tomchick, D.R., Machius, M. and Deisenhofer, J. (2004) Structure of the photolyase-like domain of cryptochrome 1 from *Arabidopsis thaliana. Proc. Natl. Acad. Sci. USA* **101**, 12142–12147.

Briggs, W.R. and Christie, J.M. (2002) Phototropins 1 and 2: versatile plant blue-light receptors. *Trend. Plant Sci.* **7**, 204–210.

Briggs, W.R. and Huala, E. (1999) Blue-light photoreceptors in higher plants. *Annu. Rev. Cell. Dev. Biol.* **15**, 33–62.

Casal, J.J., Davis, S.J., Kirchenbauer, D, *et al.* (2002) The Serine-rich N-terminal domain of oat phytochrome A helps regulate light responses and subnuclear localization of the photoreceptor. *Plant Physiol.* **129**, 1127–1137.

Chang, C., Kwok, S.F., Bleecker, A.B. and Meyerowitz, E.M. (1993) *Arabidopsis* ethylene response gene ETR1: similarity of product to two-component regulators. *Science* **262**, 539–544.

Chinkers, M. (2001) Protein phosphatases 5 in signal transduction. *Trends Endocrinol. Metab.* **12**, 28–32.

Choi, G., Kim, J.-I., Hong, S.-W, *et al.* (2005) A possible role for NDPK2 in the regulation of auxin-mediated responses for plant growth and development. *Plant Cell Physiol.* **46**, 1246–1254.

Choi, G., Yi, H., Kwon, Y.-K, *et al.* (1999) Phytochrome signalling is mediated through nucleoside diphosphate kinase 2. *Nature* **401**, 610–613.

Christie, J.M., Swartz, T.E., Bogomolni, R.A. and Briggs, W.R. (2002) Phototropin LOV domains exhibit distinct roles in regulating photoreceptor function. *Plant J.* **32**, 205–219.

Cluis, C.P., Mouchel, C.F. and Hardtke, C.S. (2004) The *Arabidopsis* transcription factor HY5 integrates light and hormone signaling pathways. *Plant J.* **38**, 332–347.

Colón-Carmona, A., Chen, D.L., Yeh, K.-C. and Abel, S. (2000) Aux/IAA proteins are phosphorylated by phytochrome in vitro. *Plant Physiol.* **124**, 1728–1738.

Daniel, X and Tobin, E. (2004) CK2 phosphorylation of CCA1 is necessary for its circadian oscillator function in *Arabidopsis. Proc. Natl. Acad. Sci. USA* **101**, 3292–3297.

del Pozo, J.C. and Estelle, M. (2000) F-box proteins and protein degradation: an emerging theme in cellular regulation. *Plant Mol. Biol.* **44**, 123–128.

Devlin, P.F. and Kay, S.A. (2000) Cryptochromes are required for phytochrome signaling to the circadian clock but not for rhythmicity. *Plant Cell* **12**, 2499–2509.

Dieterle, M., Zhou, Y.C., Schäfer, E., Funk, M. and Kretsch, T. (2001) EID1, and F-box protein involved in phytochrome A-specific light signaling. *Genes Dev.* **15**, 939–944.

Duek, P.D., Elmer, M.V., van Osten, V.R. and Fankhauser, C. (2004) The degradation of HFR1, a putative bHLH class transcription factor involved in light signalling, is regulated by phosphorylation and requires COP1. *Curr. Biol.* **14**, 2296–2301.

Esmon, C.A., Pedmale, U.V. and Liscum, E. (2005) Plant tropism: providing the power of movement to a sessile organism. *Int. J. Dev. Biol.* **49**, 665–674.

Fairchild, C.D., Schumaker, M.A. and Quail, P.H. (2000) HFR1 encodes an atypical bHLH protein that acts in phytochrome A signal transduction. *Genes Dev.* **14**, 2377–2391.

Fankhauser, C. and Chory, J. (2000) RSF1, an *Arabidopsis* locus implicated in phytochrome A signalling. *Plant Physiol.* **124**, 39–45.

Fankhauser, C. Yeh, K.-C., Lagarias, J.C., Zhang, H., Elich, T.D. and Chory, J. (1999) PKS1, a substrate phosphorylated by phytochrome that modulates light signalling in *Arabidopsis. Science* **284**, 1539–1541.

Finan, P.M., White, I.R., Redpath, S.H., Findlay, J.B. and Millner, P.A. (1994) Molecular cloning, sequence determination and heterologous expression of nucleoside diphosphate kinase from *Pisum sativum. Plant Mol Biol.* **25**, 59–67.

Giovani, B., Byrdin, M., Ahmad, M. and Brettel, K. (2003) Light-induced electron transfer in a cryptochrome blue-light photoreceptor. *Nat. Struct. Biol.* **10**, 489–490.

Hanks, S.K. and Hunter, T. (1995) The eukaryotic protein kinase superfamily: kinase (catalytic) domain structure and classification. *FASEB J.* **9**, 576–596.

Hardie, D.G. (1999) Plant protein serine/threonine kinases: classification and functions. *Ann. Rev. Plant Physiol Plant Mol. Biol* **50**, 97–131.

Hardtke, C.S., Gohda, K., Osterlund, M.T., Oyama, T., Okada, K. and Deng, X.W. (2000) HY5 stability and activity in *Arabidopsis* is regulated by phosphorylation in its COP1 binding domain. *EMBO J,* **19**, 4997–5006.

Harper, S.M. (2004) Disruption of the LOV-Jα helix interaction activates phototropin kinase activity. *Biochemistry* **43**, 16184–16192.

Harris, N., Taylor, J.E. and Roberts, J.A. (1994) Isolation of a mRNA encoding a nucleoside diphosphate kinase from tomato that is up-regulated by wounding. *Plant Mol Biol.* **25**, 739–742.

Haynes, J.G., Hartung, A.J., Hendershot, J.D., III, Passingham, R.S. and Rundle, S.J. (1999) Molecular characterization of the B' regulatory subunit gene family of *Arabidopsis* protein phosphatase 2A. *Eur. J. Biochem.* **260**, 127–136.

Hoecker, U. (2005) Regulated proteolysis in light signalling. *Curr. Opin. Plant Biol.* **8**, 469–476.

Hoecker, U., Xu, Y. and Quail, P.H. (1998) SPA: a new genetic locus involved in phytochrome A-specific signal transduction. *Plant Cell* **10**, 19–33.

Huala. E., Oeller, P.W., Liscum, E., Han, I.-S., Larsen, E. and Briggs, W.R. (1997) *Arabidopsis* NPH1: a protein kinase with a putative redox-sensing domain. *Science* **278**, 2120.

Hudson, M., Ringli, C., Boylan, M.T. and Quail, P. (1999) The FAR1 locus encodes a novel nucleic protein specific to phytochrome A signalling. *Genes Dev.* **13**, 2017–2027.

Hwang, I., Chen, H.-C. and Sheen, J. (2002) Two-component signal transduction pathways in *Arabidopsis. Plant Physiol* **129**, 500–515.

Im, Y.J., Kin, J.-I., Shen, Y, *et al.* (2004) Structural analysis of *Arabidopsis thaliana* nucleoside diphosphate kinase-2 for phytochrome-mediated light signaling. *J. Mol. Biol.* **343**, 659–670.

Kasahara, M., Swartz, T.E., Olney, M.A., *et al.* (2002) Photochemical properties of the flavin mononucleotide-binding domains of the phototropins from *Arabidoposis*, rice, and *Chalmydomonas reiinhardtii. Plant Physiol* **129**, 762–773.

Kim, D.-H., Kang, J.-G., Yang, S.-S., Chung, K.-S. and Song, P.-S. (2002) A phytochrome-associated protein phosphatase 2A modulates light signals in flowering time control in *Arabidopsis. Plant Cell* **14**, 3043–3056.

Kim, J.-I., Shen, Y, Han, Y.-J., *et al.* (2004) Phytochrome phosphorylation modulates light signaling by influencing the protein-protein interaction. *Plant Cell* **16**, 2629–2640.

Kinoshita, T., Emi, T., Tominaga, M., *et al.* (2003) Blue-light- and phosphorylation-dependent binding of a 14-3-3 protein to phototropins in stomatal guard cells of broad bean. *Plant Physiol.* **133**, 1453–1463.

Klumpp, S. and Krieglstein, J. (2002). Phosphorylation and dephosphorylation of histidine residues in proteins. *Eur. J. Biochem.* **269**, 1067–1071.

Krall, L. and Reed, J.W. (2000) The histidine kinase related domain participates in phytochrome B function but is dispensable. *Proc. Natl. Scad. Sci. USA* **97**, 8169–9174.

Lapko, V.N., Jiang, X.-Y., Smith, D.L. and Song, P.-S. (1997) Posttranslational modification of oat phytochrome A: phosphorylation of a specific serine in a multiple serine cluster. *Biochemistry* **36**, 10595–10599.

Lapko, V.N., Jiang, X.-Y., Smith, D.L. and Song, P.-S. (1999) Mass spectrometric characterization of oat phytochrome A: isoforms and posttranslational modifications. *Protein Sci.* **8**, 1032–1044.

Lariguet, P., Boccalandro, H.E., Alonso, J.M, *et al.* (2003) A growth regulatory loop that provides homeostasis to phytochrome A signalling. *Plant Cell* **15**, 2966–2978.

Lieu, Y., Loros, J. and Dunlap, J.C. (2000) Phosphorylation of the *Neurospora* clock protein FREQUENCY determines its degradation rate and strongly influences the period length of the circadian clock. *Proc. Natl. Acad. Sci. USA* **97**, 234–239.

Lin, C. and Shalitin, D. (2003) Cryptochrome structure and signal transduction. *Annu. Rev. Plant Biol.* **54**, 469–496.

Lin, C. and Todo, T. (2005) The cryptochromes. *Genome Biol.* **6**, 220.0–220.9.

Liscum, E. and Briggs, W.R. (1995) Mutations in the *NPH1* locus of *Arabidopsis* disrupt the perception of phototropic stimuli. *Plant Cell* **7**, 473–485.

Lochhead, P.A., Sibbet, G., Morrice, N. and Cleghon, V. (2005) Activation-loop autophosphorylation is mediated by a novel transitional intermediate form of DYRKs. *Cell* **121**, 925–936.

Maloof, J.N., Borevitz, J.O., Dabi, T., *et al.* (2001) Natural variation in light sensitivity of *Arabidopsis*. *Nat. Genet.* **29**, 441–446.

Matsuoka, D. and Tokutomi, S. (2005) Blue light-regulated molecular switch of Ser/Thr kinase in phototropin. *Proc. Natl. Acad. Sci. USA* **102**, 13337–13342.

McMichael, R.W. and Lagarias, J.C. (1990) Phosphopeptide mapping of Avena phytochrome phosphorylated by protein kinases in vitro. *Biochemistry* **29**, 3872–3878.

Møller, S.G., Kim, Y.-S., Kunkel, T. and Chua, N.-H. (2003) PP7 is a positive regulator of blue light signalling in *Arabidopsis*. *Plant Cell* **15**, 1111–1119.

Moon, H., Lee, B., Choi, G, *et al.* (2003) NDP kinase 2 interacts with two oxidative stress-activated MAPKs to regulate cellular redox state and enhances multiple stress tolerance in transgenic plants. *Proc. Natl. Acad. Sci. USA* **100**, 358–363.

Ni, M., Tepperman, J.M. and Quail, P. (1998) PIF3, a phytochrome-interacting factor necessary for normal photoinduced signal transduction, is a novel basic helix-loop-helix protein. *Cell* **95**, 657–667.

Nolen, B., Taylor, S. and Ghosh, G. (2004) Regulation of protein kinases: controlling activity through activation segment conformation. *Mol. Cell* **15**, 661–675.

Nomura, T., Fukui, T. and Ichikawa, A. (1991) Purification and characterization of nucleoside diphosphate kinase from spinach leaves. *Biochem. Biophys. Acta* **1077**, 47–55.

Osterlund, M.T., Hardtke, C.S., Wei, N. and Deng, X.W. (2000) Targeted destabilization of HY5 during light-regulated development of *Arabidopsis*. *Nature* **405**, 462–466.

Ota, I.M. and Varshavsky, A. (1993) A yeast protein similar to bacterial two-component regulators. *Science* **262**, 566–569.

Oyama, T., Shimura, Y. and Okada, K. (1997) The *Arabidopsis* HY5 gene encodes a bZIP protein that regulates stimulus-induced development of root and hypocotyl. *Genes Dev.* **11**, 2983–2995.

Palmer, J.M., Short, T.W., Gallagher, S. and Briggs, W.R. (1993) Blue light-induced phosphorylation of a plasma membrane-associated protein in *Zea mays* L. *Plant Physiol.* **102**, 1211–1218.

Pan, L., Kawai, M., Yano, A. and Uchimiya, H. (2000) Nucleoside diphosphate kinase required for coleoptile elongation in rice. *Plant Physiol.* **122**, 447–452.

Partch, C.L., Clarkson, M.W., Ôzgur, S., Lee, A.L. and Sancar, A. (2005) Role of structural plasticity in signal transduction by the cryptochrome blue-light photoreceptor. *Biochemistry* **44**, 3795–3805.

Reed, J.W., Elumalai, R.P. and Chory, J. (1998) Suppressors of an *Arabidopsis thaliana* phyB mutation identify genes that control light signaling and hypocotyl elongation. *Genetics* **148**, 1295–1310.

Ryu, J.S., Kim J., Kunkel, T., *et al.* (2005) Phytochrome-specific type 5 phoaphatase controls light signal flux by enhancing phytochrome stability and affinity for a signal transducer. *Cell* **120**, 395–406.

Salomon, M., Kniebl, E., von Zeppelin, T. and Rûdiger, W. (2003) Mapping of low- and high-fluence autophosphorylation sites in phototropin 1. *Biochemistry* **42**, 4217–4225.

Sang, Y., Li, Q.-H., Rubio, V., *et al.* (2005) N-terminal domain-mediated homodimerization is required for photoreceptor activity of *Arabidopsis* CRYPTOCHROME 1. *Plant Cell* **17**, 1569–1584.

Schaffer, R., Ramsey, N., Samach, A., *et al.* (1998) The late elongated hypocotyl mutant of *Arabidopsis* disrupts circadian rhythms and the photoperiodic control of flowering. *Cell* **93**, 1219–1229.

Schneider-Poetsch, H.A.W., Braun, B., Marx, S. and Schaumburg, A. (1991) Phytochromes and bacterial sensor proteins are related by structural and functional homologies. *FEBS*, **281**, 245–249.

Shalitin, D., Yang, H., Mockler, T.C., *et al.* (2002) Regulation of *Arabidopsis* cryptochrome 2 by blue-light-dependent phosphorylation. *Nature* **417**, 763–767.

Shalitin, D., Yu, X., Maymon, M., Mockler, T. and Lin, C. (2003) Blue light-dependent in vivo and in vitro phosphorylation of *Arabidopsis* cryptochrome 1. *Plant Cell* **15**, 2421–2429.

Shen, Y., Kim, J.-I. and Song, P.-S. (2005) NDPK2 as a signal transducer in the phytochrome-mediated light signalling. *J. Biol. Chem.* **280**, 5740–5749.

Short, T.W., Porst, M., Palmer, J., Fernbach, E. and Briggs, W.R. (1994) Blue light induces phosphorylation at seryl residues on pea (*Pisum sativum* L.) plasma membrane protein. *Plant Physiol.* **104**, 1317–1324.

Soh, M.S., Kim, Y.M., Han, S.J. and Song, P.-S. (2000) REP1, a basic helix-loop-helix protein, is required for a branch of phytochrome A signaling in *Arabidopsis*. *Plant Cell* **12**, 2061–2073.

Somers, D., Devlin, P.F. and Kay, S.A. (1998) Phytochromes and cryptochromes in the entrainment of the *Arabidopsis* circadian clock. *Science* **282**, 1488–1490.

Sommer, D. and Song, P.-S. (1994) A plant nucleoside diphosphate kinase homologous to the human NM23 gene-product – purification and characterization. *Biochim. Biophys. Acta Mol. Cell Res.* **1222**, 464–470.

Staiger, C.J. (2000) Signaling to the actin cytoskeleton in plants. *Annu. Rev. Plant Physiol. Plant Mol. Biol.* **51**, 257–288.

Sugano, S., Andronis, C., Green, R.M., Wang, Z.-Y. and Tobin, E. (1998) Protein kinase CK2 interacts with and phosphorylates circadian clock-associated 1 protein. *Proc. Natl. Acad. Sci. USA* **95**, 11020–11025.

Sugano, S., Andronis, C., Ong, M.S., Green, R.M. and Tobin, E. (1999) The protein kinase CK2 is involved in regulation of circadian rhythms in *Arabidopsis*. *Proc. Natl. Acad. Sci. USA* **96**, 12362–12366.

Ueno, K., Kinoshita, T., Inoue, S.-I., Takashi, E. and Shimazaki, K.-I. (2005) Biochemical characterization of plasma membrane H^+-ATPase activation in guard cell protoplasts of *Arabidopsis thaliana* in response to blue light. *Plant Cell. Physiol.* **46**, 955–963.

Von Armin, A.G. and Deng, X.W. (1994) Light inactivation of *Arabidopsis* photomorphogenic repressor COP1 involves a cell-specific regulation of its nucleocytoplasmic partitioning. *Cell* **79**, 1035–1045.

Wagner, D., Hoecker, U. and Quail, P. (1997) RED1 is necessary for phytochrome B-mediated red light-specific signal transduction in *Arabidopsis*. *Plant Cell* **9**, 731–743.

Wang, Z.-Y., Kenigsbuch, D., Sun, L., Harel, E., Ong, M.S. and Tobin, E.M. (1997) A myb-related transcription factor is involved in the phytochrome regulation of an *Arabidopsis* Lhcb gene. *Plant Cell* **9**, 491–507.

Wang, Z.-Y. and Tobin, E.M. (1998) Constitutive expression of the CIRCADIAN CLOCK ASSOCIATED 1 (CCA1) gene disrupts circadian rhythms and suppresses its own expression. *Cell* **93**, 1207–1217.

Whitelam, G.C., Johnson, E., Peng, J., *et al.* (1993) Phytochrome A null mutants of *Arabidopsis* display a wild type phenotype in white light. *Plant Cell* **5**, 757–768.

Wong, Y.-S., Cheng, H.-C., Walsh, D.A. and Lagarias, J.C. (1986) Phosphorylation of Avena phytochrome in vitro as a probe of light-induced conformational changes. *J. Biol. Chem.* **261**, 12089–12097.

Yang, H.-Q., Wu, Y.-J., Tang, R.-H., Liu, D., Liu, Y. and Cashmore, A.R. (2000) The c termini of *Arabidopsis* cryptochromes mediate a constitutive light response. *Cell* **103**, 815–827.

Yeh, K.-C. and Lagarias, J.C. (1998) Eukaryotic phytochromes: light-regulated serine/threonine protein kinases with histidine kinase ancestry. *Proc. Natl. Acad. Sci. USA* **95**, 13976–13981.

Yeh, K.-C., Wu, S.-H., Murphy, J.T. and Lagarias, J.C. (1997) A cyanobacterial phytochrome two-component light sensory system. *Science* **277**, 1505–1507.

Zeugner, A., Byrdin, M., Bouly, J.-P., *et al.* (2005) Light-induced electron transfer in *Arabidopsis* cryptochrome-1 correlates with in vivo function. *J. Biol. Chem.* **280**, 19437–19440.

Zimmermann, S., Baumann, A., Jaekel, K., Marbach, I., Engelberg, H. and Frohnmeyer, H. (1999) UV-responsive genes of *Arabidopsis* revealed by similarity to the Gcn4-mediated UV response in yeast. *J. Biol. Chem.* **274**, 17017–17024.

6 The role of ubiquitin/proteasome-mediated proteolysis in photoreceptor action

Suhua Feng and Xing Wang Deng

6.1 Introduction

Plants are sessile organisms that cannot move toward favorable or away from adverse conditions. Instead, they have evolved a high degree of developmental plasticity to cope with a changing environment, to withstand external challenges and to support growth and reproduction. Light is arguably the most important environmental factor as it influences almost all aspects of plant growth and development. Plants have evolved at least four classes of photoreceptors to perceive different wavelengths of light, including the red/far-red light absorbing phytochromes, the blue/UV-A light absorbing cryptochromes and phototropins and the uncharacterized UV-B light receptor (Sullivan and Deng, 2003).

A classic example of photoreceptor-mediated light response is the highly elaborate, yet plastic seedling development. Take, for example the model plant *Arabidopsis thaliana* where seedlings undergo photomorphogenesis (or de-etiolation) in the light and skotomorphogenesis (or etiolation) in darkness. These are two drastically different seedling developmental pathways (Deng, 1994). Mutations in photoreceptors (mainly phytochromes and cryptochromes) reduce sensitivity to light, which can lead to light-grown seedlings with etiolated characteristics. This suggests that these photoreceptors play positive roles in light-induced plant growth (Hudson, 2000; Nagy and Schafer, 2002; Lin and Shalitin, 2003). Microarray analyses suggest that genome expression profile changes (about one-third of the total *Arabidopsis* genes are differentially regulated between light and darkness) are responsible for the dramatic difference in seedling morphology grown in light versus dark (Ma *et al.*, 2001). This also indicates the complexity of the signaling pathways that a plant uses to perceive light signals and to regulate development accordingly.

Research in the past decade or so has begun to elucidate the mechanism that controls the switch between photomorphogenesis and skotomorphogenesis. It has become increasingly clear that regulated proteolysis, especially through the ubiquitin/proteasome pathway, plays a major part in controlling light signal transduction and light-induced gene expression (Strickland *et al.*, 2006; Yanagawa *et al.*, 2005). Precise and efficient removal of preexisting proteins is essential for survival and is just as important as the synthesis of new proteins. In eukaryotes, the ubiquitin/proteasome system is the major pathway to selectively degrade short-lived regulatory proteins. In short, it involves the labeling of protein targets by ubiquitin (Ub) and the subsequent degradation of multiubiquitinated proteins by the 26S proteasome

(Hershko and Ciechanover, 1998). As expected, in addition to light signaling, ubiquitin/proteasome-mediated protein degradation is also essential for many other aspects of plant development (Sullivan *et al.*, 2003; Moon *et al.*, 2004; Smalle and Vierstra, 2004).

In this chapter, we will review the current understanding of the relationship between photoreceptor action and ubiquitin/proteasome-mediated proteolysis by focusing on their functional interplay in the control of photomorphogenesis in *Arabidopsis*.

6.2 Overview of the ubiquitin/proteasome system

6.2.1 Ubiquitin conjugation and deconjugation pathways

In order for a protein to undergo proteasome-mediated degradation, the initial and most important step is its modification by Ub (so-called ubiquitination). Ub is a 76-amino acid globular protein, which is highly conserved among the eukaryotic organisms (Smalle and Vierstra, 2004). Ub is able to form a covalent isopeptide linkage with a lysine residue of its target protein through a series of ATP-dependent enzymatic reactions (Figure 6.1). The carboxyl terminus of Ub, which usually ends with two glycine residues, is first adenylated. The sulfhydryl group of a cysteine residue in an E1 Ub-activating enzyme then attacks Ub carboxyl-AMP and forms

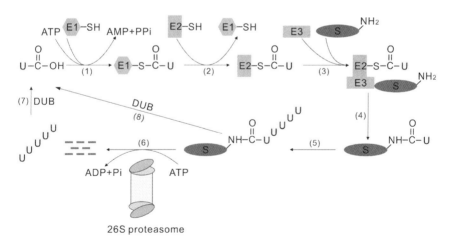

Figure 6.1 A simplified scheme of the ubiquitin/proteasome pathway. (1) Free ubiquitin (U) is activated in an ATP-dependent reaction and forms a thioester linkage with E1 Ub-activating enzyme; (2) E1 transfers activated Ub to E2 Ub-conjugating enzyme to form an E2–Ub thioester; (3) both substrate (S) and E2–Ub are bound by E3 Ub ligase; (4) E3 catalyzes the formation of an isopeptide bond between Ub and the substrate; (5) a multiubiquitin chain is formed on the substrate by sequential ubiquitination reactions; (6) the 26S proteasome recognizes and degrades the multiubiquitinated substrate; (7) deubiquitinating enzyme (DUB) regenerates free Ub by cleaving the multiubiquitin chain; (8) deubiquitination can also happen before proteasomal degradation to rescue the substrate and regenerate free Ub.

an E1–Ub thioester. In the next step, activated Ub is passed from the E1 to E2 Ub-conjugating enzyme, again through the formation of a thioester linkage between Ub carboxyl and a cysteine residue in the E2. Subsequently, the E3 Ub ligase recognizes and recruits both Ub-charged E2 and the target protein. By bringing them into close vicinity, the E3 facilitates the transfer of Ub from the E2 to the target protein. Finally, an isopeptide bond is formed between the C-terminal glycine of Ub and the ϵ-amino group of a lysine residue in the target. In order to generate a substrate that is recognizable by the proteasome, a multiubiquitin chain is usually formed, in which the carboxyl terminus of each Ub is linked to a specific lysine residue (most commonly Lys48) in the previous Ub (Hershko and Ciechanover, 1998; Sullivan *et al.*, 2003; Smalle and Vierstra, 2004). Monoubiquitination and multiubiquitination through other lysine residues in Ub (for example Lys29 and Lys63) also occur, but they do not target proteins for proteasome-mediated protein degradation (Weissman, 2001; Conaway *et al.*, 2002; Aguilar and Wendland, 2003).

Ubiquitination is a reversible process. Several families of DUBs (deubiquiti-nating enzymes) are isopeptidases that cleave the isopeptide bond between the Ub carboxyl and the lysine side chain on another protein. Distinct activities are asso-ciated with DUBs: (1) shortening of the multiubiquitin chain on a target protein from the distal end, (2) release of the multiubiquitin chain by cutting between the target protein and Ub and (3) reduction of the unanchored multiubiquitin chain into Ub monomers (Voges *et al.*, 1999). In general, the function of DUBs can involve rescuing proteins from degradation by reversing ubiquitination and maintaining an adequate cellular pool of free Ub by recycling them (Figure 6.1; Sullivan *et al.*, 2003; Smalle and Vierstra, 2004).

In addition, there is a diverse set of Ub-like proteins in eukaryotes, including SUMO (small ubiquitin-related modifier) and RUB/NEDD8 (related to ubiquitin/neural precursor cell expressed, developmentally downregulated 8). Interestingly, they both have conjugation and deconjugation systems similar to ubiquitination/deubiquitination (Hochstrasser, 2000). But unlike Ub, they do not form chains, and their functions are not to mark proteins for degradation. Recent studies have started to reveal the role of the RUB pathway in light signaling and proteasome-mediated protein degradation, a function that will be discussed in later sections.

6.2.2   Diversity of E3 Ub ligases

As mentioned earlier, E3 Ub ligases directly interact with the substrate and thus are primarily responsible for conferring specificity to the Ub pathway. Consistently, genomic analyses estimate that there are more than 1400 *Arabidopsis* proteins in-volved in the ubiquitin/proteasome system; this corresponds to more than 5% of the proteome. Among them, about 1300 are thought to be potential components of E3 Ub ligases. In contrast, only 2 E1 Ub-activating enzymes and 37 E2 Ub-conjugating enzymes have been identified in the same study (Smalle and Vierstra, 2004). There-fore, it is clear that the specificity of the Ub pathway largely resides in the large number of E3s.

The E3s in *Arabidopsis* can be categorized into two families, those containing the HECT (homologous to E6AP C-terminus) domain and those containing the RING (real interesting new gene)/U-box domain. HECT E3s differ from conventional E3s in that they form a thioester bond with Ub through a cysteine residue within the conserved HECT domain prior to transferring Ub to the substrate. *Arabidopsis* has seven HECT E3s, some of which have been studied in detail (Bates and Vierstra, 1999; Downes *et al.*, 2003). RING/U-box E3s are defined by the presence of a RING finger motif or a U-box, which are proposed to be adaptors of E2–Ub thioester. Some of the RING/U-box E3s are single polypeptides (more than 500 potential members) that contain other protein–protein interaction domains for recruiting ubiquitination targets (Smalle and Vierstra, 2004). Functional analyses for these candidates are underway. So far, the most extensively studied single subunit RING E3 related to light regulation is COP1 (constitutive photomorphogenic 1), which plays a central role in repressing photomorphogenesis in the darkness (see below).

The remaining RING/U-box E3s are multisubunit protein complexes. Their basic subunits often include a scaffold protein, a small RING finger protein and a substrate recognition unit. They can be further divided into two groups, APC (anaphase-promoting complex) or cullin-containing. The highly conserved APC contains at least 11 subunits and has important functions in cell cycle regulation. Most of the APC subunits are encoded by a single gene in *Arabidopsis* (Capron *et al.*, 2003a). As expected, mutants of several *Arabidopsis* APC subunits show defects in cell cycle progression (Blilou *et al.*, 2002; Capron *et al.*, 2003b).

Arabidopsis contains five canonical cullins (CUL1, CUL2, CUL3A, CUL3B and CUL4; Risseeuw *et al.*, 2003). Among them, CUL1 (Shen *et al.*, 2002) and CUL3 (Dieterle *et al.*, 2005; Figueroa *et al.*, 2005; Gingerich *et al.*, 2005; Thomann *et al.*, 2005; Weber *et al.*, 2005) have been characterized at the molecular and functional levels. CUL1 assembles into a so-called SCF complex with ASK (*Arabidopsis* SKP1), RBX1 (ring-box 1) and F-box protein. Within this complex, CUL1/RBX1 is the catalytic core that binds E2–Ub, and ASK/F-box protein serves as the substrate-docking site. CUL3-containing E3s share an RBX1 subunit with SCF, while they employ BTB (bric-a-brac, tramtrack and broad-complex) protein instead of ASK/F-box protein (Moon *et al.*, 2004). In *Arabidopsis*, there are 694 putative F-box proteins (Gagne *et al.*, 2002). In addition, 80 proteins containing consensus BTB domain have been identified (Gingerich *et al.*, 2005). This supports the existence of an enormous number of cullin-containing E3s and probably an equivalent number of substrates. In recent years, a number of cullin-containing E3s have been studied in many aspects of plant development, including phytohormone (auxin/gibberellin/ethylene/jasmonate) pathways, flower organogenesis, phyA (phytochrome A)-dependent far-red light signaling, circadian rhythm control, self-incompatibility responses, plant–pathogen interactions and others (Moon *et al.*, 2004; Schwechheimer and Villalobos, 2004).

6.2.3   26S proteasome

Multiubiquitinated proteins are recognized and degraded by the 26S proteasome, a 2 MDa proteolytic multisubunit complex, in an ATP-dependent manner (Figure 6.1). The 26S proteasome has 31 subunits and can be separated into two particles, a 20S core particle (CP) and a 19S regulatory particle (RP) (Voges *et al.*, 1999; Yang *et al.*, 2004). The 20S CP is a self-compartmentalized assembly that is arranged as four stacks of rings: the two end rings comprise seven α-subunits ($\alpha1$–$\alpha7$), and the two middle rings comprise seven β-subunits ($\beta1$–$\beta7$). The protease active sites at the N-termini of three different β-subunits are buried in the central CP channel (Groll *et al.*, 1997; Voges *et al.*, 1999; Unno *et al.*, 2002). Entry to the channel is blocked by the N-terminal tails of the α-subunits (Groll *et al.*, 2000).

The 20S CP is associated with one or two 19S RPs situated at either one or both ends. The RP can be further divided into a base subcomplex, which has six ATPase subunits and three non-ATPase subunits, and a lid subcomplex, which has eight non-ATPase subunits (Glickman *et al.*, 1998). The proteasome lid is evolutionarily related to the COP9 signalosome (CSN), an eight-subunit protein complex with important roles in the regulation of cullin-containing E3s and photomorphogenesis (see below). The addition of the RP confers ATP- and Ub-dependence to the proteasome holoenzyme. RP is also thought to have other regulatory functions such as gating the CP channel, activating the peptidase activity of the CP, recognizing and translocating multiubiquitinated substrates to the catalytic sites inside the CP and removing Ub from protein remnants (Coux *et al.*, 1996; Voges *et al.*, 1999; Groll *et al.*, 2000; Verma *et al.*, 2002; Hartmann-Petersen *et al.*, 2003).

The subunits of the *Arabidopsis* 26S proteasome are often encoded by gene families. This suggests that there is functional redundancy, though substrate specificities have also been demonstrated (Fu *et al.*, 1998, 1999; Yang *et al.*, 2004). In particular, the RP subunits RPN10 and RPN12a participate in ABA (abscisic acid) and cytokinin responses, respectively (Smalle *et al.*, 2002, 2003). In the case of RPN10, the ABA hypersensitivity of the *rpn10-1* mutant can be explained by the stabilization of ABI5 (ABA insensitive 5) – a positive regulator of the ABA pathway that is normally degraded by the proteasome (Lopez-Molina *et al.*, 2003).

6.3   Role of COP/DET/FUS proteins in photoreceptor-mediated signal transduction and ubiquitin/proteasome-mediated proteolysis

6.3.1   COP/DET/FUS proteins integrate divergent photoreceptor signaling pathways and downstream gene expression

Arabidopsis seedlings undergo photomorphogenesis in the light and characteristically exhibit a short hypocotyl and opened cotyledon phenotype. In contrast, when grown in the darkness, the seedlings usually have long hypocotyls and closed cotyledons instead, reflecting a different developmental program called skotomorphogenesis (Figure 6.2; Deng, 1994). Therefore, it is possible that photomorphogenesis is repressed in the darkness through certain pathways. This predicts that if these repressive pathways are disrupted by mutation, seedlings should display light-grown characteristics even when grown in the darkness. Indeed, several genetic screens selecting for such phenotypes have led to the identification of the *cop* (*constitutive photomorphogenic*) and *det* (*de-etiolated*) mutants in *Arabidopsis* (Chory *et al.*, 1989; Deng *et al.*, 1991). These mutants are all recessive and exhibit photomorphogenic features when grown in darkness, indicating that their respective wild-type genes encode

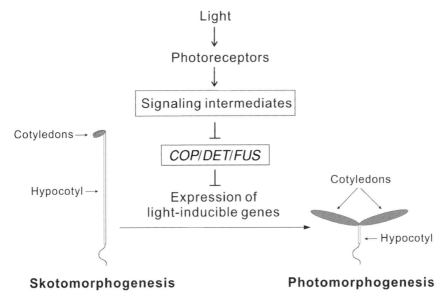

Figure 6.2 A schematic presentation of the role of *COP/DET/FUS* genes in the light control of *Arabidopsis* seedling development. The dark-grown skotomorphogenic (etiolated) seedling shown on the left has a long hypocotyl and closed cotyledons, while the light-grown photomorphogenic (de-etiolated) seedling shown on the right has a short hypocotyl and opened cotyledons. The transition from dark- to light-grown development requires the expression of a diverse array of light-inducible genes that are negatively regulated by *COP/DET/FUS* genes in the darkness. Under light conditions, light signals of different wavelengths are perceived by corresponding photoreceptors. Then the signals are transduced through the intermediates of the photoreceptor pathways and finally integrated at the *COP/DET/FUS* genes. This leads to the inhibition of *COP/DET/FUS* functions, turning on of light-inducible gene expression and proceeding on to photomorphogenic development. In the signaling cascade, the arrows indicate a positive effect and the bars indicate a negative effect.

negative regulators of photomorphogenesis. A number of the *cop* and *det* mutants are found to be allelic to the previously identified *fus* (*fusca*) mutants (Misera *et al.*, 1994). These mutants are named after the purple color of their seeds, which results from excessive accumulation of anthocyanin. In addition to the phenotypic resemblance to light-grown seedlings, dark-grown *cop/det/fus* mutants also have subcellular photomorphogenic features such as chloroplast differentiation and expression of light-inducible genes (Hardtke and Deng, 2000; Schwechheimer and Deng, 2000). This implies that light signals are integrated at the *COP/DET/FUS* loci – a conclusion supported by genetic and molecular analyses demonstrating that *COP/DET/FUS* are epistatic to various photoreceptors and light signaling intermediates (Sullivan and Deng, 2003; Wang and Deng, 2004). Further evidence also comes from genomic studies. First all photoreceptors seem to be control the expression of a similar group of genes, despite the different light signals that they perceive (Ma *et al.*, 2001). Second, genome expression profiles of the dark-grown *cop/det/fus* mutants closely resemble those of light-grown wild-type plants (Ma *et al.*, 2002, 2003). The current working model has *COP/DET/FUS* genes negatively

Table 6.1 Summary of the nine pleiotropic COP/DET/FUS proteins in *Arabidopsis*

Protein name	Corresponding *COP/DET/FUS* locus	Complex formation
COP1	*COP1/FUS1*	COP1 complex
DET1	*FUS2/DET1*	CDD complex
COP10	*COP10/FUS9*	CDD complex
CSN1	*COP11/FUS6*	COP9 signalosome
CSN2	*COP12/FUS12*	COP9 signalosome
CSN3	*COP13/FUS11*	COP9 signalosome
CSN4	*COP8/FUS4*	COP9 signalosome
CSN7	*COP15/FUS5*	COP9 signalosome
CSN8	*COP9/FUS7/FUS8*	COP9 signalosome

regulating the expression of light-inducible genes, which leads to the repression of photomorphogenesis in the darkness. Under different light conditions, corresponding photoreceptor-mediated pathways inactivate *COP/DET/FUS* in order to allow de-etiolation (Figure 6.2).

In the effort to elucidate the functional mechanism of *COP/DET/FUS* genes, nine of these loci have been cloned (Table 6.1; Serino and Deng, 2003). Their gene products turn out to exist in three protein complexes *in vivo*: the COP1 complex (Saijo *et al.*, 2003), CSN (Serino and Deng, 2003; Wei and Deng, 2003) and CDD complex (Yanagawa *et al.*, 2004). Strikingly, recent findings suggest that all three complexes are directly involved in the ubiquitin/proteasome-mediated protein degradation pathways, defining a critical linkage between regulated proteolysis and photoreceptor-mediated light signal transduction. The physiological roles and possible functional mechanisms of each COP/DET/FUS protein-based complexes will be discussed in the following sections.

6.3.2 COP1

COP1 was the first molecularly characterized *COP/DET/FUS* locus and has long remained the best understood. Full-length *COP1* mRNA encodes a protein of 658 amino acids with an approximate molecular weight of 76 kDa. Three distinct protein–protein interaction motifs are identified in COP1: an N-terminal RING finger domain (conserved in RING family E3 Ub ligases as mentioned earlier), a coiled-coil domain and a C-terminal domain containing seven WD-40 repeats (Figure 6.3; Deng *et al.*, 1992; McNellis *et al.*, 1994a). Extensive structure–function analyses suggest that COP1 may form homodimers or heterodimers through the coiled-coil domain and recruit its target proteins through the WD-40 repeats (Torii *et al.*, 1998; Holm *et al.*, 2001). Furthermore, a recent study shows that the majority of COP1 proteins in *Arabidopsis* seedlings exist as part of a large (600–700 kDa) complex (Saijo *et al.*, 2003). Presumably, COP1 is associated with multiple cofactors to regulate light-induced plant growth; so it will be of great interest to determine the subunit composition of the COP1 complex.

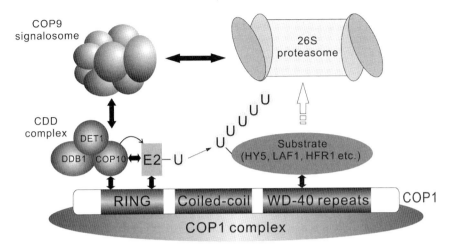

Figure 6.3 A working model of COP/DET/FUS protein functions. The COP1 complex, CDD complex and COP9 signalosome work synergistically in the repression of photomorphogenesis. In the darkness, COP1 localizes to the nucleus and therefore, is able to bind transcription factors (such as HY5, LAF1 and HFR1) through its C-terminal WD-40 repeats. These transcription factors are necessary for the expression of light-inducible genes in order for photomorphogenesis to take place. Their functions are inhibited in the darkness, because they are ubiquitinated by COP1's E3 Ub ligase activity and then degraded by the 26S proteasome. The CDD complex and E2–Ub thioester both interact with COP1's N-terminal RING finger domain and they can interact with each other as well. The COP10 subunit of the CDD complex can enhance the activity of the E2, which is required for the ubiquitination of COP1 substrates. The COP9 signalosome is essential for the nuclear localization of COP1 and the integrity of the CDD complex. Moreover, the COP9 signalosome has implicated regulatory roles in the activity and specificity of the 26S proteasome, which might be required for the proper degradation of COP1 substrates. In the diagram, lines with arrowheads on both ends indicate physical interactions, and the arrow pointing from COP10 toward the E2 indicates that COP10 can enhance E2 activity.

6.3.2.1 Light regulation of COP1 localization

A body of work has shown that COP1 acts as a switch between light signals and downstream activities; but how is this light switch regulated? Surprisingly, the expression, abundance or complex formation of COP1 does not appear to be significantly affected by light (Deng *et al.*, 1992; McNellis *et al.*, 1994b; Saijo *et al.*, 2003). Instead, light influences the nucleocytoplasmic partitioning of COP1 (von Arnim and Deng, 1994). COP1 displays nuclear enrichment in dark-grown *Arabidopsis*, but following light treatment, nuclear COP1 is rapidly depleted. Furthermore, in light-grown *Arabidopsis*, COP1 is excluded from the nucleus, while nuclear reaccumulation is observed following a subsequent shift to darkness. Different photoreceptors, including phyA, phytochrome B (phyB) and cryptochrome 1 (CRY1), are shown to mediate this localization pattern of COP1 (Osterlund and Deng, 1998), which suggests that COP1 works downstream of these light signaling pathways.

The light-regulated COP1 subcellular localization change appears to be very important for the proper control of photomorphogenesis, since it is defective in all the pleiotropic *cop/det/fus* mutants (von Arnim and Deng, 1997). Currently, all evidence suggests that the presence of COP1 in the nucleus is required for the repression of photomorphogenesis, probably because partners and/or targets essential for COP1 function are also localized in the nucleus. When light inhibits the nuclear accumulation of COP1, it physically separates COP1 from its partners and/or targets in the nucleus, allowing the plant to switch from skotomorphogenic to photomorphogenic development.

6.3.2.2   COP1 acts as an E3 Ub ligase

In order to better understand the functional mechanism of COP1, it is critical to isolate its nuclear partners and/or targets. Indeed, research efforts have been mainly focused on identifying its interacting proteins. One of the known COP1-interacting proteins is HY5 (long hypocotyl 5), a nuclear-localized bZIP (basic leucine zipper) transcription factor that binds directly to light-responsive promoters, upregulating gene expression and photomorphogenesis (Oyama *et al.*, 1997; Chattopadhyay *et al.*, 1998). COP1 interacts with HY5 through the WD-40 repeat domain in the nucleus and negatively regulates HY5 activity (Ang *et al.*, 1998; Holm *et al.*, 2001). Additionally, it has been shown that COP1–HY5 interaction leads to the degradation of the HY5 protein mediated by the ubiquitin/proteasome pathway and that the abundance of HY5 is inversely correlated with the nuclear abundance of COP1 (Osterlund *et al.*, 2000).

What could be the possible role(s) of COP1 in ubiquitin/proteasome-mediated protein degradation? On the basis of its structure, it is hypothesized that COP1 acts as an E3 Ub ligase by recruiting substrates such as HY5 via the WD-40 repeats and interacting with E2–Ub via the RING finger domain (Figure 6.3). This model is supported by several findings. First, in a substrate-independent *in vitro* assay system supplemented with Ub, E1 and E2, COP1 displays autoubiquitination activity toward itself (Seo *et al.*, 2003; Saijo *et al.*, 2003). Second, when HY5 is added to the system as a substrate, it can be ubiquitinated, which is dependent on the presence of COP1 (Saijo *et al.*, 2003). Third, in addition to HY5, several other photomorphogenesis-promoting transcription factors are also found to be the targets of COP1's *in vitro* E3 Ub ligase activity, including LAF1 (long after far-red light 1) and HFR1 (long hypocotyl in far-red 1). In both cases, genetic and physical interactions between the transcription factor and COP1 are also observed (Seo *et al.*, 2003; Jang *et al.*, 2005; Yang *et al.*, 2005). Also, it is important to note that the E2s used in these studies are from a wide variety of organisms, including mammals, *Arabidopsis* and rice. Therefore, the *in vivo* activity of COP1 may not be identical to that observed *in vitro*.

Since HY5, LAF5 and HFR1 represent three different types of transcription factors (bZIP, myb and bHLH, respectively), we can infer that COP1 is capable of negatively controlling the abundance of a wide variety of photomorphogenesis-promoting transcription factors by targeting them for ubiquitination and proteasomal degradation. Taken together with the fact that COP1 integrates different light signaling pathways, it is clear that COP1 is a master regulator acting at the junction between photoreceptor-induced signal transduction pathways and expression of light-responsive genes (Figures 6.2 and 6.3).

6.3.2.3   Interaction between photoreceptors and COP1

Through genetic screening, many signaling intermediates, possibly acting between photoreceptors and COP1, have been identified. Mutants of these components are usually defective in one or more of the photoreceptor pathways. This supports a signal cascade model: photoreceptors perceive light and generate signals that are passed along through intermediates and finally transduced to COP1 (Figure 6.2; Sullivan and Deng, 2003; Wang and Deng, 2004).

However, in some cases, direct interaction of photoreceptors and COP1 is demonstrated. The best-characterized example is the relationship between cryptochromes (CRY1 and CRY2) and COP1. Interestingly, no signal intermediate has been discovered between cryptochromes and COP1. Overexpression of the C-terminal domain of either cryptochrome (CCT1 and CCT2) leads to a constitutive light response similar to the *cop/det/fus* mutants, indicating that C-terminal domain of cryptochrome (CCT) has inhibitory effects on COP1 function (Yang *et al.*, 2000; Wang *et al.*, 2001). It has also been shown that both full-length cryptochrome and CCT can bind COP1 and that the binding is light independent (Wang *et al.*, 2001; Yang *et al.*, 2001). Therefore, it is proposed that, through direct protein–protein interaction, CCT can cause a structural modification of COP1 that antagonizes COP1's effect on its substrates such as HY5. In the context of full-length cryptochrome protein, CCT is usually folded into an inactive state. When light induces a conformation change of the cryptochrome structure, CCT is activated and becomes capable of inactivating COP1 (Chen *et al.*, 2004).

In the case of phytochromes, overexpression of the C-terminal domains of phyA and phyB does not confer a *cop/det/fus*-like phenotype (Yang *et al.*, 2000), indicating that the phytochrome signaling mechanism is different from that of the cryptochromes. As mentioned above, signal transduction cascades are defined genetically between phytochromes and COP1 (Sullivan and Deng, 2003; Wang and Deng, 2004). Nevertheless, both phyA and phyB can bind directly with COP1 (Yang *et al.*, 2001; Seo *et al.*, 2004). Little is known about how this binding affects COP1 activity.

6.3.3   CDD complex

6.3.3.1   COP10 is an E2 Ub-conjugating enzyme variant

A severe *cop10* mutant displays a similar constitutive photomorphogenic phenotype as the *cop1* mutant (Wei *et al.*, 1994), with disrupted COP1 nuclear localization and high accumulation of HY5 protein in its dark-grown seedlings (von Arnim and Deng, 1997; Osterlund *et al.*, 2000), indicating that wild-type COP10 function is required for the proper degradation of HY5 and repression of photomorphogenesis in darkness.

The *COP10* gene encodes a protein of 182 amino acids, whose sequence is highly homologous to E2 Ub-conjugating enzymes such as UBC4/UBC5 from yeast and UBC8/UBC9 from *Arabidopsis* (Suzuki *et al.*, 2002). This suggests that like COP1, which acts as an E3 Ub ligase, COP10 may also be involved in the ubiqui-tin/proteasome pathway. However, COP10 does not contain the conserved cysteine residue typical in an E2 catalytic domain. Since this cysteine is absolutely required for the conjugation of Ub, this means that COP10 is not an active E2 enzyme (Suzuki *et al.*, 2002). Instead, COP10 belongs to a family of UEV (Ub E2 variant) proteins that act in various processes, including DNA repair and cell cycle regulation (Hofmann and Pickart, 1999; Li *et al.*, 2001). For example MMS2/UEV1 works to-gether with UBC13, an active E2 enzyme, to produce noncanonical Lys63-linked Ub chains (Hofmann and Pickart, 1999).

COP10 is phylogenetically more closely related to active E2 enzymes than to other UEV family members such as TSG101 or MMS2/UEV1 (Suzuki *et al.*, 2002). Consistently, the biochemical activity of COP10 has little in common with MMS2/UEV1. It cannot form a Lys63-linked Ub chain together with UBC13 *in vitro*. Instead, it has a general enhancing effect on the activity of various E2 enzymes in the formation of either Lys48- or Lys63-linked Ub chains. It has been demonstrated that COP10 can enhance the thioester bond formation between Ub and E2, which at least partly explains the mechanism of COP10's E2 enhancement activity (Yanagawa *et al.*, 2004).

COP10 can interact with COP1 (through the RING finger domain; Figure 6.3) and E2 (Suzuki *et al.*, 2002; Yanagawa *et al.*, 2004). Taken together, the genetic and biochemical data indicate that COP10 is necessary for the COP1-mediated ubiqui-tination of photomorphogenesis-promoting transcription factors, probably through its positive effect on certain *Arabidopsis* E2s that work in cooperation with COP1 in this process (Figure 6.3).

6.3.3.2   *COP10 forms a complex with DET1 and DDB1*

The gel-filtration profile of COP10 demonstrates that most of the COP10 protein exists in a complex of approximately 300 kDa in size and that only a small fraction is in the monomeric form (Suzuki *et al.*, 2002). Biochemical purification of the COP10-containing complex from cauliflower revealed three core subunits: COP10, DET1 and DDB1 (damaged DNA binding 1) (Yanagawa *et al.*, 2004). Therefore, this complex is designated as the CDD complex.

DET1 is one of the pleiotropic *COP/DET/FUS* genes. It encodes a 62-kDa nu-clear protein with histone-binding activity (Pepper *et al.*, 1994; Benvenuto *et al.*, 2002). DDB1 was first identified in mammals on the basis of its ability to bind UV-damaged DNA (Hwang *et al.*, 1998). *Arabidopsis* has two highly conserved DDB1 proteins, DDB1a and DDB1b (Schroeder *et al.*, 2002). Both DET1 and DDB1 have been shown previously to form protein complexes. First, DDB1 copurifies with DET1 in tobacco cells (Schroeder *et al.*, 2002). In addition, human DDB1 exists in multiple CUL4A-containing complexes as a core subunit (Groisman *et al.*, 2003). More recently, DET1 and COP1 have been suggested to associate with a CUL4A-DDB1-RBX1 based E3 Ub ligase in mammals (Wertz *et al.*, 2004).

Within the CDD complex, COP10 harbors E2 enhancement activity, which is proposed to be the mechanism through which CDD complex contributes to the repression of photomorphogenesis in darkness (Figure 6.3). At the same time, complex formation is also a prerequisite for the proper *in vivo* function of COP10 (Suzuki *et al.*, 2002; Yanagawa *et al.*, 2004). The precise role of DET1 and DDB1 in the context of the CDD complex is still unclear. According to their postulated roles in DNA repair and chromosome remodeling, DET1 and DDB1 may function to guide the CDD complex to its targets or define the target specificity of COP10's E2 enhancement activity.

6.3.4   COP9 signalosome

6.3.4.1   Composition and structure of the COP9 signalosome

Six of the pleiotropic *COP/DET/FUS* genes encode subunits of the same protein complex designated the CSN. The two remaining subunits of CSN are encoded by two redundant genes that were not identified in the initial genetic screens. The eight subunits have been renamed CSN1 through CSN8, based on their molecular weight (Tables 6.1 and 6.2; Serino and Deng, 2003; Wei and Deng, 2003). A striking feature of the *Arabidopsis* COP9 signalosome is that the integrity of the complex is dependent on the presence of each subunit, which explains the nearly identical phenotype of individual CSN subunit mutants (Wei and Deng, 1999). Protein complexes homologous to *Arabidopsis* CSN are also present in many other organisms, including human (Table 6.2; Seeger *et al.*, 1998; Wei *et al.*, 1998), fission yeast (*Schizosaccharomyces pombe*; Mundt *et al.*, 1999), fruit fly (*Drosophila melanogaster*; Freilich *et al.*, 1999), budding yeast (*Saccharomyces cerevisiae*; Maytal-Kivity *et al.*, 2003; Wee *et al.*, 2002) and fungus (*Aspergillus nidulans*; Busch *et al.*, 2003).

Like COP1 and COP10, the COP9 signalosome has also been suggested to play important roles in the ubiquitin/proteasome pathway. Structurally, the signalosome appears to closely resemble the lid subcomplex of the 19S regulatory particle of the 26S proteasome. In fact, each of the eight subunits of the CSN is paralogous to a subunit of the proteasome lid subcomplex (Table 6.2; Glickman *et al.*, 1998; Wei

Table 6.2 Arabidopsis CSN subunit composition, homology with human CSN, and the relationship between CSN and the lid subcomplex of 19S proteasome regulatory particle

Subunit	Molecular weight (kDa)	Other names	Identity with human homolog (%)	Paralog in proteasome lid	Identity with lid paralog[a] (%)
CSN1	50	COP11, FUS6	45	RPN7	22
CSN2	51	FUS12	61	RPN6	21
CSN3	47	FUS11	42	RPN3	20
CSN4	45	COP8, FUS4	50	RPN5	19
CSN5	40	AJH1, AJH2	62	RPN11	28
CSN6	35	CSN6A, CSN6B	40	RPN8	22
CSN7	25	FUS5	34	RPN9	15
CSN8	22	COP9, FUS7	32	RPN12	18

[a] The identity level is calculated on the basis of the comparison between mammalian CSN and *Saccharomyces cerevisiae* proteasome lid (Wei and Deng, 1999).

et al., 1998; Wei and Deng, 1999). Additionally, physical interaction between the CSN and proteasome components has been demonstrated (Figure 6.3; Kwok *et al.*, 1999; Peng *et al.*, 2003). Therefore, it has been proposed that the CSN may be able to replace the lid subcomplex under certain conditions, bestowing distinct activities and specificities to the proteasome (Li and Deng, 2003).

6.3.4.2 Biochemical activities of the COP9 signalosome

As discussed in earlier sections, the nuclear localization of COP1 in darkness is essential for the repression of photomorphogenesis. This event is abolished in *csn* mutants, which suggests that CSN activity is required for normal COP1 function and explains the constitutive photomorphogenesis phenotype of *csn* mutants (Chamovitz *et al.*, 1996; von Arnim and Deng, 1997). At the same time, the CSN also interacts with the CDD complex and is required to maintain the stability of the CDD complex *in vivo* (Figure 6.3; Suzuki *et al.*, 2002; Yanagawa *et al.*, 2004). Therefore, the functional relationship among different COP/DET/FUS protein-based complexes appears to be quite complicated and requires further investigation.

One of the more recent breakthroughs is the discovery of the CSN's derubylation activity toward cullins (Lyapina *et al.*, 2001; Schwechheimer *et al.*, 2001; Zhou *et al.*, 2001). RUB (plant and yeast) or NEDD8 (mammals) is a small globular protein highly homologous to Ub that can undergo a ubiquitination-like enzymatic cascade (referred to as rubylation or neddylation) and form a covalent isopeptide bond with a lysine residue in the C-terminus of cullins (Hochstrasser, 1998; Hochstrasser, 2000). The CSN has been shown to cleave RUB from cullins (referred to as derubylation or deneddylation) through a novel metalloprotease activity that resides in the JAMM domain of its subunit 5, CSN5 (Cope *et al.*, 2002; Gusmaroli *et al.*, 2004). In the next section, we will discuss the function of rubylation and derubylation in ubiquitin/proteasome-mediated proteolysis and various cellular processes.

Other biochemical activities attributed to the CSN include protein phosphorylation and deubiquitination. Several protein kinases (Seeger *et al.*, 1998; Naumann *et al.*, 1999; Bech-Otschir *et al.*, 2001; Wilson *et al.*, 2001; Sun *et al.*, 2002; Uhle *et al.*, 2003) and deubiquitinating enzymes (Groisman *et al.*, 2003; Zhou *et al.*, 2003) have been suggested to associate with the CSN. Furthermore, the metalloprotease domain of CSN5 is shown to be required for the substrate deubiquitination activity (Groisman *et al.*, 2003). Most of the studies in these directions were performed in mammalian systems or in yeast, and it is not clear at this moment if the *Arabidopsis* CSN has similar functions and whether these functions contribute to the repression of photomorphogenesis.

6.3.4.3 Regulation of cullin-containing E3 Ub ligases
by the COP9 signalosome

All cullins studied so far are modified by RUB/NEDD8, including *Arabidopsis* CUL1 and CUL3 (del Pozo and Estelle, 1999; Hori *et al.*, 1999; Figueroa *et al.*, 2005). It turns out that rubylation/neddylation of cullins positively regulates the activities of cullin-containing E3 Ub ligases (Furukawa *et al.*, 2000; Podust *et al.*,

2000; Read *et al.*, 2000; Wu *et al.*, 2000; del Pozo *et al.*, 2002; Ohh *et al.*, 2002). Consistently, rubylation/neddylation is essential for most of the organisms examined, including mouse (Tateishi *et al.*, 2001), fission yeast (*S. pombe*; Osaka *et al.*, 2000), worms (*Caenorhabditis elegans*; Kurz *et al.*, 2002), fruit fly (*D. melanogaster*; Ou *et al.*, 2002) and plants (*A. thaliana*; Dharmasiri *et al.*, 2003). The mechanism underlying rubylation's stimulation of cullin-containing E3 is still under investigation. However, current hypotheses include stabilization of the interaction between RBX1 and the E2–Ub (Kawakami *et al.*, 2001; Zheng *et al.*, 2002) and facilitation of the assembly of cullin-based complexes (Schwechheimer and Villalobos, 2004).

The derubylation activity of the CSN predicts that it may act negatively on cullin-containing E3 Ub ligases. Consistent with this hypothesis, *in vitro* experiments carried out in mammalian and yeast cells show that the CSN inhibits the Ub ligase activities associated with various complexes containing CUL1 (Lyapina *et al.*, 2001; Yang *et al.*, 2002; Zhou *et al.*, 2003), CUL3 (Zhou *et al.*, 2001, 2003) and CUL4A (Groisman *et al.*, 2003). However, multiple genetic studies reveal that CSN plays the opposite role *in vivo*, i.e. reduction in CSN function leads to defects in pathways positively regulated by cullin-containing E3s (Schwechheimer *et al.*, 2001; Cope *et al.*, 2002; Doronkin *et al.*, 2003; Groisman *et al.*, 2003; Liu *et al.*, 2003; Pintard *et al.*, 2003). On the basis of these observations, it is proposed that rubylation/derubylation process is the driving force behind the dynamic cycles of assembly and disassembly of cullin-based E3 complexes, which are required to enable E3s to achieve optimal activity (Cope and Deshaies, 2003; Serino and Deng, 2003; Wei and Deng, 2003).

As expected, there is also evidence suggesting that derubylation is not the only function of CSN in the control of cullin-containing E3 Ub ligases. For example an *Arabidopsis csn* loss-of-function mutant maintains normal derubylation activity but shows aberrant floral organ formation and impaired jasmonate responses, which reflect reduced activity of known CUL1-containing SCF-type E3 Ub ligases (SCF[UFO]; Wang *et al.*, 2003; SCF[COI1]; Feng *et al.*, 2003).

How, then, does the effect of the CSN on cullin-containing E3s relate to its role in the repression of photomorphogenesis? Unfortunately this question has not been resolved. So far, the relationship between *Arabidopsis* cullins and photomorphogenesis is not fully understood, partly because null mutations of *CUL1* and *CUL3* are embryonic lethal (Shen *et al.*, 2002; Figueroa *et al.*, 2005; Gingerich *et al.*, 2005; Thomann *et al.*, 2005) and the physiological role of other cullins in *Arabidopsis* is still unclear. Nonetheless, given the enormous number of potential cullin-based complexes in *Arabidopsis*, it is possible that some of them may play negative roles in photomorphogenesis. However, as no such component has been recovered from the *cop/det/fus* mutant screens, this may mean that there is functional redundancy. It is also likely that a particular E3 is only involved in a subset of light-induced signal transduction pathways. Finally, it has been shown that human COP1, DET1 and CUL4A form a complex that has E3 Ub ligase activity toward c-Jun (Wertz *et al.*, 2004). Thus, it will be of great interest to determine if this complex is conserved in plants.

6.3.5 SPA protein family

The SPA proteins do not belong to the COP/DET/FUS group. However, they have a close relationship with COP1, both physically and functionally; therefore, their roles in photoreceptor-mediated light signal transduction will be discussed in this section.

The *SPA1* (*Suppressor of phyA-105 1*) gene was initially identified in a genetic screen for suppressors of a weak *phyA* allele (*phyA-105*), as a potential negative regulator of phyA-dependent far-red light signaling pathways (Hoecker *et al.*, 1998). Subsequently, *SPA1* was shown to encode a protein with three distinctive domains: an N-terminal kinase-like domain, a coiled-coil domain and a C-terminal domain containing four WD-40 repeats (Hoecker *et al.*, 1999). The WD-40 repeat domain, which is essential for SPA1 function, is highly homologous to the WD-40 repeat domain that is found in COP1 (44% identity in amino acid sequence). Consistently, SPA1 can also bind HY5 through the WD-40 domain, similar to the COP1–HY5 interaction (Saijo *et al.*, 2003). Furthermore, COP1 and SPA1 can interact with each other via their coiled-coil domains (Hoecker and Quail, 2001; Saijo *et al.*, 2003).

Functional analyses suggest that SPA1 indeed acts synergistically with COP1 to downregulate the protein level of HY5 in far-red light (Saijo *et al.*, 2003). More importantly, SPA1 is able to alter the *in vitro* E3 Ub ligase activity of COP1. The full-length SPA1 protein inhibits COP1's E3 Ub ligase activity toward HY5, while the coiled-coil domain of SPA1 enhances the ubiquitination of LAF1 by COP1 (Saijo *et al.*, 2003; Seo *et al.*, 2003). These apparently conflicting observations might arise from the use of different E2s and substrates in the *in vitro* assays. It is also possible that in the *in vitro* system, excessive amounts of SPA1 could sequester the HY5 from COP1, while a truncated version of SPA1 (such as the coiled-coil domain) lacking the WD-40 repeats does not have such an effect. Nonetheless, the genetic data clearly suggest that SPA1 is a negative regulator of HY5 protein accumulation in far-red light (Saijo *et al.*, 2003). Therefore, the current hypothesis is that a heterocomplex of COP1 and SPA1 recruits protein targets such as HY5 and LAF1 through the WD-40 domains and facilitates the ubiquitination and proteasomal degradation of these substrates.

There are three *SPA1* homologs in *Arabidopsis* that are named *SPA2*, *SPA3* and *SPA4* (Laubinger and Hoecker, 2003; Laubinger *et al.*, 2004). The loss of all *SPA* genes in the *spa1spa2spa3spa4* quadruple mutant leads to strong photomorphogenic phenotypes even when grown in the darkness, reminiscent of the pleiotropic *cop/det/fus* mutants. Furthermore, *spa1*, *spa2* and *spa3* mutants display hypersensitivity to light. Therefore, these SPA proteins all appear to repress photomorphogenesis. Genetic analyses show that the SPA proteins have distinct and overlapping roles in regulating light-controlled development (Laubinger *et al.*, 2004). It is tempting to speculate that like SPA1, other SPA proteins also act in concert with COP1, a hypothesis that is supported by COP1's interactions with SPA2, SPA3 and SPA4 (Laubinger and Hoecker, 2003; Laubinger *et al.*, 2004).

6.4 Other connection points between light signaling and selective proteolysis

In a simplified model, COP1 represses photomorphogenesis by promoting the ubiquitination and proteasomal degradation of transcription factors that are required for expression of light-inducible genes (Figure 6.3). However, transcription factors are not the only targets of selective proteolysis in light signal transduction pathways. In addition to COP1, there are other E3 Ub ligases responsible for protein degradation events in photomorphogenesis pathways. In this section, we will focus on several cases that define additional links between photoreceptor action and protein degradation and discuss the possible functional implications.

6.4.1 F-box proteins that are involved in light signaling

As discussed in previous sections, F-box proteins are substrate recognition units for the SCF-type E3 Ub ligases, and *Arabidopsis* has 694 putative F-box proteins (Gagne *et al.*, 2002). Among the characterized F-box proteins, several are involved in light signaling pathways, including EID1 (Empfindlicher Im Dunkelroten Licht 1) and AFR (attenuated far-red response). Therefore, these SCF complexes may represent other connections between light signaling and protein degradation in addition to the connections represented by COP/DET/FUS proteins. Nonetheless, we cannot rule out the possibility that some of these SCF complexes are subject to regulation by the COP9 signalosome.

EID1 and AFR play opposite roles in phyA-dependent signaling. The *eid1* mutant displays enhanced sensitivity to far-red light, but accumulates wild-type levels of phyA with normal degradation properties (Buche et al., 2000). Thus, SCF^{EID1} probably acts to promote degradation of positive regulators of the phyA pathway (Dieterle *et al.*, 2001). In contrast, plants in which *AFR* expression is knocked down by RNAi have impaired phyA-dependent light signaling, suggesting that the function of SCF^{AFR} is to degrade repressors of the phyA pathway. A potentially important property of *AFR* is that its mRNA level is under the control of the circadian clock, which has extensive crosstalk with the light pathways (Harmon and Kay, 2003). Finding out the identity of EID1 and AFR targets will be important to broaden our understanding of the role of protein degradation in photomorphogenesis.

Three ZTL family F-box proteins, ZTL (Zeitlupe), FKF1 (Flavin binding, Kelch repeat, F-box) and LKP2 (LOV Kelch protein 2), have well-defined functions in circadian rhythm regulation and light control of hypocotyl elongation (Nelson *et al.*, 2000; Somers *et al.*, 2000; Schultz *et al.*, 2001; Somers *et al.*, 2004). They are distinguished from the other F-box proteins by the presence of a LOV (light, oxygen or voltage) domain in the N-terminus, which is highly homologous to the FMN (flavin mononucleotide)-binding domain found in blue light photoreceptor phototropins. The LOV domain of FKF1 has indeed been shown to bind FMN, indicating that FKF1 might be able to function as a blue light receptor (Imaizumi *et al.*,

2003). TOC1 (Timing of CAB expression 1), a critical component of central circadian oscillator, is a substrate of SCF^{ZTL} (Mas *et al.*, 2003) and CDF1 (cycling Dof factor 1), a Dof transcriptional factor that represses *CO* expression, is a substrate of SCF^{FKF1} (Imaizumi *et al.*, 2005). However, these are not sufficient to explain the photomorphogenesis-related phenotypes of the mutants of *ZTL* family genes. Therefore, the implication is that there must be additional substrates of the ZTL family F-box protein containing SCF-type E3 Ub ligases.

6.4.2   Other light signaling pathway components that are targets of proteolysis

6.4.2.1   Phytochrome A

Phytochrome A is the primary photoreceptor mediating very low fluence response and high irradiance response to far-red light. It is most abundant in etiolated seedlings, and its level rapidly decreases upon exposure to light. The abundance of phyA is controlled at both the transcriptional and posttranscriptional levels (Quail *et al.*, 1995). It has also been demonstrated that the major posttranscriptional regulation mechanism is the degradation of phyA protein via the ubiquitin/proteasome pathway (Jabben *et al.*, 1989; Clough and Vierstra, 1997; Clough *et al.*, 1999). It has been known for a long time that the conversion of phyA into its activated form upon light absorption makes it more susceptible to degradation (Clough and Vierstra, 1997; Nagy and Schäfer, 2002). Recent data suggest that phyA protein stability is dependent, at least in part, on the phosphorylation/dephosphorylation status of phyA N-terminal extension (Ryu *et al.*, 2005). Collectively, these findings support a hypothesis where phyA-mediated signaling is attenuated by phosphorylation-triggered proteasomal degradation of phyA, which occurs simultaneously with phyA activation.

phyA was recently shown to be ubiquitinated by COP1 *in vitro*, and reduction of phyA abundance triggered by light treatment is impaired in weak *cop1* mutants (Seo *et al.*, 2004). These data suggest that phyA is also a target of COP1's E3 Ub ligase activity. By promoting the ubiquitination and proteasomal degradation of phyA, COP1 can desensitize phyA-dependent far-red light signaling. Determining whether phosphorylation of phyA can facilitate its recruitment by COP1 will be interesting.

It is important to note that the degradation of phyA *in vivo* is likely to require additional pathways besides COP1. Upon light exposure, phyA moves from the cytosol to the nucleus and COP1 moves in the opposite direction (von Arnim and Deng, 1994; Kircher *et al.*, 2002). Thus, COP1-mediated degradation could not be solely responsible for the rapid and dramatic loss of phyA. The identification of other factors regulating phyA degradation will shed new light on the desensitizing mechanism of phyA-dependent signaling.

6.4.2.2   Cryptochrome 2

Two well-characterized *Arabidopsis* cryptochromes (CRY1 and CRY2) have overlapping functions in controlling photomorphogenesis in blue light (Lin, 2002). While

CRY1 is more or less stable, CRY2 is highly photolabile (Ahmad *et al.*, 1998; Lin *et al.*, 1998; Guo *et al.*, 1999). When *Arabidopsis* seedlings are transferred from darkness to blue light, CRY2 protein levels decline rapidly. It is also observed that a phosphorylated form of CRY2 appears quickly after the transfer. This reaction is blue light dependent, since neither etiolated seedlings nor seedlings transferred back to darkness have phosphorylated CRY2. Moreover, after the initial accumulation, the phosphorylated CRY2 decreases with increased exposure time to blue light, together with the unphosphorylated CRY2, indicating that the phosphorylated CRY2 is degraded (Shalitin *et al.*, 2002).

It has been shown that the CCT promotes constitutive photomorphogenesis, possibly by negatively affecting COP1 function (Yang *et al.*, 2000; Wang *et al.*, 2001). CCT2 is constitutively phosphorylated *in vivo*, unlike the full-length CRY2, which is phosphorylated only in blue light (Shalitin *et al.*, 2002). Therefore, it is conceivable that blue-light-triggered phosphorylation induces a conformation change in the structure of CRY2, which in turn activates its C-terminal domain to antagonize COP1 function and promote photomorphogenesis. This is different from phyA phosphorylation, which inhibits phyA function (Kim *et al.*, 2004; Ryu *et al.*, 2005). On the other hand, phosphorylated CRY2 also becomes more prone to undergoing proteolysis, providing a way to desensitize CRY2-dependent signaling that is somewhat analogous to phyA-dependent signaling (Lin and Shalitin, 2003).

Some evidence suggests that COP1 might be involved in the degradation of CRY2 (Shalitin *et al.*, 2002). In *cop1* weak mutants, the blue-light-triggered degradation of CRY2 is partially impaired. Additionally, the relative amount of phosphorylated CRY2 accumulates at higher levels than in wild type, indicating that defects in *COP1* result in the uncoupling of CRY2 phosphorylation and degradation to some extent. This effect is specific to *COP1*, but not to the other *COP/DET/FUS* genes. A physical interaction between CRY2 and COP1 has already been demonstrated (Wang *et al.*, 2001). Further study is required to elucidate if COP1 really has E3 Ub ligase activity toward CRY2 and whether COP1 prefers phosphorylated CRY2 as a substrate. It is important to note that even in *cop1* null mutants, the degradation of CRY2 is still not completely abolished, suggesting the existence of additional pathways (Shalitin *et al.*, 2002).

6.4.2.3 *Phytochrome-interacting factor 3 and far-red elongated hypocotyl 1*

Protein–protein interaction plays a central role in many signaling processes. Biochemical approaches have led to the discovery of a number of phytochrome interacting proteins that play divergent roles in the regulation of phytochrome signaling pathways (Wang and Deng, 2004). Phytochrome-interacting factor 3 (PIF3) is a bHLH (basic helix-loop-helix) transcription factor that interacts with phyA and phyB *in vitro* and binds to the G-box motif in the promoters of various light-regulated genes (Ni *et al.*, 1998, 1999; Martinez-Garcia *et al.*, 2000). PIF3's physiological functions in phytochrome signaling are complex, including a negative role in phyB-mediated inhibition of hypocotyl elongation, a negative role in phyA- and phyB-mediated cotyledon opening and a positive role in phyA- and phyB-mediated *CHS* (*Chalcone synthase*) gene induction (Kim *et al.*, 2003). Recently, several groups have reported the rapid downregulation of PIF3 in far-red and red light, which is mediated by phytochromes. Interestingly, this is not due to the change in *PIF3* mRNA level. Rather, it has been shown that the PIF3 protein is ubiquitinated by light treatment and subsequently degraded by the proteasome (Bauer *et al.*, 2004; Park *et al.*, 2004).

Far-red elongated hypocotyl 1 (FHY1) is defined genetically as a positive regulator of photomorphogenesis, specific to phyA-dependent far-red light responses (Desnos *et al.*, 2001). Phenotypic and genomic analyses suggest that FHY1 acts early in phyA-mediated signaling, probably very close to phyA itself (Wang and Deng, 2002; Wang *et al.*, 2002). Similar to PIF3, FHY1 protein also accumulates to its highest level in darkness and drops rapidly in light conditions. While *FHY1* mRNA level decreases in light conditions (Desnos *et al.*, 2001), it is clear that proteasome-mediated degradation of FHY1 upon light exposure is the major contributor to the observed abundance change of FHY1 protein (Shen *et al.*, 2005).

Rather unexpectedly, COP1 is required for the accumulation of both PIF3 and FHY1 in the darkness; this is unlikely to be the direct result of COP1's E3 Ub ligase activity (Bauer *et al.*, 2004; Shen *et al.*, 2005). The mechanism of this regulation is still under investigation. Further, it will be of great interest to identify the E3 Ub ligases that are responsible for targeting PIF3 and FHY1 during the dark-to-light transitions.

6.5   Concluding remarks

Since the discovery of pleiotropic *COP/DET/FUS* genes as repressors of photomorphogenesis in *Arabidopsis* about 15 years ago, remarkable progress has been made in the effort to elucidate their structures and cellular roles. Genetically, *COP/DET/FUS* genes act at the intersection between photoreceptor-mediated pathways and downstream light-regulated gene expression. Biochemically, they define three protein complexes *in vivo*: the COP1 complex, CDD complex and CSN. COP1, a RING family E3 Ub ligase, is involved in the ubiquitination and degradation of various positive regulators of photomorphogenesis, from transcription factors to photoreceptors. Its activity is under the control of multiple photoreceptors, either directly or indirectly. On the other hand, the CDD complex and CSN are likely to have broader roles in regulation of the ubiquitin/proteasome pathway. The COP10 subunit of the CDD complex upregulates the activity of several E2 Ub-conjugating enzymes *in vitro*. CSN catalyzes the derubylation of cullins and is required for optimal activities of multiple cullin-containing E3 Ub ligases. In addition, other components of the ubiquitin/proteasome system (for example F-box proteins) also contribute to the fine-tuning of photomorphogenesis. In summary, these findings suggest that rapid and regulated protein destruction through the ubiquitin/proteasome pathway is a common way in which plants efficiently switch on or off developmental programs (for example photomorphogenesis and skotomorphogenesis), in order to adapt to the ever-changing environment.

While the overall picture has been established, some important details remain to be filled in. Future research could focus on the molecular mechanism of COP1 nucleocytoplasmic partitioning, purifying and analyzing the COP1 complex, searching for additional targets of COP1-mediated ubiquitination, identifying *in vivo* E2 targets of COP10, studying the function of DET1 and DDB1 in the context of the CDD complex, determining the functional relationship between the CSN and the proteasome, elucidating the precise role of the CSN in photomorphogenesis and, of course, continuing to establish more connections between light signaling and the ubiquitin/proteasome pathways. Undoubtedly, progress in these directions will enable us to better understand how light control of plant development is regulated by protein degradation.

Note

Most recently, it has been demonstrated that the *Arabidopsis* CUL4 and the CDD complex form an E3 ligase together, which physically interacts with COP1 and negatively modulates light-induced plant development. The *CUL4* reduction-of-function lines have characteristics similar to the *cop/det/fus* mutants, including constitutive photomorphogenic phenotypes and elevated accumulation level of COP1 targets such as HY5. Furthermore, like other cullins, *Arabidopsis* CUL4 undergoes CSN-regulated rubylation/derubylation cycles (Chen *et al.*, 2006). Therefore, it seems likely that the biochemical activities of the three COP/DET/FUS protein-based complexes are linked through the CUL4-containing E3 Ub ligase.

References

Aguilar, R.C. and Wendland, B. (2003) Ubiquitin: not just for proteasomes anymore. *Curr. Opin. Cell Biol.* **15**(2), 184–190.

Ahmad, M., Jarillo, J.A., Cashmore, A.R. (1998) Chimeric proteins between cry1 and cry2 *Arabidopsis* blue light photoreceptors indicate overlapping functions and varying protein stability. *Plant Cell* **10**(2), 197–207.

Ang, L.H., Chattopadhyay, S., Wei, N., *et al.* (1998) Molecular interaction between COP1 and HY5 defines a regulatory switch for light control of *Arabidopsis* development. *Mol. Cell* **1**(2), 213–222.

Bates, P.W. and Vierstra, R.D. (1999) UPL1 and 2, two 405 kDa ubiquitin-protein ligases from *Arabidopsis thaliana* related to the HECT-domain protein family. *Plant J.* **20**(2), 183–195.

Bauer, D., Viczián, A., Kircher, S., *et al.* (2004) Constitutive photomorphogenesis 1 and multiple photoreceptors control degradation of phytochrome interacting factor 3, a transcription factor required for light signaling in *Arabidopsis*. *Plant Cell* **16**(6), 1433–1445.

Bech-Otschir, D., Kraft, R., Huang, X., *et al.* (2001) COP9 signalosome-specific phosphorylation targets p53 to degradation by the ubiquitin system. *EMBO J.* **20**(7), 1630–1639.

Benvenuto, G., Formiggini, F., Laflamme, P., Malakhov, M. and Bowler, C. (2002) The photomorphogenesis regulator DET1 binds the amino-terminal tail of histone H2B in a nucleosome context. *Curr. Biol.* **12**(17), 1529–1534.

Blilou, I., Frugier, F., Folmer, S., *et al.* (2002) The *Arabidopsis HOBBIT* gene encodes a CDC27 homolog that links the plant cell cycle to progression of cell differentiation. *Genes Dev.* **16**(19), 2566–2575.

Buche, C., Poppe, C., Schäfer, E. and Kretsch, T. (2000) *eid1*: a new *Arabidopsis* mutant hypersensitive in Phytochrome A-dependent high-irradiance responses. *Plant Cell* **12**(4), 547–558.

Busch, S., Eckert, S.E., Krappmann, S. and Braus, G.H. (2003) The COP9 signalosome is an essential regulator of development in the filamentous fungus *Aspergillus nidulans*. *Mol. Microbiol.* **49**(3), 717–730.

Capron, A., Okresz, L. and Genschik, P. (2003a) First glance at the plant APC/C, a highly conserved ubiquitin-protein ligase. *Trends Plant Sci.* **8**(2), 83–89.

Capron, A., Serralbo, O., Fulop, K., *et al.* (2003b) The *Arabidopsis* anaphase-promoting complex or cyclosome: molecular and genetic characterization of the APC2 subunit. *Plant Cell* **15**(10), 2370–2382.

Chamovitz, D.A., Wei, N., Osterlund, M.T., *et al.* (1996) The COP9 complex, a novel multisubunit nuclear regulator involved in light control of a plant developmental switch. *Cell* **86**(1), 115–121.

Chattopadhyay, S., Ang, L.H., Puente, P., Deng, X.W. and Wei, N. (1998) *Arabidopsis* bZIP protein HY5 directly interacts with light-responsive promoters in mediating light control of gene expression. *Plant Cell* **10**(5), 673–683.

Chen, H., Shen, Y., Tang, X., Yu, L., Wang, J., Guo, L., Zhang, Y., Zhang, H., Feng, S., Strickland, E., Zheng, N. and Deng, X.W. (2006) *Arabidopsis* CULLIN4 forms an E3 ubiquitin ligase with RBX1 and the CDD complex in mediating light control of development. *Plant Cell* **18**(8), 1991–2004.

Chen, M., Chory, J. and Fankhauser, C. (2004) Light signal transduction in higher plants. *Annu. Rev. Genet.* **38**, 87–117.

Chory, J., Peto, C., Feinbaum, R., Pratt, L. and Ausubel, F. (1989) *Arabidopsis thaliana* mutant that develops as a light-grown plant in the absence of light. *Cell* **58**(5), 991–999.

Clough, R.C., Jordan-Beebe, E.T., Lohman, K.N., *et al.* (1999) Sequences within the N- and C-terminal domains of phytochrome A are required for PFR ubiquitination and degradation. *Plant J.* **17**(2), 155–167.

Clough, R.C. and Vierstra, R.D. (1997) Phytochrome degradation. *Plant Cell Environ.* **20**(6), 713–721.

Conaway, R.C., Brower, C.S. and Conaway, J.W. (2002) Emerging roles of ubiquitin in transcription regulation. *Science* **296**(5571), 1254–1258.

Cope, G.A. and Deshaies, R.J. (2003) COP9 signalosome: a multifunctional regulator of SCF and other cullin-based ubiquitin ligases. *Cell* **114**(6), 663–671.

Cope, G.A., Suh, G.S., Aravind, L., *et al.* (2002) Role of predicted metalloprotease motif of Jab1/Csn5 in cleavage of Nedd8 from Cul1. *Science* **298**(5593), 608–611.

Coux, O., Tanaka, K. and Goldberg, A.L. (1996) Structure and functions of the 20S and 26S proteasomes. *Annu. Rev. Biochem.* **65**, 801–847.

del Pozo, J.C., Dharmasiri, S., Hellmann, H., Walker, L., Gray, W.M. and Estelle, M. (2002) AXR1-ECR1-dependent conjugation of RUB1 to the *Arabidopsis* cullin AtCUL1 is required for auxin response. *Plant Cell* **14**(2), 421–433.

del Pozo, J.C. and Estelle, M. (1999) The *Arabidopsis* cullin AtCUL1 is modified by the ubiquitin-related protein RUB1. *Proc. Natl. Acad. Sci. USA* **96**(26), 15342–15347.

Deng, X.W. (1994) Fresh view of light signal transduction in plants. *Cell* **76**(3), 423–426.

Deng, X.W., Caspar, T. and Quail, P.H. (1991) cop1: a regulatory locus involved in light-controlled development and gene expression in *Arabidopsis*. *Genes Dev.* **5**(7), 1172–1182.

Deng, X.W., Matsui, M., Wei, N., *et al.* (1992) *COP1*, an *Arabidopsis* regulatory gene, encodes a protein with both a zinc-binding motif and a G_β homologous domain. *Cell* **71**(5), 791–801.

Desnos, T., Puente, P., Whitelam, G.C. and Harberd, N.P. (2001) FHY1: a phytochrome A-specific signal transducer. *Genes Dev.* **15**(22), 2980–2990.

Dharmasiri, S., Dharmasiri, N., Hellmann, H. and Estelle, M. (2003) The RUB/Nedd8 conjugation pathway is required for early development in *Arabidopsis*. *EMBO J.* **22**(8), 1762–1770.

Dieterle, M., Thomann, A., Renou, J.P., *et al.* (2005) Molecular and functional characterization of *Arabidopsis* Cullin 3A. *Plant J.* **41**(3), 386–399.

Dieterle, M., Zhou, Y.C., Schäfer, E., Funk, M. and Kretsch, T. (2001) EID1, an F-box protein involved in phytochrome A-specific light signaling. *Genes Dev.* **15**(8), 939–944.

Doronkin, S., Djagaeva, I. and Beckendorf, S.K. (2003) The COP9 signalosome promotes degradation of Cyclin E during early *Drosophila* oogenesis. *Dev. Cell* **4**(5), 699–710.

Downes, B.P., Stupar, R.M., Gingerich, D.J. and Vierstra, R.D. (2003) The HECT ubiquitin-protein ligase (UPL) family in *Arabidopsis*: UPL3 has a specific role in trichome development. *Plant J.* **35**(6), 729–742.

Feng, S., Ma, L., Wang, X., *et al.* (2003) The COP9 signalosome interacts physically with SCF[COI1] and modulates jasmonate responses. *Plant Cell* **15**(5), 1083–1094.

Figueroa, P., Gusmaroli, G., Serino, G., *et al.* (2005) *Arabidopsis* has two redundant Cullin3 proteins that are essential for embryo development and that interact with RBX1 and BTB proteins to form multisubunit E3 ubiquitin ligase complexes in vivo. *Plant Cell* **17**(4), 1180–1195.

Freilich, S., Oron, E., Kapp, Y., *et al.* (1999) The COP9 signalosome is essential for development of *Drosophila melanogaster.Curr. Biol.* **9**(20), 1187–1190.

Fu, H., Doelling, J.H., Arendt, C.S., Hochstrasser, M. and Vierstra, R.D. (1998) Molecular organization of the 20S proteasome gene family from *Arabidopsis thaliana. Genetics* **149**(2), 677–692.

Fu, H., Doelling, J.H., Rubin, D.M. and Vierstra, R.D. (1999) Structural and functional analysis of the six regulatory particle triple-A ATPase subunits from the *Arabidopsis* 26S proteasome. *Plant J.* **18**(5), 529–539.

Furukawa, M., Zhang, Y., McCarville, J., Ohta, T. and Xiong, Y. (2000) The CUL1 C-terminal sequence and ROC1 are required for nuclear accumulation, NEDD8 modification, and ubiquitin ligase activity of CUL1. *Mol. Cell. Biol.* **20**(21), 8185–8197.

Gagne, J.M., Downes, B.P., Shiu, S.H., Durski, A.M. and Vierstra, R.D. (2002) The F-box subunit of the SCF E3 complex is encoded by a diverse superfamily of genes in *Arabidopsis. Proc. Natl. Acad. Sci. U S A* **99**(17), 11519–11524.

Gingerich, D.J., Gagne, J.M., Salter, D.W., *et al.* (2005) Cullins 3a and 3b assemble with members of the broad complex/tramtrack/bric-a-brac (BTB) protein family to form essential ubiquitin-protein ligases (E3s) in *Arabidopsis. J. Biol. Chem.* **280**(19), 18810–18821.

Glickman, M.H., Rubin, D., Coux, O., *et al.* (1998) A subcomplex of the proteasome regulatory particle required for ubiquitin-conjugate degradation and related to the COP9-signalosome and eIF3. *Cell* **94**(5), 615–623.

Groisman, R., Polanowska, J., Kuraoka, I., *et al.* (2003) The ubiquitin ligase activity in the DDB2 and CSA complexes is differentially regulated by the COP9 signalosome in response to DNA damage. *Cell* **113**(3), 357–367.

Groll, M., Bajorek, M., Kohler, A., *et al.* (2000) A gated channel into the proteasome core particle. *Nat. Struct. Biol.* **7**(11), 1062–1067.

Groll, M., Ditzel, L., Lowe, J., *et al.* (1997) Structure of 20S proteasome from yeast at 2.4 A resolution. *Nature* **386**(6624), 463–471.

Guo, H., Duong, H., Ma, N. and Lin, C. (1999) The *Arabidopsis* blue light receptor cryptochrome 2 is a nuclear protein regulated by a blue light-dependent post-transcriptional mechanism. *Plant J.* **19**(3), 279–287.

Gusmaroli, G., Feng, S. and Deng, X.W. (2004) The *Arabidopsis* CSN5A and CSN5B subunits are present in distinct COP9 signalosome complexes, and mutations in their JAMM domains exhibit differential dominant negative effects on development. *Plant Cell* **16**(11), 2984–3001.

Hardtke, C.S. and Deng, X.W. (2000) The cell biology of the COP/DET/FUS proteins. Regulating proteolysis in photomorphogenesis and beyond? *Plant Physiol.* **124**(4), 1548–1557.

Harmon, F.G. and Kay, S.A. (2003) The F box protein AFR is a positive regulator of phytochrome A-mediated light signaling. *Curr. Biol.* **13**(23), 2091–2096.

Hartmann-Petersen, R., Seeger, M. and Gordon, C. (2003) Transferring substrates to the 26S proteasome. *Trends Biochem. Sci.* **28**(1), 26–31.

Hershko, A. and Ciechanover, A. (1998). The ubiquitin system. *Annu. Rev. Biochem.* **67**, 425–479.

Hochstrasser, M. (1998) There's the rub: a novel ubiquitin-like modification linked to cell cycle regulation. *Genes Dev.* **12**(7), 901–907.

Hochstrasser, M. (2000) Evolution and function of ubiquitin-like protein-conjugation systems. *Nat. Cell Biol.* **2**(8), E153–E157.

Hoecker, U. and Quail, P.H. (2001) The phytochrome A-specific signaling intermediate SPA1 interacts directly with COP1, a constitutive repressor of light signaling in *Arabidopsis. J. Biol. Chem.* **276**(41), 38173–38178.

Hoecker, U., Tepperman, J.M. and Quail, P.H. (1999) SPA1, a WD-repeat protein specific to phytochrome A signal transduction. *Science* **284**(5413), 496–499.

Hoecker, U., Xu, Y. and Quail, P.H. (1998) *SPA1*: a new genetic locus involved in phytochrome A-specific signal transduction. *Plant Cell* **10**(1), 19–33.

Hofmann, R.M. and Pickart, C.M. (1999) Noncanonical *MMS2*-encoded ubiquitin-conjugating enzyme functions in assembly of novel polyubiquitin chains for DNA repair. *Cell* **96**(5), 645–653.

Holm, M., Hardtke, C.S., Gaudet, R. and Deng, X.W. (2001) Identification of a structural motif that confers specific interaction with the WD40 repeat domain of *Arabidopsis* COP1. *EMBO J.* **20**(1–2), 118–127.

Hori, T., Osaka, F., Chiba, T., *et al.* (1999) Covalent modification of all members of human cullin family proteins by NEDD8. *Oncogene* **18**(48), 6829–6834.

Hudson, M.E. (2000) The genetics of phytochrome signalling in *Arabidopsis. Semin. Cell Dev. Biol.* **11**(6), 475–483.

Hwang, B.J., Toering, S., Francke, U. and Chu, G. (1998) p48 activates a UV-damaged-DNA binding factor and is defective in xeroderma pigmentosum group E cells that lack binding activity. *Mol. Cell. Biol.* **18**(7), 4391–4399.

Imaizumi, T., Schultz, T.F., Harmon, F.G., Ho, L.A. and Kay, S.A. (2005) FKF1 F-box protein mediates cyclic degradation of a repressor of *CONSTANS* in *Arabidopsis. Science* **309**(5732), 293–297.

Imaizumi, T., Tran, H.G., Swartz, T.E., Briggs, W.R. and Kay, S.A. (2003) FKF1 is essential for photoperiodic-specific light signalling in *Arabidopsis. Nature* **426**(6964), 302–306.

Jabben, M., Shanklin, J. and Vierstra, R.D. (1989) Ubiquitin-phytochrome conjugates: pool dynamics during *in vivo* phytochrome degradation. *J. Biol. Chem.* **264**(9), 4998–5005.

Jang, I.C., Yang, J.Y., Seo, H.S. and Chua, N.H. (2005) HFR1 is targeted by COP1 E3 ligase for post-translational proteolysis during phytochrome A signaling. *Genes Dev.* **19**(5), 593–602.

Kawakami, T., Chiba, T., Suzuki, T., *et al.* (2001) NEDD8 recruits E2-ubiquitin to SCF E3 ligase. *EMBO J.* **20**(15), 4003–4012.

Kim, J., Yi, H., Choi, G., Shin, B., Song, P.S. and Choi, G. (2003) Functional characterization of phytochrome interacting factor 3 in phytochrome-mediated light signal transduction. *Plant Cell* **15**(10), 2399–2407.

Kim, J.I., Shen, Y., Han, Y.J., *et al.* (2004) Phytochrome phosphorylation modulates light signaling by influencing the protein-protein interaction. *Plant Cell* **16**(10), 2629–2640.

Kircher, S., Gil, P., Kozma-Bognar, L., *et al.* (2002) Nucleocytoplasmic partitioning of the plant photoreceptors phytochrome A, B, C, D, and E is regulated differentially by light and exhibits a diurnal rhythm. *Plant Cell* **14**(7), 1541–1555.

Kurz, T., Pintard, L., Willis, J.H., *et al.* (2002) Cytoskeletal regulation by the Nedd8 ubiquitin-like protein modification pathway. *Science* **295**(5558), 1294–1298.

Kwok, S.F., Staub, J.M. and Deng, X.W. (1999) Characterization of two subunits of *Arabidopsis* 19S proteasome regulatory complex and its possible interaction with the COP9 complex. *J. Mol. Biol.* **285**(1), 85–95.

Laubinger, S., Fittinghoff, K. and Hoecker, U. (2004) The SPA quartet: a family of WD-repeat proteins with a central role in suppression of photomorphogenesis in *Arabidopsis. Plant Cell* **16**(9), 2293–2306.

Laubinger, S. and Hoecker, U. (2003) The SPA1-like proteins SPA3 and SPA4 repress photomorphogenesis in the light. *Plant J.* **35**(3), 373–385.

Li, L. and Deng, X.W. (2003) The COP9 signalosome: an alternative lid for the 26S proteasome? *Trends Cell Biol.* **13**(10), 507–509.

Li, L., Liao, J., Ruland, J., Mak, T.W. and Cohen, S.N. (2001) A TSG101/MDM2 regulatory loop modulates MDM2 degradation and MDM2/p53 feedback control. *Proc. Natl. Acad. Sci. U S A* **98**(4), 1619–1624.

Lin, C. (2002) Blue light receptors and signal transduction. *Plant Cell* **14**(Suppl.), S207–S225.

Lin, C. and Shalitin, D. (2003) Cryptochrome structure and signal transduction. *Annu. Rev. Plant Biol.* **54**, 469–496.

Lin, C., Yang, H., Guo, H., Mockler, T., Chen, J. and Cashmore, A.R. (1998) Enhancement of blue-light sensitivity of *Arabidopsis* seedlings by a blue light receptor cryptochrome 2. *Proc. Natl. Acad. Sci. U S A* **95**(5), 2686–2690.

Liu, C., Powell, K.A., Mundt, K., Wu, L., Carr, A.M. and Caspari, T. (2003) Cop9/signalosome subunits and Pcu4 regulate ribonucleotide reductase by both checkpoint-dependent and -independent mechanisms. *Genes Dev.* **17**(9), 1130–1140.

Lopez-Molina, L., Mongrand, S., Kinoshita, N. and Chua, N.H. (2003) AFP is a novel negative regulator of ABA signaling that promotes ABI5 protein degradation. *Genes Dev.* **17**(3), 410–418.

Lyapina, S., Cope, G., Shevchenko, A., *et al.* (2001) Promotion of NEDD8-CUL1 conjugate cleavage by COP9 signalosome. *Science* **292**(5520), 1382–1385.

Ma, L., Gao, Y., Qu, L., *et al.* (2002) Genomic evidence for COP1 as a repressor of light-regulated gene expression and development in *Arabidopsis*. *Plant Cell* **14**(10), 2383–2398.

Ma, L., Li, J., Qu, L., *et al.* (2001) Light control of *Arabidopsis* development entails coordinated regulation of genome expression and cellular pathways. *Plant Cell* **13**(12), 2589–2607.

Ma, L., Zhao, H. and Deng, X.W. (2003) Analysis of the mutational effects of the *COP/DET/FUS* loci on genome expression profiles reveals their overlapping yet not identical roles in regulating *Arabidopsis* seedling development. *Development* **130**(5), 969–981.

Martinez-Garcia, J.F., Huq, E. and Quail, P.H. (2000) Direct targeting of light signals to a promoter element-bound transcription factor. *Science* **288**(5467), 859–863.

Mas, P., Kim, W.Y., Somers, D.E. and Kay, S.A. (2003) Targeted degradation of TOC1 by ZTL modulates circadian function in *Arabidopsis thaliana*. *Nature* **426**(6966), 567–570.

Maytal-Kivity, V., Pick, E., Piran, R., Hofmann, K. and Glickman, M.H. (2003) The COP9 signalosome-like complex in *S. cerevisiae* and links to other PCI complexes. *Int. J. Biochem. Cell Biol.* **35**(5), 706–715.

McNellis, T.W., von Arnim, A.G., Araki, T., Komeda, Y., Miséra, S. and Deng, X.W. (1994a) Genetic and molecular analysis of an allelic series of *cop1* mutants suggests functional roles for the multiple protein domains. *Plant Cell* **6**(**4**), 487–500.

McNellis, T.W., von Arnim, A.G. and Deng, X.W. (1994b) Overexpression of *Arabidopsis* COP1 results in partial suppression of light-mediated development: evidence for a light-inactivable repressor of photomorphogenesis. Plant Cell **6**(10), 1391–1400.

Misera, S., Muller, A.J., Weiland-Heidecker, U. and Jurgens, G. (1994) The *FUSCA* genes of *Arabidopsis*: negative regulators of light responses. *Mol. Gen. Genet.* **244**(3), 242–252.

Moon, J., Parry, G. and Estelle, M. (2004) The ubiquitin-proteasome pathway and plant development. *Plant Cell* **16**(12), 3181–3195.

Mundt, K.E., Porte, J., Murray, J.M., *et al.* (1999) The COP9/signalosome complex is conserved in fission yeast and has a role in S phase. *Curr. Biol.* **9**(23), 1427–1430.

Nagy, F. and Schafer, E. (2002) Phytochromes control photomorphogenesis by differentially regulated, interacting signaling pathways in higher plants. *Annu. Rev. Plant Biol.* **53**, 329–355.

Naumann, M., Bech-Otschir, D., Huang, X., Ferrell, K. and Dubiel, W. (1999) COP9 signalosome-directed c-Jun activation/stabilization is independent of JNK. *J. Biol. Chem.* **274**(50), 35297–35300.

Nelson, D.C., Lasswell, J., Rogg, L.E., Cohen, M.A. and Bartel, B. (2000) *FKF1*, a clock-controlled gene that regulates the transition to flowering in *Arabidopsis*. *Cell* **101**(3), 331–340.

Ni, M., Tepperman, J.M. and Quail, P.H. (1998) PIF3, a phytochrome-interacting factor necessary for normal photoinduced signal transduction, is a novel basic helix-loop-helix protein. *Cell* **95**(5), 657–667.

Ni, M., Tepperman, J.M. and Quail, P.H. (1999) Binding of phytochrome B to its nuclear signalling partner PIF3 is reversibly induced by light. *Nature* **400**(6746), 781–784.

Ohh, M., Kim, W.Y., Moslehi, J.J., *et al.* (2002) An intact NEDD8 pathway is required for Cullin-dependent ubiquitylation in mammalian cells. *EMBO Rep.* **3**(2), 177–182.

Osaka, F., Saeki, M., Katayama, S., *et al.* (2000) Covalent modifier NEDD8 is essential for SCF ubiquitin-ligase in fission yeast. *EMBO J.* **19**(13), 3475–3484.

Osterlund, M.T. and Deng, X.W. (1998) Multiple photoreceptors mediate the light-induced reduction of GUS-COP1 from *Arabidopsis* hypocotyl nuclei. *Plant J.* **16**(2), 201–208.

Osterlund, M.T., Hardtke, C.S., Wei, N. and Deng, X.W. (2000) Targeted destabilization of HY5 during light-regulated development of *Arabidopsis*. *Nature* **405**(6785), 462–466.

Ou, C.Y., Lin, Y.F., Chen, Y.J. and Chien, C.T. (2002) Distinct protein degradation mechanisms mediated by Cul1 and Cul3 controlling Ci stability in *Drosophila* eye development. *Genes Dev.* **16**(18), 2403–2414.

Oyama, T., Shimura, Y. and Okada, K. (1997) The *Arabidopsis HY5* gene encodes a bZIP protein that regulates stimulus-induced development of root and hypocotyl. *Genes Dev.* **11**(22), 2983–2995.

Park, E., Kim, J., Lee, Y., *et al.* (2004) Degradation of phytochrome interacting factor 3 in phytochrome-mediated light signaling. *Plant Cell Physiol.* **45**(8), 968–975.

Peng, Z., Shen, Y., Feng, S., *et al.* (2003) Evidence for a physical association of the COP9 signalosome, the proteasome, and specific E3 ligases *in vivo*. *Curr. Biol.* **13**(13), R504–R505.

Pepper, A., Delaney, T., Washburn, T., Poole, D. and Chory, J. (1994) *DET1*, a negative regulator of light-mediated development and gene expression in *Arabidopsis*, encodes a novel nuclear-localized protein. *Cell* **78**(1), 109–116.

Pintard, L., Kurz, T., Glaser, S., Willis, J.H., Peter, M. and Bowerman, B. (2003) Neddylation and deneddylation of CUL-3 is required to target MEI-1/Katanin for degradation at the meiosis-to-mitosis transition in *C. elegans*. *Curr. Biol.* **13**(11), 911–921.

Podust, V.N., Brownell, J.E., Gladysheva, T.B., *et al.* (2000) A Nedd8 conjugation pathway is essential for proteolytic targeting of p27^{KIP1} by ubiquitination. *Proc. Natl. Acad. Sci. U S A* **97**(9), 4579–4584.

Quail, P.H., Boylan, M.T., Parks, B.M., Short, T.W., Xu, Y. and Wagner, D. (1995) Phytochromes: photosensory perception and signal transduction. *Science* **268**(5211), 675–680.

Read, M.A., Brownell, J.E., Gladysheva, T.B., *et al.* (2000) Nedd8 modification of Cul-1 activates SCF$^{\beta TrCP}$-dependent ubiquitination of IκBα. *Mol. Cell Biol.* **20**(7), 2326–2333.

Risseeuw, E.P., Daskalchuk, T.E., Banks, T.W., *et al.* (2003) Protein interaction analysis of SCF ubiquitin E3 ligase subunits from *Arabidopsis*. *Plant J.* **34**(6), 753–767.

Ryu, J.S., Kim, J.I., Kunkel, T., *et al.* (2005) Phytochrome-specific type 5 phosphatase controls light signal flux by enhancing phytochrome stability and affinity for a signal transducer. *Cell* **120**(3), 395–406.

Saijo, Y., Sullivan, J.A., Wang, H., *et al.* (2003) The COP1-SPA1 interaction defines a critical step in phytochrome A-mediated regulation of HY5 activity. *Genes Dev.* **17**(21), 2642–2647.

Schroeder, D.F., Gahrtz, M., Maxwell, B.B., *et al.* (2002) De-etiolated 1 and damaged DNA binding protein 1 interact to regulate *Arabidopsis* photomorphogenesis. *Curr. Biol.* **12**(17), 1462–1472.

Schultz, T.F., Kiyosue, T., Yanovsky, M., Wada, M. and Kay, S.A. (2001) A role for LKP2 in the circadian clock of *Arabidopsis*. *Plant Cell* **13**(12), 2659–2670.

Schwechheimer, C. and Deng, X.W. (2000) The COP/DET/FUS proteins-regulators of eukaryotic growth and development. *Semin. Cell Dev. Biol.* **11**(6), 495–503.

Schwechheimer, C., Serino, G., Callis, J., *et al.* (2001) Interactions of the COP9 signalosome with the E3 ubiquitin ligase SCF^{TIR1} in mediating auxin response. *Science* **292**(5520), 1379–1382.

Schwechheimer, C. and Villalobos, L.I. (2004) Cullin-containing E3 ubiquitin ligases in plant development. *Curr. Opin. Plant Biol.* **7**(6), 677–686.

Seeger, M., Kraft, R., Ferrell, K., *et al.* (1998) A novel protein complex involved in signal transduction possessing similarities to 26S proteasome subunits. *FASEB J.* **12**(6), 469–478.

Seo, H.S., Watanabe, E., Tokutomi, S., Nagatani, A. and Chua, N.H. (2004) Photoreceptor ubiquitination by COP1 E3 ligase desensitizes phytochrome A signaling. *Genes Dev.* **18**(6), 617–622.

Seo, H.S., Yang, J., Ishikawa, M., Bolle, B., Ballesteros, M.L. and Chua, N.H. (2003) LAF1 ubiquitination by COP1 controls photomorphogenesis and is stimulated by SPA1. *Nature* **423**(6943), 995–999.

Serino, G. and Deng, X.W. (2003) The COP9 signalosome: regulating plant development through the control of proteolysis. *Annu. Rev. Plant Biol.* **54**, 165–182.

Shalitin, D., Yang, H., Mockler, T.C., *et al.* (2002) Regulation of *Arabidopsis* cryptochrome 2 by blue-light-dependent phosphorylation. *Nature* **417**(6890), 763–767.

Shen, W.H., Parmentier, Y., Hellmann, H., *et al.* (2002) Null mutation of *AtCUL1* causes arrest in early embryogenesis in *Arabidopsis*. *Mol. Biol. Cell* **13**(6), 1916–1928.

Shen, Y., Feng, S., Ma, L., *et al.* (2005) *Arabidopsis* FHY1 protein stability is regulated by light via phytochrome A and 26S proteasome. *Plant Physiol.* **139**(3), 1234–1243.

Smalle, J., Kurepa, J., Yang, P., *et al.* (2002) Cytokinin growth responses in *Arabidopsis* involve the 26S proteasome subunit RPN12 *Plant Cell* **14**(1), 17–32.

Smalle, J., Kurepa, J., Yang, P., *et al.* (2003) The pleiotropic role of the 26S proteasome subunit RPN10 in *Arabidopsis* growth and development supports a substrate-specific function in abscisic acid signaling. *Plant Cell* **15**(4), 965–980.

Smalle, J. and Vierstra, R.D. (2004) The ubiquitin 26S proteasome proteolytic pathway. *Annu. Rev. Plant Biol.* **55**, 555–590.

Somers, D.E., Kim, W.Y. and Geng, R. (2004) The F-box protein ZEITLUPE confers dosage-dependent control on the circadian clock, photomorphogenesis, and flowering time. *Plant Cell* **16**(3), 769–782.

Somers, D.E., Schultz, T.F., Milnamow, M. and Kay, S.A. (2000) *ZEITLUPE* encodes a novel clock-associated PAS protein from *Arabidopsis*. *Cell* **101**(3), 319–329.

Strickland, E., Rubio, V. and Deng, X.W. (2006) The function of the COP/DET/FUS proteins in controlling photomorphogenesis: a role for regulated proteolysis. In: *Photomorphogenesis in Plants and Bacteria*, 3rd edn (eds Schaefer, E. and Nagy, F.), pp. 357–378. Springer, Dordrecht.

Sullivan, J.A. and Deng, X.W. (2003) From seed to seed: the role of photoreceptors in *Arabidopsis* development. *Dev. Biol.* **260**(2), 289–297.

Sullivan, J.A., Shirasu, K. and Deng, X.W. (2003) The diverse roles of ubiquitin and the 26S proteasome in the life of plants. *Nat. Rev. Genet.* **4**(12), 948–958.

Sun, Y., Wilson, M.P. and Majerus, P.W. (2002) Inositol 1,3,4-triphosphate 5/6-kinase associates with the COP9 signalosome by binding to CSN1. *J. Biol. Chem.* **277**(48), 45759–45764.

Suzuki, G., Yanagawa, Y., Kwok, S.F., Matsui, M. and Deng, X.W. (2002) *Arabidopsis* COP10 is a ubiquitin-conjugating enzyme variant that acts together with COP1 and the COP9 signalosome in repressing photomorphogenesis. *Genes Dev.* **16**(5), 554–559.

Tateishi, K., Omata, M., Tanaka, K. and Chiba, T. (2001) The NEDD8 system is essential for cell cycle progression and morphogenetic pathway in mice. *J. Cell Biol.* **155**(4), 571–579.

Thomann, A., Brukhin, V., Dieterle, M., *et al.* (2005) *Arabidopsis* CUL3A and CUL3B genes are essential for normal embryogenesis. *Plant J.* **43**(3), 437–448.

Torii, K.U., McNellis, T.W. and Deng, X.W. (1998) Functional dissection of *Arabidopsis* COP1 reveals specific roles of its three structural modules in light control of seedling development. *EMBO J.* **17**(19), 5577–5587.

Uhle, S., Medalia, O., Waldron, R., *et al.* (2003) Protein kinase CK2 and protein kinase D are associated with the COP9 signalosome. *EMBO J.* **22**(6), 1302–1312.

Unno, M., Mizushima, T., Morimoto, Y., *et al.* (2002) The structure of the mammalian 20S proteasome at 2.75 A resolution. *Structure* **10**(5), 609–618.

Verma, R., Aravind, L., Oania, R., *et al.* (2002) Role of Rpn11 metalloprotease in deubiquitination and degradation by the 26S proteasome. *Science* **298**(5593), 611–615.

Voges, D., Zwickl, P. and Baumeister, W. (1999) The 26S proteasome: a molecular machine designed for controlled proteolysis. *Annu. Rev. Biochem.* **68**, 1015–1068.

von Arnim, A.G. and Deng, X.W. (1994) Light inactivation of *Arabidopsis* photomorphogenic repressor COP1 involves a cell-specific regulation of its nucleocytoplasmic partitioning. *Cell* **79**(6), 1035–1045.

von Arnim, A.G., Osterlund, M.T., Kwok, S.F. and Deng, X.W. (1997) Genetic and developmental control of nuclear accumulation of COP1, a repressor of photomorphogenesis in *Arabidopsis*. *Plant Physiol.* **114**(3), 779–788.

Wang, H. and Deng, X.W. (2002) *Arabidopsis* FHY3 defines a key phytochrome A signaling component directly interacting with its homologous partner FAR1. *EMBO J.* **21**(6), 1339–1349.

Wang, H. and Deng, X.W. (2004) Phytochrome signaling mechanism. In: *The Arabidopsis Book* (eds Somerville, C.R. and Meyerowitz, E.M.), American Society of Plant Biologists, Rockville, MD.

Wang, X., Feng, S., Nakayama, N., *et al.* (2003) The COP9 signalosome interacts with SCF[UFO] and participates in *Arabidopsis* flower development. *Plant Cell* **15**(5), 1071–1082.

Wang, H., Ma, L., Habashi, J., Zhao, H. and Deng, X.W. (2002) Analysis of far-red light-regulated genome expression profiles of phytochrome A pathway mutants in *Arabidopsis*. *Plant J.* **32**(5), 723–733.

Wang, H., Ma, L.G., Li, J.M., Zhao, H.Y. and Deng, X.W. (2001) Direct interaction of *Arabidopsis* cryptochromes with COP1 in light control development. *Science* **294**(5540), 154–158.

Weber, H., Bernhardt, A., Dieterle, M., *et al.* (2005) *Arabidopsis* AtCUL3a and AtCUL3b form complexes with members of the BTB/POZ-MATH protein family. *Plant Physiol.* **137**(1), 83–93.

Wee, S., Hetfeld, B., Dubiel, W. and Wolf, D.A. (2002) Conservation of the COP9/signalosome in budding yeast. *BMC Genet.* **3**, 15.

Wei, N. and Deng, X.W. (1999) Making sense of the COP9 signalosome. A regulatory protein complex conserved from *Arabidopsis* to human. *Trends Genet.* **15**(3), 98–103.

Wei, N. and Deng, X.W. (2003) The COP9 signalosome. *Annu. Rev. Cell Dev. Biol.* **19**, 261–286.

Wei, N., Kwok, S.F., von Arnim, A.G., *et al.* (1994) *Arabidopsis COP8*, *COP10*, and *COP11* genes are involved in repression of photomorphogenic development in darkness. *Plant Cell* **6**(5), 629–643.

Wei, N., Tsuge, T., Serino, G., *et al.* (1998) The COP9 complex is conserved between plants and mammals and is related to the 26S proteasome regulatory complex. *Curr. Biol.* **8**(16), 919–922.

Weissman, A.M. (2001) Themes and variations on ubiquitylation. *Nat. Rev. Mol. Cell Biol.* **2**(3), 169–178.

Wertz, I.E., O'Rourke, K.M., Zhang, Z., *et al.* (2004) Human de-etiolated-1 regulates c-Jun by assembling a CUL4A ubiquitin ligase. *Science* **303**(5662), 1371–1374.

Wilson, M.P., Sun, Y., Cao, L. and Majerus, P.W. (2001) Inositol 1,3,4-triphosphate 5/6-kinase is a protein kinase that phosphorylates the transcription factors c-Jun and ATF-2. *J. Biol. Chem.* **276**(44), 40998–41004.

Wu, K., Chen, A. and Pan, Z.Q. (2000) Conjugation of Nedd8 to CUL1 enhances the ability of the ROC1-CUL1 complex to promote ubiquitin polymerization *J. Biol. Chem.* **275**(41), 32317–32324.

Yanagawa, Y., Feng, S. and Deng, X.W. (2005) Light control of plant development: a role of the ubiquitin/proteasome-mediated proteolysis. In: *Light Sensing in Plants* (eds Wada, M., Shimazaki, K. and Iino, M.), pp. 253–259. Yamada Science Foundation and Springer-Verlag, Tokyo.

Yanagawa, Y., Sullivan, J.A., Komatsu, S., *et al.* (2004) *Arabidopsis* COP10 forms a complex with DDB1 and DET1 in vivo and enhances the activity of ubiquitin conjugating enzymes. *Genes Dev.* **18**(17), 2172–2181.

Yang, H.Q., Tang, R.H. and Cashmore, A.R. (2001) The signaling mechanism of *Arabidopsis* CRY1 involves direct interaction with COP1. *Plant Cell* **13**(12), 2573–2587.

Yang, H.Q., Wu, Y.J., Tang, R.H., Liu, D., Liu, Y. and Cashmore, A.R. (2000) The C termini of *Arabidopsis* cryptochromes mediate a constitutive light response. *Cell* **103**(5), 815–827.

Yang, J., Lin, R., Sullivan, J., *et al.* (2005) Light regulates COP1-mediated degradation of HFR1, a transcription factor essential for light signaling in *Arabidopsis*. *Plant Cell* **17**(3), 804–821.

Yang, P., Fu, H., Walker, J., *et al.* (2004) Purification of the *Arabidopsis* 26 S proteasome: biochemical and molecular analyses revealed the presence of multiple isoforms. *J. Biol. Chem.* **279**(8), 6401–6413.

Yang, X., Menon, S., Lykke-Andersen, K., *et al.* (2002) The COP9 signalosome inhibits p27[KIP1] degradation and impedes G1-S phase progression via deneddylation of SCF Cul1. *Curr. Biol.* **12**(8), 667–672.

Zheng, N., Schulman, B.A., Miller, J.J., *et al.* (2002) Structure of the Cul1-Rbx1-Skp1-F box[Skp2] SCF ubiquitin ligase complex. *Nature* **416**(6882), 703–709.

Zhou, C., Seibert, V., Geyer, R., *et al.* (2001) The fission yeast COP9/signalosome is involved in cullin modification by ubiquitin-related Ned8p. *BMC Biochem.* **2**, 7.

Zhou, C., Wee, S., Rhee, E., Naumann, M., Dubiel, W. and Wolf, D.A. (2003) Fission yeast COP9/signalosome suppresses cullin activity through recruitment of the deubiquitylating enzyme Ubp12p. *Mol. Cell* **11**(4), 927–938.

7 UV-B perception and signal transduction

Gareth I. Jenkins and Bobby A. Brown

7.1 Introduction

Ultraviolet-B radiation (UV-B; 280–320 nm) is an integral component of sunlight. Most of the UV-B that reaches the earth is absorbed by the stratospheric ozone layer, and therefore UV-B wavelengths comprise only a small fraction of sunlight at the earth's surface (Pyle, 1997; McKenzie *et al.*, 2003). Nevertheless, UV-B has a major impact on the biosphere because it is the most energetic component of the daylight spectrum. UV-B can damage macromolecules such as DNA and proteins, generate reactive oxygen species (ROS) and impair cellular processes. It is well known that the levels of UV-B in sunlight are sufficient to cause damage to sensitive tissues in humans and other animals and to promote some forms of skin cancer. However, it is becoming increasingly clear that UV-B is not solely an agent of damage and has an important role as an informational signal (Paul and Gwynn-Jones, 2003; Brosché and Strid, 2003; Frohnmeyer and Staiger, 2003; Ulm and Nagy, 2005). In particular, the perception of low levels of UV-B by plants actively promotes survival because it stimulates responses that help to protect against and repair UV-damage. Furthermore, responses to UV-B modify the biochemical composition of plant tissues, influence plant morphology and help to deter pests and pathogens (Tevini and Teramura, 1989; Björn, 1996; Jansen *et al.*, 1998; Frohnmeyer and Staiger, 2003; Stratmann, 2003).

Plants are unavoidably exposed to UV-B because they need to capture sunlight for photosynthesis. The fact that plants rarely display signs of UV-damage in the natural environment demonstrates that they have evolved very effective mechanisms for UV-protection and repair. The protective mechanisms include the deposition of UV-absorbing phenolic compounds in the outer epidermal layers and the production of antioxidant systems (Björn, 1996; Rozema *et al.*, 1997; Jansen *et al.*, 1998; Frohnmeyer and Staiger, 2003). Repair involves enzymes such as DNA photolyases (Britt, 1999). UV-B exposure stimulates the expression of genes involved in both UV-protection and repair (Jenkins *et al.*, 1997; Brosché and Strid, 2003; Frohnmeyer and Staiger, 2003; Ulm and Nagy, 2005). It is therefore important to understand the cellular and molecular mechanisms of UV-B perception and signal transduction and to determine the contribution of UV-B responses to normal plant growth and development. Indeed, it will not be possible to obtain a complete understanding of the role of light in controlling plant development without knowledge of the regulatory effects of UV-B.

Despite the importance of plant responses to UV-B, remarkably little is known about the mechanisms of UV-B perception and signal transduction. Certainly much more is known about the photoreceptors and signalling components that mediate responses to other wavelengths of light. Much of the recent remarkable progress in understanding phytochrome, cryptochrome and phototropin action has resulted from the application of genetic approaches, and this strategy is now generating important new information about UV-B responses (see Section 7.5). Hence the focus of this chapter is on the most recent discoveries in UV-B perception and signalling. A broader perspective on the effects of UV-B on plants, ranging from the ecological impact to signalling processes can be obtained from several excellent recent reviews (Brosché and Strid, 2003; Caldwell *et al.*, 2003; Frohnmeyer and Staiger, 2003; Paul and Gwynn-Jones, 2003; Stratmann, 2003; Ulm and Nagy, 2005).

7.2   UV-B in the environment

Most of the UV-B and all of the UV-C radiation impinging on the earth is absorbed by the stratospheric ozone layer (Pyle, 1997; McKenzie *et al.*, 2003) and therefore UV-B wavelengths comprise less than 2% of full sunlight. Over recent decades the level of UV-B reaching the earth's surface has increased because of depletion of the ozone layer, largely as a result of the anthropogenic release of halocarbons into the atmosphere. The most severe effects have been in polar regions, but reduction in the ozone column is also evident elsewhere (McKenzie *et al.*, 2003). Fortunately, measures taken to counter the problem appear to be working (Andrady *et al.*, 2005).

Apart from ozone absorption, a variety of factors influence the amount of UV-B that plants are exposed to (McKenzie *et al.*, 2003; Paul and Gwynn-Jones, 2003). A major factor affecting UV-B levels is the solar angle, which determines the length of the light path through the atmosphere. Thus, UV-B is highest in the tropics and lowest at high latitudes. Solar elevation varies seasonally as well as diurnally, so at mid and high latitudes the UV-B irradiance is much higher in summer than in winter. Other important factors are altitude, which increases UV-B irradiance by about 5%–7% per 1000 m, and cloud cover, which typically decreases UV-B irradiance by 15%–30% (McKenzie *et al.*, 2003). Further effects on UV-B levels are caused by surface reflection and atmospheric pollution. In addition, absorption by vegetation canopies will greatly reduce the amount of UV-B reaching plants below.

It is well known that several different methods of measuring UV-B and reporting UV-B levels are used in the literature, which often makes comparison of studies difficult and confusing. Some authors report photon fluence rates (μmol photons m^{-2} s^{-1}) while others use energy levels (irradiance) per unit area (e.g. kJ m^{-2} h^{-1} or W m^{-2}). The irradiance measurements will be influenced by the wavelength distribution of the light source, as shorter wavelengths are more energetic. In addition, many authors use a weighted measure of UV-B (UV-B$_{BE}$), which incorporates the biological effectiveness of the UV-B radiation, based on the action spectrum for a selected response, such as human erythema (sunburn), or a generalised plant UV-B action spectrum (Caldwell, 1971; Caldwell *et al.*, 2003). Clearly, the different types of measurement are suited to different purposes, but for studies of UV-B photoreception and signal transduction it is usually most appropriate to measure photon fluence rates of UV-B, as in other photomorphogenic studies, because photoreception processes generally involve the detection of individual photons according to the absorption spectrum of the photoreceptor. However, if a particular response results from tissue or molecular damage, the energy of the radiation is likely to be important and an irradiance measurement is then appropriate.

7.3   Plant responses to UV-B

The scientific literature reports numerous studies of the effects of UV-B on plants. The research has involved a variety of species, different stages of development, diverse growth conditions, and various spectral qualities, amounts and durations of UV-B treatment. Inevitably these studies are difficult to compare, but nevertheless some general conclusions can be drawn. First, it is evident that UV-B has a wide-ranging impact on plants, extending from effects on gene expression, cell physiology and biosynthesis to effects on growth, morphology and development (Tevini and Teramura, 1989; Björn, 1996; Frohnmeyer and Staiger, 2003). Second, it is clear that differences in the fluence rate, duration and wavelength of UV-B produce substantial differences in response (Brosché and Strid, 2003; Frohnmeyer and Staiger, 2003; Ulm and Nagy, 2005). Furthermore, responses are dependent on the context of the UV-B treatment, that is the interaction with other environmental variables such as the amount of other light qualities, temperature, water and nutrient status (Caldwell *et al.*, 2003). In general, more modest effects of UV-B are observed under field conditions than in controlled environments (Caldwell *et al.*, 2003).

An extensive review of the diverse effects of UV-B on plants is beyond the scope of this chapter, but it is appropriate to emphasise a few key points. High fluence rates of UV-B cause damage to plant tissues and ultimately necrosis. There are reports of damage to DNA, proteins and lipids and of inhibition of photosynthetic reactions, membrane processes, etc. (Björn, 1996; Jansen *et al.*, 1998; Frohnmeyer and Staiger, 2003). Moreover, high fluence rates of UV-B generate ROS and initiate cellular stress responses (Jansen *et al.*, 1998; Brosché and Strid, 2003). However, the levels of UV-B employed in some studies are well above those experienced in nature and some of the observations may therefore have limited relevance to normal plant growth. The extent to which damage occurs in plants growing in high ambient levels of UV-B is not clear, because if the plant has acclimated to that particular light environment, the repair mechanisms are generally sufficient to prevent damage appearing. Damage is most likely to become evident when plants are exposed to high UV-B levels without acclimation.

UV-B also has regulatory, photomorphogenic effects on plants. Ambient levels of UV-B promote various changes in plant morphology and development, including the inhibition of stem extension and reduction in internode length; leaf curling, reduction in leaf surface area and increase in leaf thickness; promotion of branching; altered flowering time and reduced fertility; and reduced biomass (Tevini and Teramura,

1989; Björn, 1996; Rozema *et al.*, 1997; Jansen *et al.*, 1998; Frohnmeyer and Staiger, 2003). Several of these effects may be due to altered amounts, distribution or responsiveness to plant growth regulators such as auxins and brassinosteroids. UV-B is reported to cause photooxidation of the auxin indoleacetic acid (IAA) (Ros and Tevini, 1995), and it has been suggested that IAA degradation by specific phenolic peroxidases may be important in some UV-B responses (Jansen *et al.*, 2001). There is also evidence for the involvement of brassinosteroid signalling in UV-B responses (Savenstrand *et al.*, 2004; see Section 7.4.1.2). Nevertheless, the molecular basis of morphological responses to UV-B is poorly understood.

In addition to morphological and developmental effects on plants, UV-B regulates aspects of metabolism and hence modulates biochemical composition. It is well established that UV-B stimulates the synthesis of secondary metabolites, in particular flavonoids that accumulate in the epidermal layers (Hahlbrock and Scheel, 1989). Some flavonoids, notably the flavonols, act in conjunction with other phenolic compounds, in particular hydroxycinnamic acid esters, to provide a UV-absorbing sunscreen that limits penetration of UV-B into leaf tissues (Caldwell *et al.*, 1983; Bornman *et al.*, 1997; Jenkins *et al.*, 1997). Correlations between flavonoid content and UV sensitivity have been reported in several species (Tevini and Teramura, 1989), and it has been shown that genotypes lacking flavonoids or sinapic acid esters are much more susceptible than wild-type plants to damage by UV-B (Li *et al.*, 1993; Lois and Buchanan, 1994; Stapleton and Walbot, 1994; Landry *et al.*, 1995). UV-B also promotes the accumulation of other secondary metabolites, such as terpenoid indole alkaloids (Ouwerkerk *et al.*, 1999), but it is not clear whether these function in UV-B protection. However, the array of biochemical compounds produced in plants exposed to UV-B probably has an important role in deterring pathogens and herbivorous insects.

Many responses to UV-B involve the differential regulation of gene expression. It is clear that different fluence rates induce (or repress) the expression of different sets of genes and that different genes have characteristic kinetics of response (Brosché and Strid, 2003; Frohnmeyer *et al.*, 2003; Ulm and Nagy, 2005). The expression of stress-related genes, such as *Arabidopsis PATHOGENESIS RELATED 1 (PR-1)*, requires exposure to relatively high fluence rates of UV-B (Brosché and Strid, 2003), whereas low fluence rates are sufficient to induce a variety of genes, many of which are involved in protective responses to UV-B (Ulm *et al.*, 2004). Furthermore, expression of these latter genes may require only a very brief exposure to UV-B, as shown for the gene encoding the flavonoid biosynthesis enzyme chalcone synthase (CHS) (Frohnmeyer *et al.*, 1999; Jenkins *et al.*, 2001). As discussed further below, the gene expression responses to brief, low fluence rates of UV-B are mediated by photosensory pathways and not by the high fluence, stress response pathways.

The recent application of transcriptome analysis to UV-B responses has provided valuable information on the range of genes induced by UV-B and the nature of the response pathways. Microarray analyses of maize (Casati and Walbot 2003; 2004; Casati *et al.*, 2006), *Nicotiana longiflora* (Izaguirre *et al.*, 2003) and *Arabidopsis* (Brosché *et al.*, 2002; Ulm *et al.*, 2004; Brown *et al.*, 2005) have shown that UV-B modifies the expression of genes encoding enzymes, membrane and cytoskeletal proteins, transcription factors, signalling components and various other proteins involved in a range of cellular processes, including photosynthesis, primary and secondary metabolism, cell wall biosynthesis, stress protection, DNA-related processes, RNA processing, translation and proteolysis. Casati and Walbot (2004) found that many of the UV-B-regulated genes in maize displayed organ-specific expression, and differences were observed between seedlings and adult plants. Also, tissues not directly exposed to UV-B, including roots in soil, showed altered gene expression, implying that a signal is transmitted from UV-illuminated to non-illuminated tissues. The nature of the signal is not known but there are parallels here with the systemic acquired resistance response to pathogen attack, and it may be that similar types of signalling processes are involved. Casati and Walbot (2004) additionally reported differences in the kinetics and fluence rates required to induce particular sets of genes. Rather than showing a simple reciprocal dose–response relationship, sets of genes were stimulated above particular threshold doses of UV-B. These data provide further evidence for the existence of distinct UV-B signalling pathways operating at different fluence rates. Ulm *et al.* (2004) identified over 100 genes that showed altered expression within 1 h of giving *Arabidopsis* seedlings a brief (15 min), low irradiance UV-B treatment, and these did not include genes expressed in response to high fluence rates of UV-B. In addition, experiments using cut-off filters to produce different UV spectra provided evidence for distinct response pathways within the UV-B range, a longer wavelength UV-B pathway and a shorter wavelength UV-B pathway. The latter pathway negatively regulated the expression of a subset of genes induced by the former. Different effects of short and long wavelength UV-B have also been reported for growth responses in cucumber and other species (Shinkle *et al.*, 2004), and in this case the shorter wavelength response appears to involve DNA damage (Shinkle *et al.*, 2005).

7.4 UV-B perception and signal transduction

It is evident from the above section that plants show diverse responses to UV-B. Furthermore, there are substantial differences in the types of responses at different fluence rates and differential effects of wavelengths within the UV-B range. The challenge is to identify the cellular and molecular mechanisms that underpin the different effects of UV-B on plants. It is evident from research to date that there is no single mechanism of UV-B perception and signal transduction. Therefore it is necessary to categorise the different responses and to define the UV-B perception and signalling mechanisms responsible for each type of response.

Several authors have subdivided UV-B responses according to the UV-B fluence rate required to initiate them (e.g. Brosché and Strid, 2003; Frohnmeyer and Staiger, 2003). The resulting models are valuable in that they emphasise the existence of distinct UV-B signalling pathways. However, there is undoubtedly substantial overlap in the range of fluence rates required to initiate different types of responses. Moreover, the threshold fluence rates are likely to vary according to the developmental stage and growth conditions of the plants and, in particular, whether

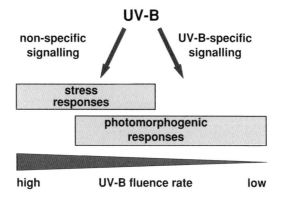

Figure 7.1   Illustration highlighting the differences between UV-B stress and UV-B photomorphogenic responses and signalling pathways. UV-B stress responses are induced at relatively high fluence rates of UV-B and are mediated by signalling processes that are not specific to UV-B. UV-B photomorphogenic responses are initiated at lower fluence rates than stress responses and are mediated by UV-B-specific signalling pathways.

they have been acclimated to UV-B. An alternative way of categorising UV-B responses is according to their function, whether they constitute a stress response or a photomorphogenic response, as illustrated in Figure 7.1. UV-B-induced stress responses may result from photodamage to molecules and/or the accumulation of ROS and are not specific to UV-B. As discussed below, several of the genes induced by UV-B stress are also induced by other stimuli because of the overlap of signalling pathways. In contrast, the signalling pathways that mediate responses to UV-B as an informational signal appear to be UV-B-specific (see Sections 7.4.2 and 7.5.2) and to result in UV-protection or morphological changes. Such responses may be defined as photosensory or photomorphogenic, using the latter term in its broadest sense – a regulatory response to light – rather than the narrow sense of only affecting morphology. There is some evidence for multiple photomorphogenic and non-photomorphogenic pathways, but it will not be possible to establish how many distinct pathways mediate UV-B responses until we know more about their components.

7.4.1   Non-photomorphogenic UV-B signalling

7.4.1.1   Damage/stress signalling

As stated above, UV-B may directly damage macromolecules such as DNA and generate ROS. The principal type of DNA damage caused by UV-B exposure in plants is the formation of cyclobutane pyrimidine dimers (CPD), with the formation of pyrimidine [6–4] pyrimidone dimers (commonly known as 6–4 photoproducts) accounting for most of the remaining damage (Britt, 1999). Unless they are repaired, these lesions will impair DNA replication and transcription. In animal cells, DNA damage initiates signalling processes that promote DNA repair and minimise the consequences of damage by arresting cell cycle progression (Sancar *et al.*, 2004). Several of the components involved in DNA damage signalling in animal cells have been shown to be present in plants. For instance, *Arabidopsis* possesses homologues of the ATM (ataxia telangiectasia mutated) and ATR (ataxia telangiectasia mutated and Rad3-related) protein kinases that act as sensors for DNA double-strand breaks and replication blocks in animal cells and initiate signalling (Garcia *et al.*, 2003; Culligan *et al.*, 2004). Moreover, the histone variant H2AX has been shown to be a target for these enzymes in *Arabidopsis* as well as in mammalian cells (Friesner *et al.*, 2005). *Arabidopsis* mutants lacking ATR are hypersensitive to UV-B, at least in a root growth assay (Culligan *et al.*, 2004), whereas those lacking ATM are not (Garcia *et al.*, 2003). It appears that signalling initiated by ATR is important in controlling cell cycle progression in situations where DNA replication is affected, including as a result of UV-B exposure (Culligan *et al.*, 2004).

Studies using a chemical assay (Dai *et al.*, 1997), EPR spectroscopy (Hideg and Vass, 1996) and detection of fluorescent ROS-sensitive reagents (Allan and Fluhr, 1997; Hideg *et al.*, 2002; Barta *et al.*, 2004) have demonstrated that UV-B generates ROS in plants. The principal form of ROS detected was the superoxide radicle, which is converted to H_2O_2 by superoxide dismutase activity. There are a number of sources of ROS in cells, including reactions in photosynthesis and respiration and the activity of enzymes such as peroxidases and oxidases (Mittler, 2002), but it is not clear which of these sources produce ROS in response to UV-B. Since UV-B causes damage to proteins involved in photosynthetic electron transport (Jansen *et al.*, 1998), it is likely that excess ROS would be generated as a result of the reduced ability to dissipate excitation energy. In support of this hypothesis, Barta *et al.* (2004) reported a correlation between the inhibition of photosynthesis by UV-B and the production of superoxide. However, UV-B is also reported to stimulate NADPH oxidase activity (Rao *et al.*, 1996; A-H-Mackerness *et al.*, 2001). Hence it is likely that UV-B produces ROS by more than one mechanism.

There is evidence that plants attempt to counteract the accumulation of ROS by enhancing ROS-scavenging systems. UV-B stimulates the expression of a number of genes concerned with antioxidant production (Casati and Walbot, 2003; Casati and Walbot, 2004; Ulm *et al.*, 2004; Brown *et al.*, 2005). In addition, the activities of some antioxidant enzymes, such as ascorbate peroxidase, increase following UV-B exposure in *Arabidopsis*, although the reported increases in wild-type plants are relatively modest and are seen at higher UV-B doses (Landry *et al.*, 1995; Rao *et al.*, 1996). Similarly, high levels of UV-B stimulate antioxidant activities in rice (Dai *et al.*, 1997).

ROS can cause oxidative damage to cellular components such as DNA, protein and lipids but they also act as signalling molecules in responses to biotic and abiotic stresses (Apel and Hirt, 2004). Differences in the nature of ROS and in their spatial and temporal production are undoubtedly important in determining which signalling pathways are activated. There is evidence that ROS are involved in the regulation of some genes by UV-B. For instance, exposure to relatively high levels of UV-B causes a strong reduction in transcript levels of the *LHCB1* gene encoding the major chlorophyll-binding protein of chloroplasts, and this response is inhibited by application of the antioxidant ascorbate (Surplus *et al.*, 1998) and of a scavenger of superoxide radicals (A-H-Mackerness *et al.*, 2001). ROS are also involved in other gene expression responses, as discussed further below.

Although it is clear that UV-B causes ROS production and that ROS initiate signalling processes that induce stress responses, more information is needed on the extent and nature of ROS production under ambient UV-B conditions. The levels of UV-B used in the above studies to measure ROS production and investigate the involvement of ROS in gene expression were sometimes well above ambient. In rice, superoxide production was only detected under above ambient UV-B (Dai *et al.*, 1997). It may be that ROS production and signalling is used by plants to modulate expression of some genes in response to naturally varying levels of UV-B, but little information is available on this point.

7.4.1.2   Overlap with defence/wound signalling

It is well established that UV-B stimulates the expression of various genes normally induced by wound and defence signalling pathways (A-H-Mackerness, 2000; Brosché and Strid, 2003; Izaguirre *et al.*, 2003; Stratmann, 2003). Examples include *PR-1*, *PR-2*, *PR-5*, the defence gene *PDF1.2* and proteinase inhibitor genes. The basis of this phenomenon is that UV-B causes the production of signalling intermediates that are components of defence/wound signalling pathways. These include jasmonic acid (JA), ethylene, salicylic acid (SA) and ROS. A-H-Mackerness *et al.* (1999) reported that UV-B exposure caused rapid increases in the levels of JA and ethylene in wild-type *Arabidopsis*. In addition, UV-B promoted a slow increase in SA (Surplus *et al.*, 1998). Transgenic plants expressing *NahG*, encoding a salicylate hydroxylase, were unable to accumulate SA and showed reduced UV-B induction of *PR-1*, *PR-2* and *PR-5* (Surplus *et al.*, 1998). The ethylene insensitive *etr-1* mutant failed to show an increase in *PR-1* and *PDF1.2* transcripts in response to UV-B whereas the JA-insensitive mutant, *jar1*, lacked *PDF1.2* induction but retained *PR-1* induction (A-H-Mackerness *et al.*, 1999). These studies provide strong evidence for the involvement of SA, JA and ethylene in the UV-B induction of defence gene expression but reveal differences in the pathways regulating particular genes.

ROS production is an early step in wound and defence signalling and, as discussed above, ROS are also produced by UV-B. It is therefore not surprising that ROS are involved in the induction of wound/defence genes by UV-B. The antioxidant ascorbate inhibited the accumulation of PR-1 protein in tobacco (Green and Fluhr, 1995) and the accumulation of transcripts of several *PR* genes in *Arabidopsis* (Surplus *et al.*, 1998). A-H-Mackerness *et al.* (2001) provided evidence that superoxide is important in mediating the induction of defence genes by UV-B, either directly or through the production of H_2O_2. Superoxide, generated by NADPH oxidase associated with the plasma membrane has a key role in defence signalling (Mittler, 2002; Apel and Hirt, 2004). Pharmacological experiments suggested that NADPH oxidase and peroxidase enzymes were likely to be responsible for the production of ROS involved in regulating defence genes by UV-B (A-H-Mackerness *et al.*, 2001). However, no information is available on the mechanism of enzyme activation.

Experiments in tomato provide mechanistic information on the overlap between UV-B and wound signalling pathways. In tomato the wound response is mediated by the small peptide systemin (Ryan, 2000). Systemin induces apoplastic alkalinisation and activation of specific MAP kinases. UV-B illumination of *Lycopersicon peru-vianum* cells was found to initiate the same signalling processes (Yalamanchili and Stratmann, 2002; Holley *et al.*, 2003). UV-B caused an increase in activity of three specific MAP kinases, two of which were also stimulated by systemin. Systemin binds to the SR160 receptor, an interaction that is blocked by the molecule suramin (Stratmann *et al.*, 2000); suramin therefore inhibits apolplastic alkalinisation and MAP kinase activation. Significantly, the UV-B induction of these processes was also blocked by suramin, suggesting that the UV-B response is mediated by activation of SR160 or a related receptor that is sensitive to suramin (Yalamanchili and Stratmann, 2002). It is therefore likely that ligand-independent receptor activation by UV-B accounts for the overlap between UV-B and wound signalling pathways. However, the mechanism of receptor activation by UV-B is unknown.

It has been reported that SR160 is identical to the tomato brassinosteroid receptor tBRI1 (Montoya *et al.*, 2002; Wang and He, 2004) and hence brassinosteroid signalling may also be activated by UV-B. This could provide a mechanism for some of the morphological effects of UV-B. It is therefore interesting that Savenstrand *et al.* (2004) have reported that *Arabidopsis* brassinosteroid biosynthesis mutants and the *bri1* receptor mutant show reduced expression of several genes in response to UV-B.

The overlap between defence/wound and UV-B pathways appears to have important consequences for plants growing in the natural environment. A number of studies have shown that UV-B illumination promotes increased tolerance of plants to insect herbivory (Caldwell *et al.*, 2003; Stratmann, 2003). Izaguirre *et al.* (2003) reported that UV-B illumination regulated a substantial number of genes that were also insect-responsive in field grown *N. longiflora*. It is likely that UV-B stimulates the expression of genes that are involved in producing toxic secondary metabolites, proteinase inhibitors and other compounds that deter herbivory in a range of species, as the protective effect of UV-B appears widespread (Caldwell *et al.*, 2003). Furthermore, UV-B and wounding by herbivorous insects act synergistically in some instances to amplify the level of response (Stratmann, 2003).

7.4.2 Photomorphogenic UV-B signalling

As stated previously, there are a range of UV-B responses that are evidently not stress responses and can be considered photomorphogenic. Examples are the suppression of hypocotyl extension by low fluence rates of UV-B (Kim *et al.*, 1998; Boccalandro *et al.*, 2001; Suesslin and Frohnmeyer, 2003) and gene expression responses that provide UV protection, the best studied being the UV-B induction of *CHS* and other genes involved in flavonoid biosynthesis (Jenkins *et al.*, 1997, 2001). Several lines of evidence indicate that photomorphogenic UV-B responses are not mediated by stress/wound/defence signalling pathways. First, the threshold UV-B doses that initiate photomorphogenic responses are substantially lower than those needed to induce stress/defence/wound gene expression. For instance, less than 0.1 μmol m^{-2} s^{-1} UV-B (less than 1/30 to 1/40 of the fluence rate of UV-B in sunlight) induces hypocotyl growth suppression in *Arabidopsis* (Kim *et al.*, 1998; Boccalandro *et al.*, 2001) and 5 min ambient UV-B exposure is sufficient to induce the accumulation of *CHS* transcripts (Jenkins *et al.*, 2001). Indeed, UV-B pulses of less than a second are reported to stimulate transcription from the *CHS* promoter in parsley cells (Frohnmeyer *et al.*, 1999). Second, the UV-B induction of low fluence UV-B responses is not mediated by wound/defence signalling molecules. In contrast to the results for defence genes (A-H-Mackerness *et al.*, 2001), no reduction was observed in the UV-B stimulation of *CHS* expression in JA and ethylene signalling mutants, including *etr1* and *jar1* (C.M. Pidgeon and G.I. Jenkins, unpublished). Furthermore, compounds that generate ROS did not induce *CHS* expression in *Arabidopsis* cells and compounds that remove ROS did not prevent the UV-B response (Jenkins *et al.*, 2001), again in contrast to the situation with defence genes (Green and Fluhr, 1995; Surplus *et al.*, 1998). A-H-Mackerness *et al.* (2001) presented pharmacological data suggesting that the UV-B induction of *CHS* in *Arabidopsis* might be mediated by nitric oxide, but contrary results were obtained in experiments with an *Arabidopsis* cell culture (C.M. Pidgeon and G.I. Jenkins, unpublished). Hence there is strong evidence that distinct signalling pathways mediate photomorphogenic and non-photomorphogenic UV-B responses.

7.4.2.1 UV-B perception

The nature of photomorphogenic UV-B perception remains a mystery. There is strong evidence that it is not mediated by the known photoreceptors. Although cryptochrome (cry) and phytochrome (phy) photoreceptors absorb UV-B wavelengths to some extent and could in principle be UV-B photoreceptors, several papers report that photomorphogenic UV-B responses are retained in mutants lacking these photoreceptors. In *Arabidopsis* leaves, the induction of *CHS* transcripts by UV-B was undiminished in a *cry1cry2* double mutant (Wade *et al.*, 2001) and in mutants lacking one or more of phyA, phyB, phyD and phyE (Wade *et al.*, 2001; Brosché and Strid, 2003). Similarly, Ulm *et al.* (2004) reported that the UV-B induction of several genes in *Arabidopsis* seedlings was unaltered in *phyAphyB* and *cry1cry2* mutant plants. The UV-B-induced promotion of cotyledon opening in response to red light, detected by phyB, is not mediated by phytochromes or cryptochromes (Boccalandro *et al.*, 2001). In addition, the suppression of hypocotyl extension by UV-B is retained in *phyAphyB* seedlings (Boccalandro *et al.*, 2001; Suesslin and Frohnmeyer, 2003). Kim *et al.* (1998) had previously reported that the UV-B suppression of hypocotyl extension was diminished in *phyAphyB* seedlings, but the reason for this discrepancy is not clear.

For many years authors have speculated that there may be a UV-B photoreceptor but no such molecule has ever been identified. The principal reasons for proposing such a photoreceptor are that photomorphogenic UV-B responses cannot be explained by known UV-B absorbing molecules, as discussed above, whereas the UV-B specificity of these responses suggests a specific 'receptor'. Identification of the putative photoreceptor has been hampered by lack of information about its cellular location and lack of a unique biochemical or photobiological property to aid its isolation, bearing in mind that numerous compounds in cells have some capacity for UV-B absorption. The action spectra for UV-B responses suggest that a UV-B 'photoreceptor' would have maximal absorption in the 295–300 nm wavelength range (Ensminger, 1993). Possible chromophores would include pterins or flavins, most likely in the reduced form, with the absorption spectrum dependent on the protein environment. Compounds that antagonise flavins and pterins impaired the UV-B suppression of tomato hypocotyl extension (Ballaré *et al.*, 1995), and similar experiments suggested that flavin was involved in the UV-B induction of anthocyanin synthesis in maize (Khare and Guruprasad, 1993). Moreover, Ensminger and Schäfer (1992) found that feeding riboflavin to parsley cells enhanced CHS protein and flavonoid accumulation in response to UV-B, but not blue light, and obtained evidence that flavin could bind to cell membranes. However, there is still no direct evidence for the existence of a specific UV-B photoreceptor comparable to the known photoreceptors.

As mentioned above, in animal cells UV-B initiates some responses via DNA damage signalling pathways. However, there is compelling evidence that low fluence UV-B responses in plants are not mediated by DNA damage signalling. Action spectra for the responses show maxima around 295–300 nm, whereas damage responses are maximal at shorter wavelengths corresponding to DNA absorption (Ensminger, 1993; Ballaré *et al.*, 1995). In addition, photomorphogenic UV-B responses are initiated by amounts of UV-B that do not cause detectable damage, and mutants that are defective in DNA repair, which might be expected to show increased levels of response at a given fluence rate, do not show altered responses. For instance, the suppression of hypocotyl extension and promotion of cotyledon opening by low fluence rates of UV-B in *Arabidopsis* were unaltered in *uvr1*, *uvr2* and *uvr3* mutant backgrounds defective in DNA repair activities (Kim *et al.*, 1998; Boccalandro *et al.*, 2001) and, similarly, the induction of several genes by low fluence UV-B was unaltered in *uvr2* (Ulm *et al.*, 2004). Moreover, Frohnmeyer *et al.* (1999) found that subsecond UV-B exposure, sufficient to stimulate *CHS* promoter activity in parsley protoplasts, did not produce detectable CPD formation. In addition, CHS protein expression was most strongly stimulated by UV-B above 305 nm whereas CPD formation was maximal at shorter wavelengths. Similarly, Kalbin *et al.* (2001) found no correlation between the levels of CPD formation following UV-B exposure and the expression of several genes in pea. If the photomorphogenic pathway were mediated by DNA damage formation, one would expect that light qualities that promote photorepair of damaged nucleotides would reduce the UV-B response. However, this is not observed; in fact when blue light is given together with UV-B, the induction of *CHS* expression is not reduced but is enhanced (Ohl *et al.*, 1989; Fuglevand *et al.*, 1996). In contrast, induction of a β-1,3-glucanase by high levels of UV-B in French bean was negated by light that stimulated photorepair, indicating a response to DNA damage (Kucera *et al.*, 2003).

Another possible mechanism of UV-B perception is the activation of plasma membrane receptor kinases, analogous to the initiation of inflammatory responses by UV-B in animal cells. The inflammatory response involves cytokines that interact with receptors, including receptor tyrosine kinases at the plasma membrane. UV-B activates these receptors independently of the ligand (Bender *et al.*, 1997; Herrlich and Bohmer, 2000). ROS generated by UV-B inactivate tyrosine phosphatases, leading to activation of the receptor either by autophosphorylation or phosphorylation by a separate kinase. The signalling pathway initiated by receptor activation, whether by cytokines or UV-B, leads to transcription of various genes in the inflammatory response. Plants do not possess receptor tyrosine kinases but do have a large number of other receptor kinases (Johnson and Ingram, 2005) including, as mentioned above, the brassinosteroid/systemin receptor. It is therefore possible that UV-B could activate such receptors in plants. If activation occurs via oxidative stress, this may preclude such a mechanism mediating responses to low fluence rates of UV-B and the responses showing UV-B specificity, but some other form of activation is conceivable. Hence it is intriguing that the *bri1* mutant has reduced induction of several genes regulated by low doses of UV-B (Savenstrand *et al.*, 2004). However, expression was reduced and not eliminated and, moreover, it is not clear whether the reduction in *bri1* was specific to UV-B. Nevertheless, the possible involvement of receptor kinases in plant UV-B perception merits further investigation.

7.4.2.2 Signal transduction

Cell physiological and pharmacological approaches have provided some information on photomorphogenic UV-B signalling processes (Jenkins *et al.*, 2001). Christie and Jenkins (1996) provided evidence that the UV-B induction of *CHS* expression in *Arabidopsis* suspension culture cells required calcium ions, calmodulin and protein phosphorylation. Experiments with various calcium channel antagonists and Ca^{2+}-ATPase inhibitors suggested the involvement of an internal calcium pool rather than flux across the plasma membrane (Christie and Jenkins, 1996; Long and Jenkins, 1998). The UV-B pathway was both kinetically distinct from the UV-A/blue light (cry1) pathway inducing *CHS* in the *Arabidopsis* cells (Jenkins *et al.*, 2001) and pharmacologically distinct, in that it was inhibited by the calmodulin antagonist W-7 (Christie and Jenkins, 1996). Moreover, the UV-B signalling pathway was different to the phytochrome signalling pathway inducing *CHS* expression in tomato hypocotyls and soybean cell cultures (Bowler and Chua, 1994). Similar findings were reported for parsley (Frohnmeyer *et al.*, 1997) and soybean cells (Frohnmeyer *et al.*, 1998). Subsequently, Long and Jenkins (1998) concluded that redox processes at the plasma membrane were involved in UV-B signal transduction because the UV-B induction of *CHS* expression in *Arabidopsis* cells was inhibited both by the cell impermeable electron acceptor ferricyanide and the flavoprotein antagonist diphenylene iodonoium.

Although the above inhibitor studies demonstrate that the UV-B signalling pathway regulating *CHS* expression is distinct from the phytochrome and cryptochrome signalling pathways, they do not provide direct evidence for cell physiological events coupled to UV-B perception. It is necessary to obtain measurements of calcium fluxes, electron transport or other activities that complement the pharmacological data. Unfortunately, little evidence of this nature has been obtained. Attempts in the author's laboratory to show a UV-B-induced calcium flux in *Arabidopsis* cells were unsuccessful (T.N. Bibikova, S. Kennington and G.I. Jenkins, unpublished data) and no direct evidence of redox processes associated with photomorphogenic UV-B perception has been presented. UV-B was reported to promote an increase in the cytosolic calcium concentration in parsley cells (Frohnmeyer *et al.*, 1999), but the effect was relatively small and gradual and unlike most other calcium responses to external stimuli. Thus, further experiments are required to test whether calcium fluxes and redox processes are involved in photomorphogenic UV-B signalling.

The conclusion to be drawn from this section is that physiological and pharmacological studies have told us quite a lot about what is *not* involved in photomorphogenic UV-B perception and signal transduction but little about the physiological processes and molecular components that are involved. The implication is that low fluence UV-B signalling is mediated by a novel pathway distinct from the known abiotic stress and light signalling pathways. The evidence discussed in the following section shows that this is the case.

7.5   Genetic approach

7.5.1   Screens for UV-B signalling mutants

The application of genetic approaches has been responsible for much of the recent impressive progress in photomorphogenesis research, including the identification of the cryptochrome and phototropin photoreceptors and of components involved in phytochrome signalling. Hence, the use of these approaches promises to generate important new insights into UV-B perception and signal transduction. The key is to design genetic screens that will identify components that are regulatory and specific to the UV-B response. As discussed below, this approach is proving to be successful.

In principle, several different types of genetic approach can be used to identify genes involved in UV-B perception and signalling. One possibility is to exploit natural genetic variation. Genotypic differences in UV-B sensitivity have been reported in a variety of species ranging from members of natural populations to cultivars of important crops (Tevini and Teramura, 1989; Sullivan *et al.*, 1992; Correia *et al.*, 1999; Sato *et al.*, 2003). In addition, it is known that various ecotypes of *Arabidopsis* differ in their responses to UV-B radiation (Torabinejad and Caldwell, 2000; Cooley *et al.*, 2001; Kalbina and Strid, 2006). Studies to determine the genetic basis of such variation in UV-B tolerance could lead to the identification of novel regulatory components. For instance, Sato *et al.* (2003) have mapped quantitative trait loci associated with resistance to UV-B in rice, providing the basis for future studies to identify genes responsible for resistance. Using a different approach, namely transcriptome analysis, Casati *et al.* (2006) have obtained novel information about the genetic basis of UV-B tolerance in maize. Cultivars growing at high altitudes in the Andes, which are naturally exposed to elevated levels of UV-B and display increased UV-tolerance, showed increased expression of a number of genes encoding putative chromatin remodelling proteins. Transgenic plants in which expression of some of these components was reduced using RNAi were less tolerant of UV-B. Thus, it appears that adaptation to elevated UV-B in maize involves modifications to the regulation of chromatin organisation. These changes may help to counter the genomic instability promoted by elevated levels of UV-B (Ries *et al.*, 2000).

Several laboratories have screened directly for mutants with altered tolerance of UV-B by examining levels of tissue damage and necrosis. As might be expected, alterations in the production of UV-B absorbing sunscreen compounds and in the ability to repair damaged DNA account for many of the phenotypes. For instance, Lois and Buchanan (1994) isolated a *UV sensitive* (*uvs*) mutant that lacked UV-absorbing flavonoids, whereas a UV-B resistant *UV tolerant1* (*uvt1*) mutant contained higher levels of UV-absorbing compounds than the wild type (Bieza and Lois, 2001). However, the increased survival of the *UV-B insensitive 1* (*uvi1*) mutant was unrelated to sunscreen accumulation but instead was correlated with enhanced repair of DNA damage (Tanaka *et al.*, 2002). The mutant showed more rapid repair of 6-4 photoproducts in darkness and CPDs in the light, and had increased expression of the *PHR1* gene encoding the type II CPD photolyase. In contrast, specific defects in DNA repair were identified in a number of mutants showing hypersensitivity to UV-B. The *UV hypersensitive 1* (*uvh1*) mutant (Harlow *et al.*, 1994) was found to be defective in a subunit of the nucleotide excision repair endonuclease involved in dark repair of damaged DNA (Gallego *et al.*, 2000; Liu *et al.*, 2000). The *UV resistance locus 2* (*uvr2*) mutant was altered in the gene encoding the type II photolyase PHR1 that repairs CPDs (Ahmad *et al.*, 1997; Jiang *et al.*, 1997a; Landry *et al.*, 1997) and *uvr3* (Jiang *et al.*, 1997a) was defective in the photolyase that repairs 6-4 photoproducts (Nakajima *et al.*, 1998). The genes corresponding to several other hypersensitive mutants (Britt *et al.*, 1993; Jenkins *et al.*, 1995; Jiang *et al.*, 1997b) have not been identified.

Kliebenstein *et al.* (2002) isolated a further UV-B hypersensitive mutant, *uvr8*. Like the other *uvr* mutants, *uvr8* grew normally in light lacking UV-B but showed severe necrosis or died following UV-B exposure. However, in contrast to the other *uvr* mutants, *uvr8* plants exposed to UV-B had reduced levels of flavonoids and anthocyanin. Consistent with the reduction in flavonoid accumulation, the mutant had much reduced *CHS* expression in response to UV-B. In contrast, UV-B illumination of the mutant led to greatly increased accumulation of PR1 and PR5 proteins, probably because the plants were subject to high levels of stress. Kliebenstein *et al.* (2002) cloned the *UVR8* gene and found that it encoded a protein similar in sequence to human Regulator of Chromatin Condensation 1 (RCC1). The authors speculated that UVR8 is involved in UV-B signal transduction and, as discussed in Section 7.5.2, subsequent research has proved this to be correct.

While screens for altered UV-B sensitivity have the potential to identify mutants defective in UV-B perception or signal transduction, it is evident that most of the screens undertaken to date have identified gene products concerned with DNA repair or the synthesis of sunscreen pigments. Although these studies have been important, not least in the characterisation of several DNA repair enzymes, it appears that the best approach for isolating UV-B perception or signalling mutants is to focus on a primary photomorphogenic UV-B response that does not depend on altered survival.

Thus, Suesslin and Frohnmeyer (2003) screened for mutants altered in the suppression of hypocotyl extension by UV-B. A complication in undertaking this screen is that the cryptochrome and phytochrome photoreceptors absorb UV-B to some extent and are active in suppressing hypocotyl growth. Thus, to screen for UV-B perception and signalling mutants it is essential to select conditions that produce a UV-B-specific response. In addition, it is necessary to choose conditions that give a photomorphogenic UV-B suppression of hypocotyl extension rather than a damage response (Kim *et al.*, 1998). Thus, the authors exposed dark-grown seedlings to short pulses of UV-B radiation that produced a small reduction in hypocotyl elongation but did not cause visible damage. They showed that the response was specific to a UV-B pathway, as it was unaltered in several *cry* and *phy* mutants. Using these conditions they isolated several *UV light insensitive* (*uli*) mutants with longer hypocotyls than wild type in UV-B, but normal hypocotyl lengths in other light qualities. Thus the mutants appeared to be specific to the UV-B pathway suppressing hypocotyl extension.

Further experiments focused on *uli3*, and the authors attempted to establish whether the mutant was altered in other UV-B responses by assaying UV-B-induced gene expression. Transcript levels of several genes were measured in seedlings grown for 3 days in either UV-A light or UV-A supplemented with UV-B, compared to controls grown in darkness. *PR1* transcripts were not expressed in UV-A alone in either the mutant or wild type but were reduced in UV-A+UV-B in the mutant compared to wild type. Levels of *CHS* and *NDPK1a* transcripts were reduced in both UV-A+UV-B and UV-A alone in *uli3* compared to wild type. Therefore the mutant may be altered in the UV-B regulation of *PR1* but the altered regulation of other transcripts was not specific to UV-B. A complication is that *CHS* transcript levels are synergistically enhanced by exposure to UV-B and UV-A together (Fuglevand *et al.*, 1996). Hence it would be interesting to know whether the mutant has an altered *CHS* gene expression response to UV-B alone and to employ a brief illumination sufficient to induce *CHS* transcripts rather than 3 days exposure.

The *ULI3* gene was found to encode a protein that shared limited sequence identity with a human diacylglycerol kinase. The transcript was induced by UV-B and to a lesser extent by UV-A illumination of dark-grown seedlings. Microscopical imaging of a ULI3–GFP fusion indicated that the protein was localised predominantly in the cytoplasm. Unfortunately, none of these observations provide evidence for a molecular function of ULI3. Thus, further studies are needed to establish the role of ULI3 in the UV-B-induced suppression of hypocotyl extension.

An alternative to using hypocotyl extension to identify mutants in a photomorphogenic UV-B response is to screen for altered gene expression. A mutant screen can be developed using a transgenic line in which a UV-B-induced promoter drives expression of a suitable reporter coding sequence. As already discussed, genes such as *CHS* and several encoding transcription factors are induced by the photomorphogenic UV-B pathway and their promoters are therefore suitable for use in this approach (Jenkins *et al.*, 2001; Ulm *et al.*, 2004). Jackson *et al.* (1995) isolated several mutants with hyperinduction of a *CHS* promoter::GUS fusion, but these *increased chalcone synthase expression* (*icx*) mutants are not specific to the UV-B

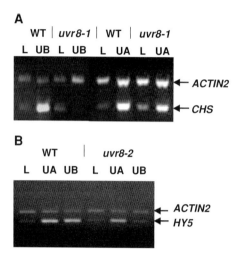

Figure 7.2 UVR8 acts specifically in the UV-B induction of *CHS* and *HY5* transcripts. Wild type (WT) and *uvr8* plants were grown for 21 days in a low fluence rate of white light (L; 25 μmol m^{-2} s^{-1}) and transferred to UV-B (UB; 3 μmol m^{-2} s^{-1}) for 4 h or UV-A (UA; 100 μmol m^{-2} s^{-1}) for 6 h. *CHS*, *HY5* and *ACTIN2* transcripts were assayed by RT-PCR as described by Brown *et al.* (2005).

pathway (Wade *et al.*, 2003). Brown *et al.* (2005) used a transgenic *Arabidopsis* line expressing a *CHS* promoter::luciferase fusion to screen for UV-B response mutants. Plants grown from mutagenised seed were given a UV-B treatment to induce *CHS::LUC* expression, and putative mutants that lacked the response were identified using a photon-counting camera. *CHS* transcripts were assayed to confirm the phenotype (Figure 7.2A). To identify mutants defective only in their response to UV-B, plants were given a UV-A light treatment, which induces *CHS* expression via cry1 (Wade *et al.*, 2001). The screen, which involved over 50 000 mutagenised plants, identified four independent mutants lacking specifically the UV-B induction of *CHS* expression. Subsequent genetic analysis showed that all these mutants were allelic with the *uvr8-1* mutant isolated by Kliebenstein *et al.* (2002). The lack of other classes of mutants isolated in the screen suggests that the UV-B-specific pathway inducing *CHS* may have relatively few components. Furthermore, it is clear that UVR8 is an important component of the photomorphogenic UV-B pathway inducing *CHS* expression.

7.5.2   UVR8

Brown *et al.* (2005) further examined the role of UVR8 in the regulation of gene expression by UV-B. Reverse transcriptase-polymerase chain reaction (RT-PCR) analyses showed that the *uvr8* mutant retains induction of *CHS* transcripts by both cry1 (UV-A illumination; Figure 7.2A) and phyA (FR illumination of dark-grown seedlings). In addition, *uvr8* is unaltered in the stimulation of *CHS* expression by non-light stimuli, including low temperature and sucrose. These observations provide strong evidence that UVR8 acts in a UV-B-specific pathway. Hence UVR8 is the first UV-B-specific signalling component to be identified in photomorphogenic pathways regulating gene expression.

Further experiments showed that the UVR8 pathway controls expression of a range of UV-B-induced genes, including a number that are crucial for UV-protection. Brown *et al.* (2005) used whole-genome microarrays to study gene expression in *uvr8* compared to wild type. Plants grown in a low fluence rate of white light lacking UV-B were given a UV-B treatment that activated the photomorphogenic UV-B pathway but did not induce the stress-responsive genes expressed at high levels of UV-B. Statistical analysis of the microarray data identified 72 UV-B-induced genes regulated by UVR8 with a 5% estimated frequency of false positives (the FDR value; Breitling *et al.*, 2004). This represents the minimum number of UVR8-regulated genes because more are included in the list if a higher percentage FDR is applied; e.g. 113 genes at 10% FDR. RT-PCR studies on a number of these genes have been used to validate the microarray data and indicate that the results are reliable (B.A. Brown and G.I. Jenkins, unpublished data). Table 7.1 shows a selection of the UVR8-regulated genes.

The microarray data show that UVR8 regulates most of the flavonoid biosynthesis genes, consistent with the biochemical analysis presented by Kliebenstein *et al.* (2002). The role of flavonoids in UV-protection is well established (see Section 7.3). UVR8 also regulates several genes concerned with other secondary metabolic

Table 7.1 A sample of genes regulated by UVR8

Gene	Name	Function	HY5
At5g13930	Chalcone synthase	Flavonoid biosynthesis	Yes
At3g55120	Chalcone isomerase	Flavonoid biosynthesis	Yes
At3g51240	Flavanone 3-hydroxylase	Flavonoid biosynthesis	Yes
At5g08640	Flavonol synthase 1	Flavonol biosynthesis	Yes
At5g42800	Dihydroflavonol 4-reductase	Anthocyanin biosynthesis	Yes
At1g65060	4-Coumarate-CoA ligase 3	Phenylpropanoid pathway	Yes
At3g57020	Strictosidine synthase	Alkaloid biosynthesis	No
At1g78510	Solanesyl diphosphate synthase	Prenylquinone biosynthesis	No
At4g31870	Glutathione peroxidase	Oxidative stress protection	No
At3g22840	Early light-induced protein (ELIP1)	Photoprotection	Yes
At1g12370	PHR1	Type II DNA photolyase	Yes
At5g24850	CryD	Blue light photoreceptor	Yes
At5g11260	HY5	Transcription factor	—
At3g17610	HYH	Transcription factor	No
At5g24120	RNA polymerase Sigma subunit E	Transcription (putatively plastid genome)	No
At1g06430	FtsH8 protease	Proteolysis	No
At5g02270	ABC transporter	Transport	No

A selection of the minimum set of 72 genes shown by microarray analysis to be regulated by UVR8, i.e. stimulated by UV-B in wild type but not in *uvr8-1* (see Brown *et al.*, 2005, for the full list). The genes shown are all within the 0.1% false discovery rate. The HY5 column identifies genes additionally regulated by HY5 on the basis of microarray analysis, i.e. those not induced by UV-B in *hy5-1* within the 0.75% false discovery rate.

pathways, including the gene encoding strictosidine synthase, which is a key enzyme in terpenoid indole alkaloid biosynthesis. UV-B stimulates strictosidine synthase expression and the accumulation of these alkaloids in *Catharanthus roseus* (Ouwerkerk *et al.*, 1999). Since these compounds absorb UV-B, it is tempting to speculate that they could contribute to UV-protection in some species. In addition, UVR8 regulates expression of the type II photolyase PHR1. As stated above, the *uvr2* mutant lacking this enzyme is highly sensitive to UV-B (Jiang *et al.*, 1997a; Landry *et al.*, 1997). Therefore UVR8 has an important regulatory role in the repair of DNA damage. Further UVR8-regulated genes are concerned with protection against oxidative stress (e.g. glutathione peroxidases; Milla *et al.*, 2003) and photooxidative damage (ELIP proteins; Hutin *et al.*, 2003). It is well established that some photosynthetic components, such as the D1 and D2 polypeptides of photosystem II are particularly susceptible to damage by UV-B (Jansen *et al.*, 1998; Booij-James *et al.*, 2000). Hence the significance of the UV-B induction of ELIP proteins and several other chloroplast proteins by UVR8 may be to help maintain photosynthetic activity in sunlight. The microarray data therefore demonstrate that UVR8 regulates expression of a range of components with vital functions in protecting plants against UV-B.

As noted above, UVR8 has sequence similarity and predicted structural similarity to human RCC1 (Kliebenstein *et al.*, 2002). RCC1 and its homologues in other eukaryotes are guanine nucleotide exchange factors (GEFs) for the small GTP-binding protein Ran (Renault *et al.*, 2001; Dasso, 2002). RCC1 is a nuclear protein that associates with chromatin and its activity produces a Ran-GTP/Ran-GDP gradient across the nuclear envelope that drives nucleocytoplasmic transport. In addition, RCC1 and Ran-GTP are involved in controlling progression of the cell cycle and mitosis. Unsurprisingly, human and yeast mutants lacking RCC1 fail to grow. It is very unlikely that UVR8 is involved in nucleocytoplasmic transport and cell cycle control because the *uvr8* mutant grows normally under non-UV-B illumination. Conversely, there is no evidence that RCC1 in other eukaryotes mediates UV-B responses or the regulation of gene expression. Brown *et al.* (2005) obtained direct evidence that UVR8 and RCC1 function differently in that they found that UVR8 had very little Ran GEF activity. Thus, although UVR8 is similar in sequence to RCC1 it is unlikely to be a functional homologue of RCC1. Moreover, Ran GEF activity is unlikely to be the basis of UVR8 activity.

Nonetheless, UVR8 does share some features with RCC1: a GFP–UVR8 fusion was present in the nucleus of transgenic plants and associated with chromatin (Brown *et al.*, 2005). In addition, GST-UVR8, expressed in *E. coli* bound strongly to a histone-agarose column *in vitro*. Interestingly, GFP–UVR8 is not exclusively localised in the nucleus, as GFP fluorescence is observed in the cytoplasm (Brown *et al.*, 2005). This is not because of aberrant overexpression from the *35S* promoter as the same pattern of localisation is observed in lines with very low expression. Hence the localisation of UVR8 contrasts with that of other RCC1-family proteins, which are exclusively nuclear. The localisation in the cytoplasm raises the possibility that UVR8 may move into the nucleus in response to UV-B rather like the phytochromes, but this has yet to be established.

7.5.3 HY5

HY5 is a bZIP transcription factor that plays a key role in the regulation of seedling photomorphogenesis as a downstream effector of several light signalling pathways (Osterlund *et al.*, 2000; Chen *et al.*, 2004). Recent evidence indicates that HY5 also has an important function in mediating photomorphogenic UV-B responses. Ulm *et al.* (2004) found that *HY5* was among the genes induced by UV-B in their microarray analysis of wild-type *Arabidopsis* seedlings. Moreover, the induction of *HY5* transcripts by UV-B was retained in *phy* and *cry* mutants, demonstrating the independence of the response from the known photoreceptors. Ulm *et al.* (2004) further demonstrated that *HY5* regulation was at the transcriptional level by using a *HY5* promoter::luciferase fusion. Northern hybridisations (Ulm *et al.*, 2004) and microarray analysis (Oravecz *et al.*, 2006) with wild-type and *hy5* mutant seedlings exposed to UV-B showed that HY5 was required for the UV-B-induced expression of a range of genes. Thus HY5 has an important role in UV-B responses.

In their microarray analysis, Brown *et al.* (2005) found that *HY5* was induced by UV-B in wild-type plants and was among the genes not expressed following UV-B exposure in the *uvr8* mutant. RT-PCR experiments confirmed this observation, demonstrating that UVR8 controls *HY5* expression (Figure 7.2B). Nevertheless, the *uvr8* mutant retained the induction of *HY5* transcripts by UV-A (Figure 7.2B) and far-red light, indicating that UVR8 controls *HY5* transcript accumulation specifically in response to UV-B.

Brown *et al.* (2005) used microarrays to compare genes reduced in expression in response to UV-B in the *hy5* mutant with those reduced in expression in *uvr8*. They found that approximately half of the genes regulated by UVR8 were also regulated by HY5. These data show that HY5 acts downstream of the UVR8 pathway to control transcription of a substantial number of UVR8-regulated genes. Some of these genes are shown in Table 7.1. Given the importance of these genes in UV protection, the *hy5* mutant would be expected to be very sensitive to UV-B and this was found to be the case (Brown *et al.*, 2005; Oravecz *et al.*, 2006). The implication is that HY5 is required for survival of plants under UV-B radiation. These findings extend previous conclusions regarding the function of HY5: in addition to promoting photomorphogenesis in seedlings, HY5 has a vital role in established plants in protecting against UV-B damage and maintaining photosynthetic competence.

The hypothesis developed by Brown *et al.* (2005) is that the association of UVR8 with chromatin facilitates the activation or binding of transcription factors that regulate target genes such as *HY5* (Figure 7.3). Brown *et al.* (2005) tested this model by using chromatin immunoprecipitation to see whether UVR8 associated with chromatin in the region of the *HY5* gene promoter. In this technique, plants are treated with formaldehyde to cross-link proteins to chromatin and then the chromatin is isolated and fragmented by sonication. An antibody is then used to immunoprecipitate chromatin bound to a selected protein, and the resultant DNA is analysed by PCR to test for the presence of particular gene sequences. This experiment was undertaken with transgenic plants expressing GFP–UVR8, and an anti-GFP antibody was used to obtain chromatin fragments. It was found that GFP–UVR8 associated

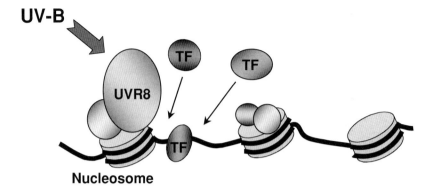

Figure 7.3   A model of the regulation of transcription by UVR8. UV-B activates UVR8 by unknown mechanisms. UVR8 associates with chromatin by binding to histones in the nucleosomes. UVR8 is proposed to facilitate the binding and/or activation of transcription factors (TF) that regulate transcription of target genes.

with a chromatin fragment containing the *HY5* promoter (sequences –331 to +23), but not with a control gene. No such association was found with chromatin from control *35S::GFP* plants. These results indicate that UVR8 is involved in the regulation of *HY5* transcription through its association with chromatin. Furthermore, it is likely that UVR8 regulates other genes, although these have yet to be identified. It is unknown how the association of UVR8 with chromatin promotes transcription, although it is likely that other proteins are involved.

7.5.4   Other transcription factors involved in UV-B responses

Apart from HY5, several other transcription factors appear to be involved in regulating gene expression in response to UV-B. HYH is similar in sequence to HY5, and its expression is induced by UV-B (Ulm *et al.*, 2004; Brown *et al.*, 2005) and regulated by UVR8 (Brown *et al.*, 2005). However, it is not yet known which genes HYH regulates. It is possible that it regulates some of the UVR8 pathway genes that are not controlled by HY5, but alterations in UV-B-induced gene expression in the *hyh* mutant have not been reported. The microarray studies of Ulm *et al.* (2004) in *Arabidopsis* and Casati and Walbot (2003, 2004) in maize identified several additional transcription factors induced by UV-B, but their roles in UV-B responses are unknown.

The best studied UV-B response at the molecular level is the regulation of flavonoid biosynthesis genes, and several transcription factors that regulate expression of these genes by UV-B have been identified. Among these is HY5 which, as mentioned above, is required for the UV-B induction of several flavonoid biosynthesis genes (see Table 7.1). In addition, a number of bHLH and MYB transcription factors act as positive regulators of flavonoid biosynthesis genes, and expression of several of these transcription factors is induced by UV-B in maize (Piazza *et al.*, 2002). In contrast, Jin *et al.* (2000) reported that expression of the AtMYB4 transcription factor, which represses transcription of the cinnamate 4-hydroxylase gene involved in sinapate ester biosynthesis, is switched off by UV-B. In consequence,

UV-protective sinapate esters accumulate following UV-B illumination of *Arabidopsis* seedlings. An *AtMYB4* mutant was found to be resistant to UV-B because of its elevated levels of sinapate esters. It may be that AtMYB4 functions principally in seedlings, because RT-PCR experiments indicate that *AtMYB4* expression is not downregulated by UV-B in mature plants (B.A. Brown and G.I. Jenkins, unpublished).

7.5.5   COP1

In seedlings, the accumulation of HY5 and other several components is regulated by the CONSTITUTIVELY PHOTOMORPHOGENIC1 (COP1) protein. In darkness, COP1 catalyses the ubiquitination of HY5 in the nucleus, which targets it for proteolytic degradation by the proteasome (Osterlund *et al.,* 2000). COP1 therefore prevents HY5 from inducing gene expression in darkness and hence represses photomorphogenesis. Following illumination, COP1 is inactivated and transferred out of the nucleus permitting HY5 accumulation and expression of its target genes. In *cop1* mutants, HY5 accumulates in darkness and seedlings have a constitutively photomorphogenic phenotype.

In contrast to its role as a negative regulator of photomorphogenesis, COP1 is a positive regulator of UV-B responses. Oravecz *et al.* (2006) reported that *cop1* seedlings lack the UV-B induction of *HY5, CHS* and a range of other genes; transcriptome analysis showed that approximately 75% of the genes induced in response to a photomorphogenic UV-B treatment in wild-type were reduced in expression in the *cop1-4* allele. About half of the genes regulated by COP1 were additionally controlled by HY5, indicating that COP1 acts through HY5 and other, as yet unidentified transcription factor(s). As expected from the gene expression results, the *cop1-4* mutant showed increased susceptibility to damage by UV-B. However, it was less sensitive than *hy5*, presumably because it has a higher residual level of expression of genes conferring tolerance in the *cop1-4* allele. Oravecz *et al.* (2006) found that UV-B promoted the nuclear accumulation of COP1 tagged with yellow fluorescent protein, in contrast to the nuclear exclusion observed in fluorescent white light. However, COP1 enrichment occurs much more slowly than the induction of gene expression by UV-B, indicating that it is a secondary process, perhaps reinforcing the response.

Clearly the role of COP1 in UV-B signalling differs from that in photomorphogenesis. Whereas COP1 acts as a negative regulator of HY5 in photomorphogenesis, degrading it in darkness, it acts together with HY5 in UV-B responses. One possibility is that COP1 directs the proteolysis of a negative regulator of the UV-B pathway that activates HY5, although alternatively it may act via a novel mechanism not involving proteolysis (Oravecz *et al.,* 2006). Further investigation of the role and regulation of COP1 should provide novel insights into the mechanism of UV-B signalling.

7.6   Concluding remarks

It is now well established that in plants UV-B mediates photomorphogenic responses distinct from stress responses. The importance of photomorphogenic UV-B responses should not be underestimated, as one of their key functions is to regulate the expression of genes that provide UV protection and the repair of UV damage, both of which are essential for survival in sunlight. Protection and repair are evidently very effective as plants rarely display signs of UV-damage in the natural environment. This leads to a surprising conclusion that is contrary to many perceptions of UV-B responses: rather than being an agent of damage, ambient UV-B actively promotes plant survival. Furthermore, photomorphogenic UV-B responses substantially influence morphogenesis and biochemical composition, which in turn affect the susceptibility of plants to attack by pathogens and herbivorous insects. Thus photomorphogenic UV-B responses have a major impact on many aspects of plant growth, metabolism and development.

Recent research has revealed that UVR8 has a key role in plant responses to UV-B, as it orchestrates the expression of a range of genes with vital functions in UV-protection and damage repair (Brown *et al.*, 2005). Further research is required to identify other components of the UVR8 pathway, but the transcription factor HY5 is evidently an important downstream effector. UVR8 regulates transcript levels of HY5 specifically in response to UV-B. COP1 also regulates *HY5* expression and promotes the expression of a range of UV-B induced genes. It is therefore important to determine the functional relationship of COP1 and UVR8 in mediating UV-B responses.

The hypersensitivity of the *uvr8*, *cop1* and *hy5* mutants to UV-B demonstrates the importance of the UVR8/COP1/HY5 pathway(s) for survival under UV-B radiation. Nevertheless, we do not yet know the functions of a substantial number of the genes regulated by these components. Studies of these genes may provide new insights into the ways plants protect against and ameliorate the damaging effects of UV-B. Indeed, microarray analyses of wild-type *Arabidopsis* (Ulm *et al.*, 2004) and maize (Casati and Walbot, 2004; Casati *et al.*, 2006) have already generated important new information on the strategies plants have evolved to cope with UV-B exposure.

There is very little mechanistic information on the regulation of morphological responses to UV-B, apart from a few studies showing the involvement of particular plant growth regulators. A potentially valuable approach will be to undertake genetic screens for mutants altered specifically in morphological responses to UV-B, similar to that performed by Suesslin and Frohnmeyer (2003). In addition, our understanding of the regulation of metabolism by UV-B is limited. Flavonoid biosynthesis has been studied extensively, but it is evident from the microarray studies undertaken to date that UV-B affects a much wider range of metabolic processes. Further research is needed to extend this work and to establish how particular metabolic processes contribute to the acclimation of plants to UV-B.

Despite the importance of photomorphogenic UV-B responses, the underlying mechanisms of UV-B perception and signal transduction remain poorly understood. However, there is now a prospect of making significant progress because of the identification of UVR8 as the first component that acts specifically to mediate UV-B responses. A number of important questions need to be addressed. In particular, how does UV-B activate UVR8? How does UVR8 promote transcription of target genes? There is no direct evidence to suggest that UVR8 is a photoreceptor, although it is premature to completely rule out this possibility. Hence, the puzzle of UV-B photoreception is still not solved. Moreover, it is still not known where UV-B perception takes place, as UVR8 is present in both the nucleus and the cytoplasm. Resolution of this point is very important as it will help to exclude some candidate photoreception processes and lend support to others. For instance, at present it is not clear how the information obtained from cell physiological and pharmacological studies, which points to signalling events in the cytoplasm (Christie and Jenkins, 1996; Frohnmeyer *et al.*, 1997, 1999) and possibly at the plasma membrane (Long and Jenkins, 1998), ties in with UVR8 activation.

The model developed by Brown *et al.* (2005) proposes that the association of UVR8 with chromatin facilitates transcriptional initiation of target genes such as *HY5* (Figure 7.3) Research is needed to test and extend this model, in particular to understand how UVR8 interacts with chromatin and to identify components that are involved in transcriptional regulation via UVR8. Indeed, our understanding of the effects of UV-B on processes associated with chromatin in plants is limited, yet these processes are undoubtedly very important. As discussed above, there is evidence that DNA damage signalling initiated by UV-B regulates cell cycle progression (Culligan *et al.*, 2004) and that chromatin modification is involved in adaptation to UV-B (Casati *et al.*, 2006). Hence further research into chromatin-related processes promises to provide valuable insights into the regulation of plant gene expression and development by UV-B.

Acknowledgements

We thank Virginia Walbot and coworkers for allowing us to refer to data in press. We are grateful to John Christie and members of GIJ's laboratory for critical comments on the manuscript.

References

A-H-Mackerness, S. (2000) Plant responses to ultraviolet-B (UV-B : 280–320 nm) stress: what are the key regulators? Invited review. *Plant Growth Regul.* **32**(1), 27–39.

A-H-Mackerness, S., John, C.F., Jordan, B. and Thomas, B. (2001) Early signaling components in ultraviolet-B responses: distinct roles for different reactive oxygen species and nitric oxide. *FEBS Lett.* **489**(2–3), 237–242.

A-H-Mackerness, S., Surplus, S.L., Blake, P., *et al.* (1999) Ultraviolet-B-induced stress and changes in gene expression in *Arabidopsis thaliana*: role of signalling pathways controlled by jasmonic acid, ethylene and reactive oxygen species. *Plant Cell Environ.* **22**(11), 1413–1423.

Ahmad, M., Jarillo, J.A., Klimczak, L.J., *et al.* (1997) An enzyme similar to animal type II photolyases mediates photoreactivation in *Arabidopsis*. *Plant Cell* **9**(2), 199–207.

Allan, A.C. and Fluhr, R. (1997) Two distinct sources of elicited reactive oxygen species in tobacco epidermal cells. *Plant Cell* **9**(9), 1559–1572.

Andrady, A., Aucamp, P.J., Bais, A.F., *et al.* (2005) Environmental effects of ozone depletion and its interactions with climate change: progress report, 2004. *Photochem. Photobiol. Sci.* **4**(2), 177–184.

Apel, K. and Hirt, H. (2004) Reactive oxygen species: Metabolism, oxidative stress, and signal transduction. *Annu. Rev. Plant Biol.* **55**, 373–399.

Ballaré, C.L., Barnes, P.W. and Flint, S.D. (1995) Inhibition of hypocotyl elongation by ultraviolet-B radiation in de-etiolating tomato seedlings. 1. The Photoreceptor. *Physiol. Plant.* **93**(4), 584–592.

Barta, C., Kalai, T., Hideg, K., Vass, I. and Hideg, E. (2004) Differences in the ROS-generating efficacy of various ultraviolet wavelengths in detached spinach leaves. *Funct. Plant Biol.* **31**(1), 23–28.

Bender, K., Blattner, C., Knebel, A., Iordanov, M., Herrlich, P. and Rahmsdorf, H.J. (1997) UV-induced signal transduction. *J. Photochem. Photobiol. B* **37**(1–2), 1–17.

Bieza, K. and Lois, R. (2001) An *Arabidopsis* mutant tolerant to lethal ultraviolet-B levels shows constitutively elevated accumulation of flavonoids and other phenolics. *Plant Physiol.* **126**(3), 1105–1115.

Bjorn, L.O. (1996) Effects of ozone depletion and increased UV-B on terrestrial ecosystems. *Int. J. Environ. Stud.* **51**, 217–243.

Boccalandro, H.E., Mazza, C.A., Mazzella, M.A., Casal, J.J. and Ballare, C.L. (2001) Ultraviolet B radiation enhances a phytochrome-B-mediated photomorphogenic response in *Arabidopsis*. *Plant Physiol.* **126**(2), 780–788.

Booij-James, I.S., Dube, S.K., Jansen, M.A.K., Edelman, M. and Mattoo, A.K. (2000) Ultraviolet-B radiation impacts light-mediated turnover of the photosystem II reaction center heterodimer in *Arabidopsis* mutants altered in phenolic metabolism. *Plant Physiol.* **124**(3), 1275–1283.

Bornman, J.F., Reuber, S., Cen, Y.-P. and Weissenböck, G. (1997) Ultraviolet radiation as a stress factor and the role of protective pigments. In: *Plants and UV-B: Responses to Environmental Change* (ed. Lumsden, P.J.), pp. 157–168. Cambridge University Press, Cambridge, UK.

Bowler, C. and Chua, N.H. (1994) Emerging themes of plant signal-transduction. *Plant Cell* **6**(11), 1529–1541.

Breitling, R., Armengaud, P., Amtmann, A. and Herzyk, P. (2004) Rank products: a simple, yet powerful, new method to detect differentially regulated genes in replicated microarray experiments. *FEBS Lett.* **573**(1–3), 83–92.

Britt, A.B. (1999) Molecular genetics of DNA repair in higher plants. *Trends Plant Sci.* **4**(1), 20–25.

Britt, A.B., Chen, J.J., Wykoff, D. and Mitchell, D. (1993) A UV-sensitive mutant of *Arabidopsis* defective in the repair of pyrimidine-pyrimidinone (6-4) dimers. *Science* **261**(5128), 1571–1574.

Brosché, M., Schuler, M.A., Kalbina, I., Connor, L. and Strid, A. (2002) Gene regulation by low level UV-B radiation: identification by DNA array analysis. *Photochem. Photobiol. Sci.* **1**(9), 656–664.

Brosché, N. and Strid, A. (2003) Molecular events following perception of ultraviolet-B radiation by plants. *Physiol. Plant.* **117**(1), 1–10.

Brown, B.A., Cloix, C., Jiang, G.H., *et al.* (2005) A UV-B-specific signaling component orchestrates plant UV protection. *Proc. Natl. Acad. Sci. USA* **102**(50), 18225–18230.

Caldwell, M.M. (1971) Solar UV irradiation and the growth and development of higher plants. In: *Photophysiology*, Vol. 6 (ed. Giese, A.C.), pp. 131–177. Academic Press, New York.

Caldwell, M.M., Ballaré, C.L., Bornman, J.F., *et al.* (2003) Terrestrial ecosystems increased solar ultraviolet radiation and interactions with other climatic change factors. *Photochem. Photobiol. Sci.* **2**(1), 29–38.

Caldwell, M.M., Robberecht, R. and Flint, S.D. (1983) Internal filters – prospects for UV-acclimation in higher-plants. *Physiol. Plant.* **58**(3), 445–450.

Casati, P., Stapleton, A.E., Blum, J.E. and Walbot, V. (2006) Genome-wide analysis of high-altitude maize and gene knockdown stocks implicates chromatin remodeling proteins in response to UV-B. *Plant J.* **46**, 613–627.

Casati, P. and Walbot, V. (2003) Gene expression profiling in response to ultraviolet radiation in maize genotypes with varying flavonoid content. *Plant Physiol.* **132**(4), 1739–1754.

Casati, P. and Walbot, V. (2004) Rapid transcriptome responses of maize (Zea mays) to UV-B in irradiated and shielded tissues. *Genome Biol.* **5**(3), R16.

Chen, M., Chory, J. and Fankhauser, C. (2004) Light signal transduction in higher plants. *Annu. Rev. Genet.* **38**, 87–117.

Christie, J.M. and Jenkins, G.I. (1996) Distinct UV-B and UV-A blue light signal transduction pathways induce chalcone synthase gene expression in *Arabidopsis* cells. *Plant Cell* **8**(9), 1555–1567.

Cooley, N.M., Higgins, J.T., Holmes, M.G. and Attridge, T.H. (2001) Ecotypic differences in responses of *Arabidopsis thaliana* L. to elevated polychromatic UV-A and UVB+A radiation in the natural environment: a positive correlation between UV-B+A inhibition and growth rate. *J. Photochem. Photobiol. B* **60**(2–3), 143–150.

Correia, C.M., Areal, E.L.V., Torres-Pereira, M.S. and Torres-Pereira, J.M.G. (1999) Intraspecific variation in sensitivity to ultraviolet-B radiation in maize grown under field conditions – II. Physiological and biochemical aspects. *Field Crops Res.* **62**(2–3), 97–105.

Culligan, K., Tissier, A. and Britt, A. (2004) ATR regulates a G2-phase cell-cycle checkpoint in *Arabidopsis thaliana. Plant Cell* **16**(5), 1091–1104.

Dai, Q.J., Yan, B., Huang, S.B., *et al.*(1997) Response of oxidative stress defense systems in rice (Oryza sativa) leaves with supplemental UV-B radiation. *Physiol. Plant.* **101**(2), 301–308.

Dasso, M. (2002) The Ran GTPase: theme and variations. *Curr. Biol.* **12**(14), R502–R508.

Ensminger, P.A. (1993) Control of development in plants and fungi by far-UV radiation. *Physiol. Plant.* **88**(3), 501–508.

Ensminger, P.A. and Schäfer, E. (1992) Blue and ultraviolet-B light photoreceptors in parsley cells. *Photochem. Photobiol.* **55**(3), 437–447.

Friesner, J.D., Liu, B., Culligan, K. and Britt, A.B. (2005) Ionizing radiation-dependent gamma-H2AX focus formation requires ataxia telangiectasia mutated and ataxia telangiectasia mutated and Rad3-related. *Mol. Biol. Cell* **16**(5), 2566–2576.

Frohnmeyer, H., Bowler, C. and Schäfer, E. (1997) Evidence for some signal transduction elements involved in UV- light-dependent responses in parsley protoplasts. *J. Exp. Botany* **48**(308), 739–750.

Frohnmeyer, H., Bowler, C., Zhu, J.K., Yamagata, H., Schäfer, E. and Chua, N.H. (1998) Different roles for calcium and calmodulin in phytochrome- and UV-regulated expression of chalcone synthase. *Plant J.* **13**(6), 763–772.

Frohnmeyer, H., Loyall, L., Blatt, M.R. and Grabov, A. (1999) Millisecond UV-B irradiation evokes prolonged elevation of cytosolic-free Ca^{2+} and stimulates gene expression in transgenic parsley cell cultures. *Plant J.* **20**(1), 109–117.

Frohnmeyer, H. and Staiger, D. (2003) Ultraviolet-B radiation-mediated responses in plants. Balancing damage and protection. *Plant Physiol.* **133**(4), 1420–1428.

Fuglevand, G., Jackson, J.A. and Jenkins, G.I. (1996) UV-B, UV-A, and blue light signal transduction pathways interact synergistically to regulate chalcone synthase gene expression in *Arabidopsis. Plant Cell* **8**(12), 2347–2357.

Gallego, F., Fleck, O., Li, A., Wyrzykowska, J. and Tinland, B. (2000) AtRAD1, a plant homologue of human and yeast nucleotide excision repair endonucleases, is involved in dark repair of UV damages and recombination. *Plant J.* **21**(6), 507–518.

Garcia, V., Bruchet, H., Camescasse, D., Granier, F., Bouchez, D. and Tissier, A. (2003) AtATM is essential for meiosis and the somatic response to DNA damage in plants. *Plant Cell* **15**(1), 119–132.

Green, R. and Fluhr, R. (1995) UV-B-induced PR-1 accumulation is mediated by active oxygen species. *Plant Cell* **7**(2), 203–212.

Hahlbrock, K. and Scheel, D. (1989) Physiology and molecular-biology of phenylpropanoid metabolism. *Annu. Rev. Plant Physiol. Plant Mol. Biol.* **40**, 347–369.

Harlow, G.R., Jenkins, M.E., Pittalwala, T.S. and Mount, D.W. (1994) Isolation of Uvh1, an *Arabidopsis* mutant hypersensitive to ultraviolet-light and ionizing-radiation. *Plant Cell* **6**(2), 227–235.

Herrlich, P. and Bohmer, F.D. (2000) Redox regulation of signal transduction in mammalian cells. *Biochem. Pharmacol.* **59**(1), 35–41.

Hideg, E., Barta, C., Kalai, T., Vass, I., Hideg, K. and Asada, K. (2002) Detection of singlet oxygen and superoxide with fluorescent sensors in leaves under stress by photoinhibition or UV radiation. *Plant Cell Physiol.* **43**(10), 1154–1164.

Hideg, E. and Vass, I. (1996) UV-B induced free radical production in plant leaves and isolated thylakoid membranes. *Plant Sci.* **115**(2), 251–260.

Holley, S.R., Yalamanchili, R.D., Moura, D.S., Ryan, C.A. and Stratmann, J.W. (2003) Convergence of signaling pathways induced by systemin, oligosaccharide elicitors, and ultraviolet-B radiation at the level of mitogen-activated protein kinases in Lycopersicon peruvianum suspension-cultured cells. *Plant Physiol.* **132**(4), 1728–1738.

Hutin, C., Nussaume, L., Moise, N., Moya, I., Kloppstech, K. and Havaux, M. (2003) Early light-induced proteins protect *Arabidopsis* from photooxidative stress. *Proc. Natl. Acad. Sci. USA* **100**(8), 4921–4926.

Izaguirre, M.M., Scopel, A.L., Baldwin, I.T. and Ballaré, C.L. (2003) Convergent responses to stress. Solar ultraviolet-B radiation and Manduca sexta herbivory elicit overlapping transcriptional responses in field-grown plants of Nicotiana longiflora. *Plant Physiol.* **132**(4), 1755–1767.

Jackson, J.A., Fuglevand, G., Brown, B.A., Shaw, M.J. and Jenkins, G.I. (1995) Isolation of *Arabidopsis* mutants altered in the light- regulation of chalcone synthase gene expression using a transgenic screening approach. *Plant J.* **8**(3), 369–380.

Jansen, M.A.K., Gaba, V. and Greenberg, B.M. (1998) Higher plants and UV-B radiation: Balancing damage, repair and acclimation. *Trends Plant Sci.* **3**(4), 131–135.

Jansen, M.A.K., van den Noort, R.E., Tan, M.Y.A., Prinsen, E., Lagrimini, L.M. and Thorneley, R.N.F. (2001) Phenol-oxidizing peroxidases contribute to the protection of plants from ultraviolet radiation stress. *Plant Physiol.* **126**(3), 1012–1023.

Jenkins, G.I., Fuglevand, G. and Christie, J.M. (1997) UV-B perception and signal transduction. In: *Plants and UV-B: Responses to Environmental Change* (ed. Lumsden, P.J.), pp. 135–156. Cambridge University Press, Cambridge, UK.

Jenkins, G.I., Long, J.C., Wade, HK., Shenton, M.R. and Bibikova, T.N. (2001) UV and blue light signalling: pathways regulating chalcone synthase gene expression in *Arabidopsis*. *New Phytol.* **151**(1), 121–131.

Jenkins, M.E., Harlow, G.R., Liu, Z.R., Shotwell, M.A., Ma, J. and Mount, D.W. (1995) radiation-sensitive mutants of *Arabidopsis thaliana*. *Genetics* **140**(2), 725–732.

Jiang, C.Z., Yee, J., Mitchell, D.L. and Britt, A.B. (1997a) Photorepair mutants of *Arabidopsis*. *Proc. Natl. Acad. Sci. USA* **94**(14), 7441–7445.

Jiang, C.Z., Yen, C.N., Cronin, K., Mitchell, D. and Britt, A.B. (1997b) UV- and gamma-radiation sensitive mutants of *Arabidopsis thaliana*. *Genetics* **147**(3), 1401–1409.

Jin, H.L., Cominelli, E., Bailey, P., *et al.* (2000) Transcriptional repression by AtMYB4 controls production of UV– protecting sunscreens in *Arabidopsis*. *EMBO J.* **19**(22), 6150–6161.

Johnson, K.L. and Ingram, G.C. (2005) Sending the right signals: regulating receptor kinase activity. *Curr. Opin. Plant Biol.* **8**(6), 648–656.

Kalbin, G., Hidema, J., Brosché, M., Kumagai, T., Bornman, J.F. and Strid, A. (2001) UV-B-induced DNA damage and expression of defence genes under UV-B stress: tissue-specific molecular marker analysis in leaves. *Plant Cell Environ.* **24**(9), 983–990.

Kalbina, I. and Strid, A. (2006) Supplementary ultraviolet-B irradiation reveals differences in stress responses between *Arabidopsis thaliana* ecotypes. *Plant Cell Environ.* **29**(5), 754–763.

Khare, M. and Guruprasad, K.N. (1993) UV-B-induced anthocyanin synthesis in maize regulated by fmn and inhibitors of FMN photoreactions. *Plant Sci.* **91**(1), 1–5.

Kim, B.C., Tennessen, D.J. and Last, R.L. (1998) UV-B-induced photomorphogenesis in *Arabidopsis thaliana*. *Plant J.* **15**(5), 667–674.

Kliebenstein, D.J., Lim, J.E., Landry, LG. and Last, R.L. (2002) *Arabidopsis* UVR8 regulates ultraviolet-B signal transduction and tolerance and contains sequence similarity to human regulator of chromatin condensation 1. *Plant Physiol.* **130**(1), 234–243.

Kucera, B., Leubner-Metzger, G. and Wellmann, E. (2003) Distinct ultraviolet-signaling pathways in bean leaves. DNA damage is associated with β-1,3-glucanase gene induction, but not with flavonoid formation. *Plant Physiol.* **133**(4), 1445–1452.

Landry, L.G., Chapple, C.C.S. and Last, R.L. (1995) *Arabidopsis* mutants lacking phenolic sunscreens exhibit enhanced ultraviolet-B injury and oxidative damage. *Plant Physiol.* **109**(4), 1159–1166.

Landry, L.G., Stapleton, A.E., Lim, J., Hoffman, P., Hays, J.B., Walbot, V. and Last, R.L. (1997) An *Arabidopsis* photolyase mutant is hypersensitive to ultraviolet-B radiation. *Proc. Natl. Acad. Sci. USA* **94**(1), 328–332.

Li, J.Y., Oulee, T.M., Raba, R., Amundson, R.G. and Last, R.L. (1993) *Arabidopsis* flavonoid mutants are hypersensitive to UV-B irradiation. *Plant Cell* **5**(2), 171–179.

Liu, Z.R., Hossain, G.H., Islas-Osuna, M.A., Mitchell, D.L. and Mount, D.W. (2000) Repair of UV damage in plants by nucleotide excision repair: *Arabidopsis* UVH1 DNA repair gene is a homolog of *Saccharomyces cerevisiae* RAD1. *Plant J*. **21**(6), 519–528.

Lois, R. and Buchanan, B.B. (1994) Severe sensitivity to ultraviolet-radiation in an *Arabidopsis* mutant deficient in flavonoid accumulation. 2. Mechanisms of UV-resistance in *Arabidopsis*. *Planta* **194**(4), 504–509.

Long, J.C. and Jenkins, G.I. (1998) Involvement of plasma membrane redox activity and calcium homeostasis in the UV-B and UV-A blue light induction of gene expression in *Arabidopsis*. *Plant Cell* **10**(12), 2077–2086.

McKenzie, R.L., Bjorn, L.O., Bais, A. and Ilyasd, M. (2003) Changes in biologically active ultraviolet radiation reaching the Earth's surface. *Photochem. Photobiol. Sci*. **2**(1), 5–15.

Milla, M.A.R., Maurer, A., Huete, A.R. and Gustafson, J.P. (2003) Glutathione peroxidase genes in *Arabidopsis* are ubiquitous and regulated by abiotic stresses through diverse signaling pathways. *Plant J*. **36**(5), 602–615.

Mittler, R. (2002) Oxidative stress, antioxidants and stress tolerance. *Trends Plant Sci*. **7**(9), 405–410.

Montoya, T., Nomura, T., Farrar, K., Kaneta, T., Yokota, T. and Bishop, G.J. (2002) Cloning the tomato CURL3 gene highlights the putative dual role of the leucine-rich repeat receptor kinase tBRI1/SR160 in plant steroid hormone and peptide hormone signaling. *Plant Cell* **14**(12), 3163–3176.

Nakajima, S., Sugiyama, M., Iwai, S., *et al.* (1998) Cloning and characterization of a gene (UVR3) required for photorepair of 6-4 photoproducts in *Arabidopsis thaliana*. *Nucleic Acids Res*. **26**(2), 638–644.

Ohl, S., Hahlbrock, K. and Schafer, E. (1989) A stable blue-light-derived signal modulates ultraviolet-light-induced activation of the chalcone-synthase gene in cultured parsley cells. *Planta* **177**(2), 228–236.

Oravecz, A., Baumann, A., Máté, Z., *et al.* (2006) Constitutively Photomorphogenic1 is required for the UV-B response in *Arabidopsis*. *Plant Cell* **18**(8), 1975–1990.

Osterlund, M.T., Wei, N. and Deng, X.-W. (2000) The roles of photoreceptor systems and the COP1-targeted destabilization of HY5 in light control of *Arabidopsis* seedling development. *Plant Physiol*. **124**(4), 1520–1524.

Ouwerkerk, P.B.F., Hallard, D., Verpoorte, R. and Memelink, J. (1999) Identification of UV-B light-responsive regions in the promoter of the tryptophan decarboxylase gene from *Catharanthus roseus*. *Plant Mol. Biol*. **41**(4), 491–503.

Paul, N.D. and Gwynn-Jones, D. (2003) Ecological roles of solar UV radiation: towards an integrated approach. *Trends Ecol. Evol*. **18**(1), 48–55.

Piazza, P., Procissi, A., Jenkins, G.I. and Tonelli, C. (2002) Members of the c1/pl1 regulatory gene family mediate the response of maize aleurone and mesocotyl to different light qualities and cytokinins. *Plant Physiol*. **128**(3), 1077–1086.

Pyle, J.A. (1997) Global ozone depletion. In: *Plants and UV-B : Responses to Environmental Change* (ed. Lumsden, P.J.), pp. 3–11. Cambridge University Press, Cambridge, UK.

Rao, M.V., Paliyath, C. and Ormrod, D.P. (1996) Ultraviolet-B- and ozone-induced biochemical changes in antioxidant enzymes of *Arabidopsis thaliana*. *Plant Physiol*. **110**(1), 125–136.

Renault, L., Kuhlmann, J., Henkel, A. and Wittinghofer, A. (2001) Structural basis for guanine nucleotide exchange on Ran by the regulator of chromosome condensation (RCC1). *Cell* **105**(2), 245–255.

Ries, G., Heller, W., Puchta, H., Sandermann, H., Seidlitz, H.K. and Hohn, B. (2000) Elevated UV-B radiation reduces genome stability in plants. *Nature* **406**(6791), 98–101.

Ros, J. and Tevini, M. (1995) Interaction of UV-radiation and IAA during growth of seedlings and hypocotyl segments of sunflower. *J. Plant Physiol*. **146**(3), 295–302.

Rozema, J., vandeStaaij, J., Bjorn, L.O. and Caldwell, M. (1997) UV-B as an environmental factor in plant life: stress and regulation. *Trends Ecol. Evol*. **12**(1), 22–28.

Ryan, C.A. (2000) The systemin signaling pathway: differential activation of plant defensive genes. *Biochim. Biophys. Acta*. **1477**(1–2), 112–121.

Sancar, A., Lindsey-Boltz, L.A., Unsal-Kacmaz, K. and Linn, S. (2004) Molecular mechanisms of mammalian DNA repair and the DNA damage checkpoints. *Annu. Rev. Biochem.* **73**, 39–85.

Sato, T., Ueda, T., Fukuta, Y., Kumagai, T. and Yano, M. (2003) Mapping of quantitative trait loci associated with ultraviolet- B resistance in rice (*Oryza sativa* L.). *Theor. Appl. Genet.* **107**(6), 1003–1008.

Savenstrand, H., Brosché, M. and Strid, A. (2004) Ultraviolet-B signalling: *Arabidopsis* brassinosteroid mutants are defective in UV-B regulated defence gene expression. *Plant Physiol. Biochem.* **42**(9), 687–694.

Shinkle, J.R., Atkins, A.K., Humphrey, E.E., Rodgers, C.W., Wheeler, S.L. and Barnes, P.W. (2004) Growth and morphological responses to different UV wavebands in cucumber (*Cucumis sativum*) and other dicotyledonous seedlings. *Physiol. Plant.* **120**(2), 240–248.

Shinkle, J.R., Derickson, D.L. and Barnes, P.W. (2005) Comparative photobiology of growth responses to two UV-B wavebands and UV-C in dim-red-light- and white-light-grown cucumber (*Cucumis sativus*) seedlings: physiological evidence for photoreactivation. *Photochem. Photobiol.* **81**(5), 1069–1074.

Stapleton, A.E. and Walbot, V. (1994) Flavonoids can protect maize DNA from the induction of ultraviolet-radiation damage. *Plant Physiol.* **105**(3), 881–889.

Stratmann, J. (2003) Ultraviolet-B radiation co-opts defense signaling pathways. *Trends Plant Sci.* **8**(11), 526–533.

Stratmann, J., Scheer, J. and Ryan, C.A. (2000) Suramin inhibits initiation of defense signaling by systemin, chitosan, and a beta-glucan elicitor in suspension-cultured *Lycopersicon peruvianum* cells. *Proc. Natl. Acad. Sci. USA* **97**(16), 8862–8867.

Suesslin, C. and Frohnmeyer, H. (2003) An *Arabidopsis* mutant defective in UV-B light-mediated responses. *Plant J.* **33**(3), 591–601.

Sullivan, J.H., Teramura, A.H. and Ziska, L.H. (1992) Variation in UV-B sensitivity in plants from a 3,000-m elevational gradient in Hawaii. *Am. J. Bot.* **79**(7), 737–743.

Surplus, S.L., Jordan, B.R., Murphy, A.M., Carr, J.P., Thomas, B. and A-H-Mackerness, S. (1998) Ultraviolet-B-induced responses in *Arabidopsis thaliana*: role of salicylic acid and reactive oxygen species in the regulation of transcripts encoding photosynthetic and acidic pathogenesis-related proteins. *Plant Cell Environ.* **21**(7), 685–694.

Tanaka, A., Sakamoto, A., Ishigaki, Y., *et al.* (2002) An ultraviolet-B-resistant mutant with enhanced DNA repair in *Arabidopsis*. *Plant Physiol.* **129**(1), 64–71.

Tevini, M. and Teramura, A.H. (1989) UV-B effects on terrestrial plants. *Photochem. Photobiol.* **50**(4), 479–487.

Torabinejad, J. and Caldwell, M.M. (2000) Inheritance of UV-B tolerance in seven ecotypes of *Arabidopsis thaliana* L. Heynh. and their F-1 hybrids. *J. Hered.* **91**(3), 228–233.

Ulm, R., Baumann, A., Oravecz, A., *et al.* (2004) Genome-wide analysis of gene expression reveals function of the bZIP transcription factor HY5 in the UV-B response of *Arabidopsis*. *Proc. Natl. Acad. Sci. USA* **101**(5), 1397–1402.

Ulm, R. and Nagy, F. (2005) Signalling and gene regulation in response to ultraviolet light. *Curr. Opin. Plant Biol.* **8**(5), 477–482.

Wade, H.K., Bibikova, T.N., Valentine, W.J. and Jenkins, G.I. (2001) Interactions within a network of phytochrome, cryptochrome and UV-B phototransduction pathways regulate chalcone synthase gene expression in *Arabidopsis* leaf tissue. *Plant J.* **25**(6), 675–685.

Wade, H.K., Sohal, A.K. and Jenkins, G.I. (2003) *Arabidopsis* ICX1 is a negative regulator of several pathways regulating flavonoid biosynthesis genes. *Plant Physiol.* **131**(2), 707–715.

Wang, Z.Y. and He, J.X. (2004) Brassinosteroid signal transduction – choices of signals and receptors. *Trends Plant Sci.* **9**(2), 91–96.

Yalamanchili, R.D. and Stratmann, J.W. (2002) Ultraviolet-B activates components of the systemin signaling pathway in *Lycopersicon peruvianum* suspension-cultured cells. *J. Biol. Chem.* **277**(32), 28424–28430.

Part III   Physiological responses

8 Photocontrol of flowering

Paul Devlin

8.1 Introduction

The precise control of the timing of flowering is crucial to the fitness of the individual plant and of the species as a whole. The fittest plants are those that produce the highest number of offspring that survive, grow to maturity and reproduce in turn, thus passing on their genes (Darwin, 1859). Control on flowering time optimises the chances that a given plant will flower when all factors are optimum, when reserves are sufficient for a high yield, when suitable pollinators are available and when environmental conditions are favourable. For the species it can furthermore allow coordination of flowering to maximise the efficiency of cross-pollination that creates new diversity (West *et al.*, 1999). An inbuilt flexibility in the developmental programme is essential if the plant is to take full advantage of beneficial conditions, and also to allow adaptation to adverse conditions. For example, under extreme stress, reproduction becomes a priority and flowering must occur despite suboptimal circumstances.

The timing of flowering is an important issue for humans too. Much of our arable produce is a result of fruit production following flowering. Many crops, such as cabbage, must be harvested prior to flowering. The times of harvest are strictly governed by when flowering occurs. An understanding of the regulation of flowering, therefore, has huge potential for economic benefit.

Flowering plants can be grouped into three categories according to the nature of their regulation of flowering. Those showing *autonomous* regulation respond only to internal cues such as age and size and are not dependent on favourable environmental conditions to trigger the transition to flowering. *Obligate* responders, by contrast, respond only to external cues such as day length or temperature. *Facultative* responders can respond to either internal or external cues, ensuring that flowering will occur after a prolonged growth period in which no environmental cues are received.

The transition to flowering, itself, is a fascinating transformation. The same meristem that, up until that point, was producing vegetative structures such as leaves and stems switches to produce flowers. Many plants go through three phases of development: a *juvenile* phase during which time the plant becomes established, a *vegetative* phase when the plant produces leaves and accumulates biomass following establishment and then a *reproductive* phase (McDaniel *et al.*, 1992). A shift from juvenile to vegetative is generally essential before a subsequent transition to flowering can occur even in obligate responders, with the juvenile and vegetative phases often being distinguished by a change in form of the leaves or even of the whole architecture of the plant. The length of the juvenile period can last anything from a few weeks in an annual plant to several years in some trees.

The process of floral evocation also comprises three steps. The shift from juvenile to vegetative state is often referred to as a gaining of *competence* to flower. The subsequent *determination* to flower involves the triggering of the programme involved in the transition to flowering. At this point, the transition will still occur even after removal of the trigger itself. The final step is the *expression* of flowering. This is the point at which the flowers, themselves, are finally produced (McDaniel *et al.*, 1992).

This chapter focuses on a specific set of external cues from the light environment, leading to the triggering of flowering. Many exquisitely sensitive information gathering systems combine to maximise photosynthetic light harvest. Light information plays a central role in the regulation of plant growth: germination, seedling establishment, leaf production, branching, elongation and production of light-harvesting pigments, to name but a few aspects. Photocontrol of flowering uses many of the same light-detecting components as these systems but differs from these other mechanisms, which are well described in Chapters 9 and 10, as it concerns, not the optimisation of light harvesting, but the optimisation of reproduction. An important factor for all life in latitudes outside of the tropics is the time of the year. In many plants, flowering is timed to coincide with favourable seasons, avoiding potentially damaging extremes of climate and ensuring the availability of potential pollinators that may also be limiting under those climatic extremes. Such synchronisation is often achieved by a response to the changing day length that heralds the approaching season – a response that is termed *photoperiodism*. However, light signals can also trigger flowering in many plants as part of a phenomenon known as the shade-avoidance response (see Chapter 9). This is a response to imminent shading by neighbouring plants, and in extreme cases where the plant is unable to avoid such competitive stress, flowering is initiated despite suboptimal conditions to ensure production of offspring. The photocontrol of flowering will be also considered in the context of other cues, however. No single regulatory mechanism in the flowering pathway acts in isolation. Signals involved in the photocontrol of flowering converge upon the same targets as responses to both internal cues and a number of other external cues, and each of these cues and the points of their convergence will be briefly discussed.

This chapter will also focus on our understanding of the flowering regulatory pathway at a molecular level. Much of our knowledge of this pathway comes from the study of the model plant *Arabidopsis thaliana*. *Arabidopsis* is a facultative responder with a very short life history and has, thus, proved an excellent model in dissecting the pathways involved in both internal and external cues. The additional amenability of *Arabidopsis* for genetic analysis has, over the past few years, allowed several leaps forward in our understanding of the regulation of flowering and hence the bulk of the evidence presented will come from this species. *Arabidopsis* is a long-day plant. Under the control of a *photoperiodic* pathway, it flowers as the days are lengthening with the approach of summer. It must reach a certain stage of vegetative development before flowering can occur, regulated by an *autonomous* pathway, and many varieties of *Arabidopsis*, known as winter annuals, will not flower until they have experienced a prolonged exposure to low temperatures, a process known as *vernalisation*. A number of flowering time mutants have been isolated in *Arabidopsis*. The precise nature of their phenotypes has allowed the disrupted genes to be placed in one or other flowering pathway. For example, mutants in the photoperiodic pathway show an insensitivity to day length, whilst mutants in the autonomous pathway show a generally perturbed flowering time irrespective of environmental conditions. It does appear, however, that *Arabidopsis* forms an excellent model for other species too. Indeed, many of the components acting in *Arabidopsis* have already been shown to be present and involved in the regulation of flowering in other species including crop species.

8.2 Internal cues

The achievement of competence to flower is the result of an internal cue generally referred to as the autonomous pathway. One further internal cue is also important in *Arabidopsis*: Gibberellins have been shown to be essential for flowering to occur in non-inductive photoperiods leading to the conclusion that a *gibberellin-dependent* pathway also acts.

8.2.1 The autonomous pathway

A key component of the autonomous pathway is FLOWERING LOCUS C (FLC). FLC is a MADS-box transcription factor that acts as a suppressor of flowering (Michaels and Amasino, 1999). Loss of function of FLC leads to an early flowering phenotype that is, nonetheless, still responsive to external cues (Michaels and Amasino, 1999). For example, like a wild-type seedling, an *flc* mutant will flower earlier in long days than in short days but in each case the *flc* mutant will be earlier flowering than a wild type grown alongside.

FLC levels are tightly regulated by a number of other components in the autonomous pathway. FCA, FY, FPA, FVE, LD, FLD and FLK all act in the autonomous pathway as promoters of flowering. Recessive mutants of *fca, fy, fpa, fve, ld, fld* or *flk* all show a very much delayed transition to flowering relative to wild-type seedlings (Quesada *et al.*, 2005). However, like *flc*, these mutants are also still responsive to photoperiod and vernalisation (Koornneef *et al.*, 1991). Such mutations result in increased levels of expression of the *FLC* gene (Michaels and Amasino, 1999), leading to the conclusion that FCA, FY, FPA, FVE, LD, FLD and FLK all act as repressors of *FLC* expression. Conversely, one further factor, FRIGIDA (FRI), has been shown to be required for high expression of FLC. *fri* mutants, like *flc* mutants, show an early flowering yet environmentally responsive phenotype (Johanson *et al.*, 2000).

FLC can, thus, be considered the key player in the autonomous pathway, with the balance of all of these other factors acting to regulate its expression (Figure 8.1). In *Arabidopsis*, gaining the competence to flower manifests itself as an increase

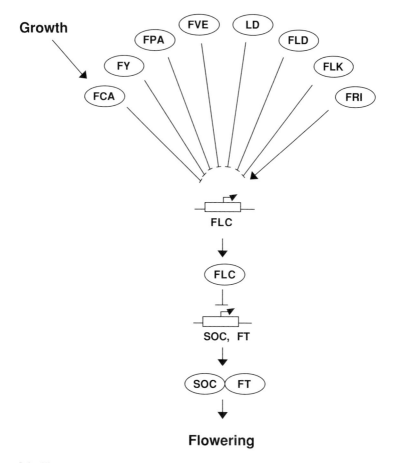

Figure 8.1   The autonomous pathway in *Arabidopsis thaliana*. FLOWERING LOCUS C (FLC) acts to inhibit flowering by repressing the floral meristem identity genes, *SUPPRESSOR OF CO* (*SOC*) and *FLOWERING LOCUS T* (*FT*). In turn, a balance of positive and negative factors regulates the level of *FLC* expression. FCA, FY, FPA, FVE, LD, FLD and FLK act to repress expression of *FLC*. FRI acts to promote *FLC* expression. *FCA* expression increases with age and eventually tips the balance in favour of repression of *FLC*, allowing flowering to proceed.

in responsiveness to external cues over time, and it is believed that this is a result of a decrease in the level of the floral inhibitor, FLC. This is consistent with the recent demonstration in *Arabidopsis* of the temporal regulation of the expression of one of the repressors of *FLC* expression, FCA (Quesada *et al.*, 2003). FCA is an RNA-binding protein (Macknight *et al.*, 2002) and is able to autoregulate its own expression by promoting cleavage and polyadenylation of its own third intron (Quesada *et al.*, 2003). The resulting truncated form of FCA protein is non-functional, and thus the level of functional FCA is kept low until a specific stage in development when this autoregulatory negative feedback becomes less effective and functional FCA protein then accumulates in regions of active cell proliferation

(Quesada *et al.*, 2003). At such a time the expression of the floral repressor *FLC* will, itself, be repressed and flowering will be promoted. Interestingly, FCA, FY, FPA, FVE, FLD and FLK are all involved in either RNA processing or histone modification. Consistent with this, *FLC* expression seems particularly sensitive to disruption of either of these processes (Simpson, 2004).

FLC has been shown to act by repressing a class of floral integrators known as the floral meristem identity genes. Activation of the genes *FLOWERING LOCUS T (FT)* (Kardailsky *et al.*, 1999) and *SUPPRESSOR OF OVEREXPRESSION OF CONSTANS 1 (SOC1)* (also known as *AGAMOUS-LIKE 20*) (Borner *et al.*, 2000; Lee *et al.*, 2000; Samach *et al.*, 2000) occurs upon removal of FLC (also discussed in Chapter 10). These genes are crucial to the transitions to flowering, their activation providing the ultimate trigger for flowering. They directly regulate a further class of factors known as the floral homeotic genes that regulate the development of the flowers themselves (Figure 8.1).

8.2.2 The gibberellin-dependent pathway

Arabidopsis is normally a *facultative* long-day plant, meaning that it will eventually flower under short days in the absence of a promoting external stimulus. However, an absolute requirement for gibberellic acid (GA) for flowering under short days has been demonstrated in *Arabidopsis* by the fact that the *ga1-3* mutant fails to flower under short days (Wilson *et al.*, 1992). *GA1* encodes *ent*-kaurine synthase A, which catalyses the first committed step of GA biosynthesis (Sun and Kamiya, 1994) and, thus, the *ga1-3* mutant is completely devoid of GA.

Exogenous application of GA induces flowering in many species (Bernier, 1988). *Arabidopsis* is no exception: exogenous treatment of *Arabidopsis* with GA accelerates flowering under both long and short days, though more so under short days (Langridge, 1957). Likewise, the *spindly (spy)* mutation, resulting in constitutively active GA signalling, causes an early flowering phenotype under both long and short days (Jacobsen and Olszewski, 1993), also demonstrating that GA can promote flowering under both conditions in *Arabidopsis*. As a result, a gibberellin-dependent pathway has also been added to the network of stimuli regulating flowering in *Arabidopsis*.

The gibberellin-dependent pathway has been shown to regulate *SOC1* (Moon *et al.*, 2003) and another floral meristem identity gene, *LEAFY (LFY)* (Blazquez *et al.*, 1998). As with *SOC1*, activation of *LFY* triggers the transition to flowering via activation of floral homeotic genes (Weigel and Nilsson, 1995).

8.3 External cues

8.3.1 Photoperiodism

One of the most conspicuous indicators of the season is the flowering of common species of plants around us. Many of these plants respond specifically to the day length to ensure timing of flowering. Photoperiodic plants can be placed into two categories: *Short-day plants* flower when the days are shorter than a certain minimum. *Long-day plants* flower when the day length is longer than a certain minimum. For example, Xanthium, a short-day plant, flowers as the day length declines towards autumn (Salisbury, 1963). Fuschia, a long-day plant, flowers as the days lengthen in spring (Vince-Prue, 1975).

8.3.1.1   Long days or short nights?

Short days effectively consist of a short day and a long night and vice versa. Consequently, questions were raised as to whether day length or night length was the crucial factor. If short-day plants are maintained in days of only 16 h in total such that they are given both short days and short nights, they will not flower despite the short day length. This clearly indicates that long nights are the key, and this has led to short-day plants sometimes being referred to as *dark dominant*. Similar experiments back up this conclusion: When short-day plants growing in short days are given a brief 10-min 'night break', a pulse of light, in the middle of each long night, the otherwise inductive effect of the short-day conditions is lost and flowering is inhibited, the conclusion being that a certain minimum length of uninterrupted darkness is the key trigger for a short-day plant (Vince-Prue, 1975).

The precise timing of this night break is crucial, however. In a very revealing experiment, Coulter and Hamner (1964) grew plants of soya bean (*Glycine max*), a short-day plant, in cycles of 8 h of light and 64 h of darkness. Night breaks were given at a particular point in the inductive dark period during each 72-h cycle. Night breaks given between 0 and 11 h after transfer to darkness completely inhibited flowering. Similarly, night breaks given between 28 and 38 h after transfer to darkness or between 49 and 60 h after transfer to darkness also inhibited flowering. Night breaks given at other times were ineffective at inhibiting flowering. This suggested that there were 'light-sensitive' periods during a long night. The pattern repeats with a period length of about 24 h, a classic indicator of the involvement of a circadian rhythm. Circadian rhythms are common to almost all life on earth allowing synchronisation of physiology and metabolism to the day/night cycle of the earth. The circadian rhythm is generated by an internal oscillator that continues to run with a 24-h period even in constant environmental conditions. Such an oscillator appears to measure the duration of darkness, and flowering will occur in short-day plants only if this oscillator has moved the plant through its light-sensitive phase without it perceiving light.

In long-day plants, by contrast, it is the duration of the day that is important, and this has led to long-day plants sometimes being referred to as *light dominant*. Night breaks are also effective in long-day plants. However, in long-day plants, night breaks act to trigger flowering in otherwise non-inductive conditions rather than inhibit flowering in otherwise inductive conditions. Prolonged night breaks of the order of at least 2 h are required to cause flowering in long-day plants but a rhythm of responsivity to night breaks is still evident, nonetheless: in *Arabidopsis*, for example a clear circadian rhythm of responsivity to 4-h night breaks was demonstrated by Carré (1998). In long-day plants, as in short-day plants, a circadian oscillator appears to measure the duration of darkness. However, in long-day plants, flowering will occur only if light is present during the sensitive phase of this oscillator.

8.3.1.2   The circadian clock

An endogenous circadian oscillator appears to be a common feature of most of the eukaryotic organisms (Dunlap, 1999). The advantage of being able to synchronise physiological processes with the day/night cycle is easy to conceive. Not only does this allow anticipation of the dramatic changes in the environment occurring at dawn and dusk but also allows a coordination of steps in metabolic pathways to an optimal time of day. The clock regulates a huge range of biochemical processes, ultimately even regulating development and behaviour. In many cases, the resulting overt rhythms are easy to observe. Commonly studied circadian rhythms include leaf movement in plants, eclosion in insects and even our own sleep–wake cycle.

One feature common to all circadian rhythms is that they persist in the absence of external stimuli, evidence of an internal rather than an external driver. However, when an organism is maintained in constant environmental conditions, it becomes apparent that whilst the period of the rhythm is consistent, the period length of the internal rhythm is generally not exactly 24 h. Depending on the organism, or even the individual, the rhythm can be either a little faster or a little slower than the actual day/night cycle. Consequently, the circadian clock requires a daily resetting by environmental cues to keep it entrained to the right time. As such, light forms the most important clock-resetting cue. The clock is particularly sensitive to light around dawn and dusk when a pulse of light in an otherwise dark setting can cause a 'phase shift', an advance or a delay in the rhythm. Light in advance of 'expected' dawn will result in a phase advance, whilst light later than 'expected' dusk will cause a phase delay.

Furthermore, circadian rhythms are temperature compensated. They continue to run with approximately the same period over a wide range of external temperatures. Unlike other biochemical reactions that show an approximate doubling of reaction time for every 10°C rise in temperature, the mechanism responsible for generation of the circadian rhythm is buffered against temperature-induced changes in rate.

The mechanism of the circadian clock has, to a large extent, been elucidated in animals and good progress is being made towards this in plants and fungi. Clocks in all of these kingdoms share a common *modus operandi*. A transcriptional feedback loop forms the basis of the rhythm in each case. As the quantity of protein of one or more key 'state variables' reaches a certain threshold, the protein feeds back to switch off transcription of its own message. A delay in the process ensures that the reaction never reaches equilibrium and a perpetual oscillation results. Various positively and negatively acting elements acting in the oscillator loop have been identified in model plants, animals and fungi. However, whilst the mechanism used to generate a rhythm is consistent, the elements of the loop themselves are not conserved. It appears that the clock has evolved separately in each phylum.

In *Arabidopsis*, the pseudo-response regulator protein, TOC1, gradually accumulates during the evening and early night. TOC1 positively regulates the transcription of two single-myb transcription factors, CCA1 and LHY. These accumulate around

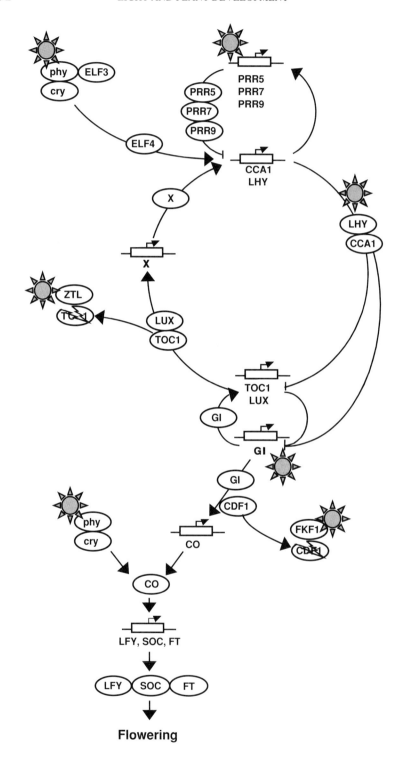

dawn and, in turn, act to negatively regulate *TOC1* expression so that TOC1 levels fall to a minimum around dawn. Without TOC1, *CCA1* and *LHY* transcription is no longer stimulated and the levels of CCA1 and LHY then fall again allowing TOC1 to begin to re-accumulate. Thus the cycle repeats (Figure 8.2). CCA1 and LHY inhibit *TOC1* expression by binding to an 'evening element' (AAAATATCT) in the TOC1 promoter. They also act to regulate other circadian 'output' genes either positively through the 'CCA1 binding site' (AAAAATCT), in which case these will peak around dawn, or negatively through evening element, in which case these will peak around dusk.

The way in which TOC1 promotes *CCA1* and *LHY* expression is not understood, however. TOC1 shows no homology to a transcription factor and thus another factor has been invoked to fill this gap in the loop. An elegant mathematical modelling-based approach also showed that no such model involving the factors CCA1, LHY and TOC1 alone could replicate the rhythm observed *in vivo*. TOC1 protein levels are minimal around dawn, at which time TOC1 would be expected to be maximally activating expression of *CCA1* and *LHY*. This would also appear to predict the existence of an additional, TOC1-dependent component as the direct activator of *LHY* and *CCA1*, a factor termed 'Factor X' (Locke *et al.*, 2005) (Figure 8.2).

Resetting of the clock at dawn is proposed to occur through light activation of *CCA1* and *LHY* expression. In response to red light, the photoreceptor, phytochrome, binds to and inactivates PIF3, a negatively acting transcription factor, bound to a

---

Figure 8.2 The photoperiodic pathway in the long-day plant, *Arabidopsis thaliana*. The CONSTANS (CO) protein acts to promote flowering by activating the floral meristem identity genes, *LEAFY* (*LFY*), *SUPPRESSOR OF CO* (*SOC*) and *FLOWERING LOCUS T* (*FT*). *CO* expression is under the control of the circadian clock, peaking in the evening. The CO protein will only accumulate at this time if it is stabilised by light and so activation of the floral meristem identity genes can occur only in long days. The circadian clock regulating *CO* expression consists of a central loop involving the single-myb domain proteins CIRCADIAN CLOCK ASSOCIATED 1 (CCA1) and LATE ELONGATED HYPOCOTYL (LHY) that act redundantly to suppress expression of the pseudo response regulator *TIMING OF CAB1* (*TOC1*) and the myb protein LUX ARRHYTHMO (LUX). TOC1 and LUX act in turn to promote expression of *CCA1* and *LHY* thus generating a sustained rhythm. Due to the early expression peak of *TOC1* and *LUX*, another factor, FACTOR X, is proposed to act in this loop between TOC1/LUX and CCA1/LHY. Other interlocked subsidiary loops are necessary to maintain a robust rhythm. Expression of the pseudo response regulators, *PRR1*, *PRR5* and *PRR7*, is regulated by CCA1/LHY. These in turn act to negatively regulate expression of *CCA1/LHY*. CCA1, LHY and TOC1 all negatively regulate expression of *GIGANTEA* (*GI*), whilst GI positively regulates TOC1 expression. Light input to keep the clock entrained to the day/night cycle occurs primarily via light-induced acute stimulation of *CCA1/LHY* expression involving the phytochrome and cryptochrome photoreceptors and the EARLY FLOWERING 4 (ELF4) protein. Additional light-induced acute stimulation of *PRR1*, *PRR5*, *PRR7* and *GI* also contributes as does a light regulation of LHY and TOC1 protein stability, the latter involving the action of the F-box protein, ZEITLUPE (ZTL). The gating factor, ELF3, periodically inhibits the action of phytochrome in this system. Regulation of *CO* expression occurs via the circadian clock component GI that promotes *CO* expression. A Dof transcription factor, *CYCLING DOF FACTOR* (*CDF*), which binds to the *CO* promoter also negatively regulates *CO* expression. CDF is regulated by FLAVIN BINDING KELCH REPEAT F-BOX PROTEIN 1 (FKF1). FKF1 may target CDF1 for degradation in response to blue light. The circadian control of FKF1 expression and the light regulation of FKF1 function appear to coincide to control the daytime CO waveform.

G-box sequence (CACGTG) present in the *CCA1* and *LHY* promoters. This negates the repressive effect of PIF3, stimulating a rapid and transient peak in expression of *CCA1* and *LHY* (Martinez-Garcia *et al.*, 2000). Light also appears to be able to promote translation of LHY (Kim *et al.*, 2003). A sudden increase in the level of *CCA1* and *LHY* transcript will shift the phase of the clock to a point in the cycle at which these transcripts are normally highly expressed and the cycle will continue from that point having been reset to a new 'time'.

Genetic analysis has revealed a number of factors involved in this process. The red light photoreceptor family, the phytochromes, and the blue light photoreceptor family, the cryptochromes, play important roles in clock resetting (Millar *et al.*, 1995; Somers *et al.*, 1998a). In constant light, the resetting effect of light results in a shortening of the period length of the clock as light intensity increases. Phytochrome and cryptochrome mutants display reduced sensitivity to light in this respect. The combination of multiple phytochromes and cryptochromes allows a plasticity in recruitment of different photoreceptors under different environmental conditions (Devlin and Kay, 2000). One further mutant, *ztl*, also shows a reduced sensitivity to light intensity in the regulation of period length. The *ztl* mutant is less sensitive to both red and blue light implying that ZTL acts down stream of both groups of photoreceptors (Somers *et al.*, 2000). ZTL is a member of the FLAVIN-BINDING, KELCH REPEAT, F-BOX (FKF) family. F-box proteins are involved in targeting a substrate for ubiquitination, a prelude to degradation by the proteasome machinery, and ZTL has been shown to bind to TOC1 and to specifically target it for proteasome-mediated degradation (Mas *et al.*, 2003) (Figure 8.2). In the *ztl* mutant, rather than dipping during the late subjective night/early subjective day, the TOC1 protein maintains a constitutively high level both in light/dark cycles and in constant light (Mas *et al.*, 2003). Such a high level of TOC1 expression is consistent with a lengthening of the period in constant light (Mas *et al.*, 2003).

Another component involved in light resetting is EARLY FLOWERING 4 (ELF4). ELF4 is light and clock regulated. ELF4 positively regulates light induction of *CCA1* and *LHY* and, in turn, CCA1 and LHY positively regulate expression of *ELF4* (Kikis *et al.*, 2005). This reciprocal relationship has been suggested to be evidence of an interlocking, autoregulatory transcriptional feedback loop working in conjunction with, or parallel to, that previously described for CCA1, LHY and TOC1. However, *ELF4* transcript continues to oscillate in a *ccal lhy* double mutant and vice versa suggesting that the situation is more complex than this. An alternative possibility is that ELF4 may act in conjunction with TOC1 in regulating *CCA1*/*LHY* transcription possibly as part of a multi-protein complex (Figure 8.2).

The clock is particularly sensitive to light for resetting around dawn and dusk. During the daytime, a 'dead-zone' of reduced response to light is observed. This periodic reduction of light sensitivity is known as 'gating'. The insensitivity to light allows the clock to continue to run through the day rather than continually being reset. Two components of the gating mechanism have been identified in *Arabidopsis*, EARLY FLOWERING 3 (ELF3) (McWatters *et al.*, 2000; Covington *et al.*, 2001) and TIME FOR COFFEE (TIC) (Hall *et al.*, 2003). These components act at the times when the clock is most sensitive to light, around dawn and dusk, to prevent excessive sensitivity to light. ELF3 is necessary for the plant to pass through the light to dark transition at dusk without a loss of rhythmicity, whilst TIC allows the plant to pass through the dark to light transition. The molecular nature of the action of these components is uncertain, although ELF3 has been shown to directly interact with the phytochrome, phyB (Figure 8.2). Nonetheless, lesions in either component result in arrhythmicity in constant light and a failure of the clock to anticipate the light/dark transitions of a light/dark cycle.

One additional factor was recently added to the clock loop. LUX ARRHYTHMO (LUX) is another single-myb transcription factor necessary for oscillation of the central clock components. In wild-type seedlings, *LUX* transcript levels oscillate in phase with those of *TOC1*. In a *lux* mutant, as in a *toc1* null mutant, levels of *CCA1* and *LHY* expression are clamped low and, conversely, in a *cca1 lhy* double mutant, *LUX* expression, like *TOC1* expression, is constitutively high. LUX, therefore, appears to act in the same way as TOC1 in the clock loop. Indeed, *LUX* expression, like *TOC1* expression, is suppressed by CCA1 and LHY, which bind to an evening element in the *LUX* promoter (Onai *et al.*, 1998; Hazen *et al.*, 2005). It is unclear whether LUX acts completely independent of TOC1, forming a second interlocked loop, equally necessary for circadian rhythmicity, or whether LUX may act with TOC1, possibly as part of a complex (Figure 8.2).

A number of such additional interlocked loops have been proposed. In each case, it is unclear whether their components do, in fact, form a second loop, fine-tuning the LHY/CCA1/TOC1 loop, or whether they may fit into the LHY/CCA1/TOC1 loop itself. TOC1 is part of a five-member PSEUDO RESPONSE REGULATOR (PRR) gene family (Matsushika *et al.*, 2000). Levels of *PRR9*, *PRR7*, *PRR5*, *PRR3* and *PRR1* (*TOC1*) transcripts each oscillate with a circadian rhythm, the levels of each transcript peaking in that order at approximately 2–3 h intervals beginning at dawn. Mutations in any one of the five PRRs result in mild perturbation of the period of the circadian rhythm but the additive effect of these mutations is more dramatic (Mizuno and Nakamichi, 2005). For example, the *prr9* mutant shows a slightly long period length in constant light, whilst the *prr9 prr7* double mutant shows a very long period and the *prr9 prr7 prr5* triple mutant is arrhythmic in both light and dark (Farre *et al.*, 2005; Nakamichi *et al.*, 2005). The expression of *CCA1* is constitutively derepressed in the *prr9 prr7 prr5* triple mutant, whereas the expression of TOC1 is severely attenuated (Nakamichi *et al.*, 2005). Farre *et al.* (2005) recently demonstrated that CCA1 and LHY had a positive effect on *PRR7* and *PRR9* expression levels. Furthermore, CCA1 binds to the promoters of *PRR9* and *PRR7* highlighting a direct link to the LHY/CCA1/TOC1 loop (Figure 8.2). Curiously, PRR9 and PRR7 appear to be involved in the transmission of light signals to the clock as well as in the regulation of the central oscillator. The phenotypes of the *prr9* and *prr7* monogenic mutants are light dependent (Farre *et al.*, 2005). Furthermore, *PRR9* expression is rapidly and transiently induced by light, dependent on phytochrome action (Makino *et al.*, 2001; Ito *et al.*, 2005). The PRR9/PRR7/PRR5 circuitry might serve as a pacemaker that finely tunes the periods of rhythms by either shortening or lengthening depending on certain conditions (Mizuno and Nakamichi, 2005).

Another interlocked feedback loop capable of maintaining circadian expression of TOC1 was also predicted in order to explain the persistence of a rhythm in the *cca1 lhy* double mutant. In this loop, it is proposed that expression of a 'Factor Y' is repressed by TOC1 and that in turn 'Factor Y' would act to promote transcription of *TOC1*. In order for the model to fit experimental data, expression of 'Factor Y' has been proposed to be suppressed by CCA1/LHY (Locke *et al.*, 2005). A candidate for 'Factor Y' has emerged in GIGANTEA (GI) (Locke *et al.*, 2005). Its pattern of expression fits well to that proposed for 'Factor Y' and expression of *GI* has been demonstrated to be suppressed to a constant low level by overexpression of either TOC1 (Makino *et al.*, 2002) or LHY (Fowler *et al.*, 1999) (Figure 8.2). Furthermore, loss-of-function *gi* mutants show severely disrupted circadian rhythmicity (Locke *et al.*, 2005) (Figure 8.2). Such secondary feedback loops have already been shown to exist in animals and are believed to improve the robustness of the circadian system.

8.3.1.3   The coincidence model

The rhythm of responsiveness to night breaks in both long- and short-day plants suggests a circadian clock regulation of night break sensitivity. A key component in this process in *Arabidopsis* is CONSTANS (CO). In the light, CO has been shown to directly induce expression of the floral meristem identity genes, *FT*, *SOC* and *LFY*. The *co* mutant shows an insensitivity to long days for the induction of flowering. Conversely, a CO overexpressor shows a constitutively early flowering phenotype. However, it is not merely the presence of CO protein that stimulates flowering. Expression of *CO* and accumulation of CO protein shows a clear circadian rhythm peaking during the subjective night (Suarez-Lopez *et al.*, 2001). This rhythm can be observed in both long and short days demonstrating that some additional signal specific to long days must also play a role. It was proposed that CO could form a 'coincidence' mechanism whereby light incident upon CO at a time of high *CO* expression could activate it allowing it to induce flowering. The phase of the rhythm of *CO* expression is such that it is high at times of sensitivity to light for the regulation of flowering. In long days, the light will still be incident upon the plant at the time when the level of CO protein is rising. Conversely, in short days the plant will already be in darkness at this time. As a consequence long days would activate CO but short days would not. Such a system is perhaps more correctly called the external coincidence model (Bünning, 1960; Pittendrigh and Minis, 1964; Thomas and Vince-Prue, 1997). It relies on a coincidence between an internal factor and an external stimulus.

The identification of the photoreceptors involved in the photoperiodic response in *Arabidopsis* paved the way for the verification of this model. Two photoreceptors, phytochrome A (phyA) and cryptochrome 2 (cry2), are responsible for the detection of long days. Mutants deficient in either phyA or cry2 show a late-flowering phenotype specifically in long days. The *cry2* mutant is almost completely day-length insensitive when grown under white light (Guo *et al.*, 1998). The *phyA* mutant shows a more subtle phenotype under these conditions but shows a pronounced deficiency in the detection of low-intensity incandescent day extensions where short days of standard white light are extended using low-intensity incandescent light. These extensions are sufficiently low intensity to have no significant photosynthetic impact but the red and far-red rich wavelengths strongly activate phyA. In wild-type plants this leads to a pronounced acceleration in flowering time (Johnson *et al.*, 1994). The involvement of phyA in the perception of red/far-red light in the photoperiodic induction of flowering in *Arabidopsis* explains the requirement for prolonged night breaks (Carre, 1998). PhyA commonly acts via a response mode known as the high irradiance response characterised by a requirement for prolonged irradiation (see Chapter 1).

Yanovsky and Kay (2002) demonstrated that phyA and cry2 activate expression of the floral meristem identity gene, *FT*, under long days but not short days. Furthermore, they showed that this effect required high levels of *CO* expression verifying the external coincidence model for the induction of flowering in *Arabidopsis*. Valverde *et al.* (2004) subsequently demonstrated a mechanism by which light signals interact with CO. They showed that the CO protein is unstable in the dark and is consequently degraded under these conditions. The action of light through phyA and cry2 is capable of stabilising the CO protein allowing it to accumulate to sufficient levels for induction of *FT* expression.

FKF1, another member of the FKF family that also contains ZTL, is crucial in controlling the precise waveform of *CO* expression and CO protein abundance. FKF1 is necessary for the accumulation of *CO* transcripts during the afternoon (Imaizumi *et al.*, 2003). FKF1 interacts with a Dof transcription factor, CYCLING DOF FACTOR 1 (CDF1), which binds to the *CO* promoter. Plants with elevated levels of CDF1 flower late and have reduced expression of *CO* demonstrating that CDF1 acts as a suppressor of *CO* expression. CDF1 protein is more stable in *fkf1* mutants indicating that FKF1 controls its stability (Imaizumi *et al.*, 2003). Expression of *FKF1* peaks in the evening consistent with its demonstrated role at this time of day. An additional level of complexity is provided by the fact that light also regulates the expression of the *CO* gene in itself. The action requires blue light and it has been demonstrated that FKF1, itself, is the photoreceptor in this response. The eponymous flavin-binding domain of FKF1 has been demonstrated to directly bind a flavin mononucleotide chromophore, causing the protein to act as a blue-light photoreceptor. It is proposed that the F-box domain of FKF1 may target CDF1 for degradation in response to blue light. (Imaizumi *et al.*, 2003). Hence, the circadian control of *FKF1* expression and the light regulation of FKF1 function appear to coincide to control the daytime CO waveform (Figure 8.2).

Curiously, ZTL itself appears to have the opposite effect on *CO* expression. It down-regulates transcription of *CO* to delay flowering (Somers *et al.*, 2004). The *ztl-1* mutant has only a modest effect on flowering (Somers *et al.*, 2000) but the effect of ZTL on *CO* expression is observed in plants overexpressing ZTL. ZTL overexpression results in down-regulation of *CO* transcript and flowering time is delayed in direct proportion to the level of ZTL (Somers *et al.*, 2004).

8.3.1.4   *Flowering time mutants of* Arabidopsis

As with the defects in the light responsive regulators of CO, mutations affecting the circadian clock-associated regulators of CO often result in a reduction in photoperiod sensitivity. In fact, a number of clock components were originally identified courtesy of a mutation that resulted in a flowering time defect. The *lhy-1* mutation in *Arabidopsis*, which results in a constitutively high level of expression of *LHY* causing arrhythmicity, is one example. As the gene name suggests, *lhy-1* was identified as having a photoperiod insensitive late-flowering phenotype (Schaffer *et al.*, 1998).

Loss-of-function mutants of *LHY* or *CCA1* both exhibit a shortening of the circadian period length (Green and Tobin, 1999; Mizoguchi *et al.*, 2002). Both also show an early flowering phenotype in short days (Mizoguchi *et al.*, 2002). CCA1 and LHY act redundantly in the circadian clock and, as predicted, the *lhy cca1* double loss-of-function mutant is arrhythmic in constant light. However, under light/dark cycles, these *lhy cca1* plants do show a diurnal rhythm with dramatically earlier phases of expression of the clock-associated genes *GI* and *TOC1*. The combination of these mutations also has an additive effect on photoperiodic induction of flowering, causing very early flowering specific to short days (Mizoguchi *et al.*, 2002). *CO* expression also shows an early phase in the *cca1 lhy* double-null mutant (Mizoguchi *et al.*, 2005). This early phase means that the rise in *CO* expression now coincides with the presence of light towards the end of a short day triggering flowering.

By contrast, loss of the clock component *gi* results in a day-length-insensitive late-flowering phenotype (Fowler *et al.*, 1999). The late-flowering phenotype of the *gi-3* mutation is epistatic to the early flowering phenotype observed in the *lhy cca1* double-null mutant under short days. The absence of GI causes a reduction in *CO* expression in this *lhy cca1* double-null mutant background. Furthermore, overexpression of GI results in a dramatic early flowering phenotype under all conditions and enhanced expression of *CO* and *FT*, despite delaying circadian phase (Mizoguchi *et al.*, 2005). It is concluded that the role of GI in the photoperiodic regulation of flowering is not restricted to its role in the circadian clock. GI appears to be more-directly involved in regulating the expression of *CO*, seemingly providing a link between the clock and *CO* expression (Figure 8.2). It is interesting to note that *GI* expression is also shifted to an earlier phase in *lhy cca1* double-null mutants under short days and is required for the expression of *CO*, thus supporting this model. GI does not promote CO expression and flowering by activating *FKF1* transcription, however. GI is not required to activate *FKF1* expression, and in GI overexpressing plants, *FKF1* mRNA expression is not increased (Mizoguchi *et al.*, 2005).

It should also be noted that the delay in flowering of *lhy cca1* double-null mutants caused by loss of CO is weaker than that caused by the loss of GI, suggesting that besides promoting flowering by activating CO and FT, GI can promote flowering independently of these genes. Similarly, loss of CO or FT only partially suppresses the early flowering phenotype resulting from overexpression of GI (Mizoguchi *et al.*, 2005).

The *toc1-1* mutation results in a short-period phenotype of about 21 h. The mutant displays a severely diminished responsivity to photoperiod, being early flowering in short days consistent with an early onset of CO transcription (Somers *et al.*, 1998b). In an elegant experiment, Strayer *et al.* (2000) demonstrated that *toc1-1* regains photoperiod sensitivity when the total day length is shortened to 21 h to match the endogenous period length of the *toc1-1* mutant. Thus, *toc1-1* can distinguish short-day/long-night cycles made up of 14-h light/7-h darkness from long-day/short-night cycles of 7-h light/14-h darkness.

There is not always such an obvious correlation between the period length of the clock and the resulting flowering phenotype, however. Both the long-period *prr9 prr7* double mutant and the short-period *prr7 prr5* double mutant show a late-flowering phenotype in long days. The PRR family proteins have also been shown to be involved in light signalling and it is possible that they may in some way affect a more direct light-regulated regulation of flowering through the shade-avoidance pathway (see below).

The *elf3* gating mutant is arrhythmic in constant white light but is constitutively early flowering. It is proposed that the clock is stopped at a phase at which transcription is activated in this mutant (McWatters *et al.*, 2000). ELF3 has been shown to negatively regulate *GI*, *CO* and *FT* transcript levels as the expression of all three genes is increased in the *elf3* mutant. Like the *gi* mutant, the *elf3* mutant also points to a CO-independent mechanism of regulation of flowering. The *elf3 co* double mutant flowers much earlier in long days than the *co* monogenic mutant, although *FT* message levels remain very low (Kim *et al.*, 2005). It is possible that this represents the action of GI discussed earlier.

8.3.1.5   Application to other species

The involvement of CO in an external coincidence model for the regulation of flowering appears to be conserved in other species (Simpson, 2003). Orthologues of the phytochrome and cryptochrome photoreceptors and the clock-associated genes, as well as *GI*, *CO* and *FT*, have been found in a number of species (Liu *et al.*, 2001; Hayama *et al.*, 2003; Nemoto *et al.*, 2003; Boxall *et al.*, 2005; Hecht *et al.*, 2005; Lariguet and Dunand, 2005). In rice, promotion of flowering occurs in response to short days. Nonetheless, the same network of factors appear to be involved in controlling this response (Yano *et al.*, 2000; Hayama *et al.*, 2002; Kojima *et al.*, 2002), although one key element in the way that the components interact is reversed to confer inhibition of flowering in long days rather than promotion. The rice orthologue of CO has been demonstrated to be required for the suppression of flowering under long-day conditions (Yano *et al.*, 2000), as opposed to the promotion of flowering seen in these conditions in long-day plants. Other components appear to act in exactly the same way as in *Arabidopsis*. The rice orthologue of FT has been shown to activate flowering in rice (Kojima *et al.*, 2002). Hayama *et al.* (2002) demonstrated that expression of the rice *GI* (*OsGI*) is circadian controlled and that its temporal expression pattern is very similar to that of *Arabidopsis GI* under both short-day and long-day conditions. In wild-type rice plants, *GI* expression peaks just before the dark period. The levels of rice *CO* mRNA also show circadian rhythms under short-day and long-day conditions, expression being high at night and low during the day just as in *Arabidopsis*. Under long-day conditions, the rice *CO* mRNA level is relatively high at dawn. In contrast, under short-day conditions, the rice *CO* mRNA level at dawn is much lower than that under long-day conditions. As a consequence, light would only be incident upon CO in long days (Hayama *et al.*, 2003). Just as in *Arabidopsis*, rice *FT* mRNA shows striking differences between long-day and short-day conditions in wild-type rice plants. It is strongly suppressed under long-day conditions but shows a diurnal rhythm under short-day conditions; its level is high during the day, becoming low during the night. Finally, when *OsGI* is overexpressed in rice, the result is late flowering under both long-day and short-day conditions. Overexpression of *OsGI* in rice results in an increase in expression of the rice *CO* orthologue, just as in *Arabidopsis*. However, expression of the rice *FT* orthologue is suppressed in these plants. It appears that there is a simple reversal in the way that CO regulates FT in rice: CO acts to suppress FT expression (Hayama *et al.*, 2003).

GI has even been shown to be involved in the short day-induced photoperiod-regulation of tuberisation in potato (Martinez-Garcia *et al.*, 2002) demonstrating that this mechanism appears to be involved in other photoperiodic responses in addition to flowering.

There is also a slight difference in the photoreceptors detecting light in the external coincidence model observed in short-day plants. Light-stable phytochromes (as opposed to light labile phyA) play a key role in the perception of the night break. A short, red light night break is sufficient to inhibit flowering in a short-day plant such as *Xanthium strumarium*. A clear red/far-red reversibility is also exhibited here, with far-red light acting to negate the effect of a red light night break if given immediately afterwards (Hendricks and Siegelman, 2006). This requirement for a relatively short pulse of light and the demonstration of red/far-red reversibility is indicative of a classical 'low fluence response' mediated by the light-stable phytochromes (see Chapter 1). Red light causes the production of an active Pfr form of phytochrome that inhibits flowering, presumably by stabilising the CO protein, whilst far-red light causes the reversion of this Pfr to the inactive Pr form. Although light stable, these phytochromes exhibit a gradual dark reversion of Pfr to Pr (Eichenberg *et al.*, 2000), hence the Pfr formed during the light period in a short day is not able to act to inhibit flowering during the crucial Pfr-sensitive phase of the clock during the subsequent long night. The involvement of light-stable phytochromes in the inhibition of flowering by photoperiod in short-day plants is supported by recent genetic evidence from the study of phytochrome-deficient mutant of rice (Izawa *et al.*, 2000).

8.3.1.6   Site of perception of photoperiodic stimulus

Treatment of a single leaf with inductive photoperiods can induce flowering in the meristem leading to the proposal that a floral stimulus named 'florigen' can be perceived in the leaf and subsequently transmitted through the plant to the meristem (Chailakhyan, 1936). Furthermore, this florigen is graft transmissible and can even promote flowering when an induced leaf is grafted onto another plant of a different species. Transmission has, in fact, been observed between long-day and short-day plant species (Zeevaart, 1976). Two pieces of evidence demonstrate that the signal is transmitted through the phloem. Girdling experiments in which the phloem is removed prevent the transmission of the signal. Likewise, the movement of florigen has been shown to correlate closely with the movement of radiolabelled photosynthetic assimilates from donor to recipient (Zeevaart, 1976).

It took nearly 70 years for any aspect of this florigen signal to be identified. Numerous detailed biochemical analyses of phloem exudate failed to provide a likely candidate molecule until Huang *et al.* (Huang *et al.*, 2005) identified at least one component as the mRNA of the floral meristem identity gene, *FT*. Several key pieces of evidence preceded this discovery. In *Arabidopsis*, *CO* has been demonstrated to be expressed mainly in the leaf where it activates *FT* expression in the leaf phloem. Likewise, the *FT* gene is expressed strongly in the leaf under inductive conditions but is not expressed in the meristem (Takada and Goto, 2003; An *et al.*, 2004). Ectopic expression of *CO* in the leaf or in the phloem tissue was demonstrated to be able to induce *FT* expression in the phloem and to rescue a *co* mutant. However, expression of *CO* specifically in the meristem is unable to trigger *FT* expression in the phloem and does not induce flowering (An *et al.*, 2004; Ayre and Turgeon, 2004). Finally, ectopic expression of *FT* in the meristem is sufficient to induce flowering (An *et al.*, 2004). Huang *et al.* (2005) reasoned that the transmissible floral signal may be either the mRNA or the protein of *FT*. They tested plants in which expression of an introduced *FT* gene was specifically induced in the leaf and looked for the accumulation of *FT* mRNA in the shoot apical meristem. They found that *FT* mRNA did indeed accumulate in the shoot apex a few hours after induction of *FT* expression in the leaf and that flowering was induced as a result of this treatment. Furthermore, courtesy of a slight difference between their leaf-induced FT transcript and the endogenous *FT* mRNA, they were able to demonstrate the *FT* mRNA accumulating in the apex originated from their introduced, leaf-expressed gene. Interestingly, they also demonstrated that this increase in *FT* expression formed a part of a self-propagating stimulus triggering expression of the endogenous *FT* gene. This is consistent with a number of observations from other species showing that the florigen stimulus is self-propagating such that a plant induced to flower as a result of a graft can itself be used as a donor of the florigen signal in a subsequent graft (Zeevaart, 1976). It remains to be seen whether FT protein also plays a part in the signal or, indeed, whether other factors may also contribute.

8.3.2   Shade avoidance

Discussion of the photocontrol of flowering is often limited to photoperiodism. However, a second way in which light regulates flowering can be seen in the phenomenon of shade avoidance. Shade avoidance is discussed in Chapter 9. The response is mediated as a result of changes in light quality due to the reflection of light from neighbouring plants in close proximity. A preferential absorption of red and blue light by chlorophyll leads to a depletion in these wavelengths in light reflected from or transmitted through green vegetation. The reflected light that we see is rich in green wavelengths but is also very much enriched in the far-red region of the spectrum. This low red:far-red light has a dramatic effect on the phytochrome photoequilibrium within a plant, converting phytochrome from the active Pfr form to the inactive Pr form, and this results in a series of 'avoidance' effects including increased elongation growth, increased apical dominance and, eventually, acceleration of flowering. It is proposed that the acceleration of flowering as a result of prolonged shade is, in effect, a stress response triggering production of seeds as a way of allowing offspring to survive in a harsh environment until conditions improve. In terms of the magnitude of the response, the acceleration of flowering due to shade is far more dramatic than that due to photoperiod.

The shade-avoidance response is primarily mediated by the light-stable phytochromes: phyB plays the major role in *Arabidopsis*, whilst phyD and phyE have been shown to affect specific aspects of the response, often acting redundantly with phyB (Devlin *et al.*, 1998, 1999). PhyA has a moderating function (Yanovsky *et al.*, 1995; Devlin *et al.*, 1996). All of these phytochromes affect the flowering response. PhyB Pfr exerts an inhibitory effect on elongation growth and flowering and it is the loss of this Pfr in low red:far-red light that triggers shade avoidance. The inhibitory effect of phyB Pfr on flowering is nicely demonstrated when *Arabidopsis* seedlings are germinated and grown in constant blue light. In the absence of red light, very little Pfr is formed and as a result flowering is very rapid (Guo *et al.*, 1998). In constant red light, in contrast, *Arabidopsis* seedlings flower much later. The *phyB* mutant displays a constitutive shade-avoiding phenotype (Nagatani *et al.*, 1991). It flowers equally fast if it is grown in constant red or constant blue light (Guo *et al.*, 1998). The *cry2* mutant flowers late in constant red light but in constant blue light it behaves just as wild type, flowering early due to the lack of Pfr. The *phyB cry2* double mutant flowers as early as the *phyB* monogenic mutant in both red and blue light, demonstrating that the constitutively shade-avoiding phenotype of the *phyB* mutation is epistatic to a photoperiod-insensitive late-flowering phenotype of the *cry2* mutation (Mockler *et al.*, 2006). Similar experiments where seedlings were grown in high versus low red:far-red ratio white light also demonstrate the fact that the shade-avoidance response is epistatic to the late flowering of the *cry2* mutant. The *cry2* mutant is no longer late flowering under low red:far-red ratio white light (Mas *et al.*, 2000). The autonomous pathway mutant, *fca*, can also be rescued by shade treatment demonstrating that the shade avoidance is also able to supersede regulation of flowering by this pathway too (Bagnall, 1993).

The regulation of flowering by shade also acts through the floral meristem identity genes, *FT* and *LFY*. Expression of *FT* and *LFY* is up-regulated in response to treatments that simulate vegetative shade (Hempel *et al.*, 1997; Devlin *et al.*, 2003). In addition, expression of *FT* and *LFY* is constitutively high in the early-flowering *phyB* mutant (Blazquez and Weigel, 1999; Cerdan and Chory, 2003; Halliday *et al.*, 2003). This regulation of *FT* expression by phyB does not involve CO since *CO* mRNA levels do not correlate with flowering time in the *phyB* mutant (Cerdan and Chory, 2003; Halliday *et al.*, 2003). *pft1* is the one mutant that has been identified as specifically acting in this pathway. The *pft1* mutation rescues the early flowering phenotype of the *phyB* mutation in white light. Furthermore, seedlings of *pft1* also show no acceleration of flowering or increase in *FT* transcript levels in response to simulated shade implicating PFT1 as a positive regulator of *FT*, acting in response to removal of phyB Pfr (Cerdan and Chory, 2003).

Curiously, the action of phyB in the regulation of flowering time by shade shows a marked temperature dependency (see Chapter 10). The early flowering phenotype of the *phyB* mutant that is apparent at 22°C is not observed in plants grown at

16°C (Halliday *et al.*, 2003). However, wild-type plants still show a pronounced acceleration of flowering in response to shade at 16°C demonstrating that other phytochromes continue to act at this temperature. PhyE appears to play a much more prominent role in the regulation of flowering by shade at lower temperatures. This phyE action was also shown to occur via the regulation of *FT* expression (Halliday and Whitelam, 2003).

The effectiveness of low red:far-red ratio light in inducing flowering is also dependent on the time of the day. Induction of flowering in *Arabidopsis* by daily 4-h pulses of low red:far-red ratio light is much more effective if pulses are given during the late part of the subjective night (Deitzer, 1984). This appears to be part of a wider circadian regulation of shade-avoidance responses as recently demonstrated by Salter and coworkers (Salter *et al.*, 2003) that is consistent with the gating of photoreceptor action by the clock discussed earlier (Millar and Kay, 1996).

8.3.3 Vernalisation

Vernalisation refers to the process by which winter annual or biennial species are induced to flower following prolonged exposure to low temperature (Henderson and Dean, 2004). Many such species have an absolute requirement for vernalisation in order for flowering to be triggered. This treatment mimics the conditions experienced during winter and this requirement is essentially part of a timing mechanism that ensures that a plant flowers in the spring, avoiding flowering during harsh conditions. Vernalisation acts to cause a permanent epigenetic repression of the floral repressor, *FLC* (Michaels and Amasino, 1999; Sheldon *et al.*, 1999). FLC represses expression of the floral meristem identity genes, *FT* and *SOC1*, as part of the autonomous pathway discussed earlier (Kardailsky *et al.*, 1999; Borner *et al.*, 2000) (Figure 8.1). Indeed, the autonomous pathway mutant, *fca*, can be rescued by vernalisation treatment (Bagnall, 1993).

8.4 Convergence of the flowering pathways

The floral meristem identity genes, *LFY*, *FT* and *SOC1*, form the key triggers of flowering induced by the floral promotion pathways. *LFY* is induced by the photoperiodic, gibberellin-dependent and the shade-avoidance pathways. *FT* is induced by the photoperiodic, the autonomous/vernalisation and the shade-avoidance pathways. *SOC1* is induced by the photoperiodic and the autonomous/vernalisation pathways (Figure 8.3). The fact that these genes are common to more than one pathway implicates them as points of convergence of the floral promotion pathways. Indeed, key pieces of work involving *Arabidopsis* have indicated that, at least in two cases, the site of these convergences is the promoter of the floral meristem identity gene. The convergence of the autonomous/vernalisation and photoperiodic pathways was investigated by Samach *et al.* (2000). Given the fact that CO shows strong homology to a transcription factor (Putterill *et al.*, 1995), Samach *et al.* looked for primary targets of CO. They engineered plants to express a fusion protein composed of CO

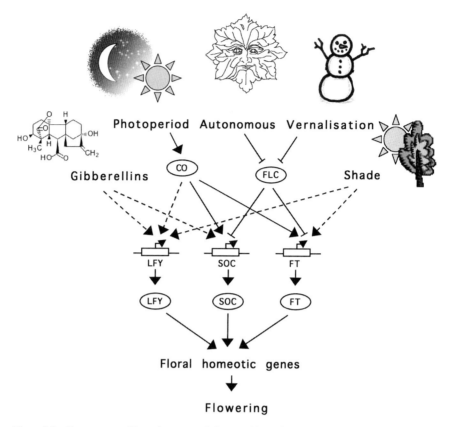

Figure 8.3   Convergence of the pathways regulating transition to flowering in *Arabidopsis*. Flowering is a result of signals from a number of internal and external cues that converge on the promoter regions of the floral meristem identity genes, *LEAFY* (*LFY*), *SUPPRESSOR OF CO* (*SOC*) and *FLOWERING LOCUS T* (*FT*). The gibberellin-dependent pathway regulates expression of *LFY* and *SOC*. The photoperiodic pathway regulates expression of *LFY*, *SOC* and *FT*. The autonomous/vernalisation pathway regulates expression of *SOC* and *FT*. The shade-avoidance pathway regulates expression of *LFY* and *FT*. Direct transcriptional activity of the factors CO and FLC is represented by solid lines. Indirect regulation by these factors is represented by dotted lines. LFY, SOC and FT regulate floral homeotic genes that directly control the production of flowers.

and the ligand-binding domain of the glucocorticoid receptor. Treatment with the steroid dexamethasone was used to direct the fusion protein to the nucleus where it activated the transcription of genes downstream of *CO* in the flowering pathway. By combining the fusion protein approach with the application of the protein synthesis inhibitor cycloheximide, Samach and co-workers were able to identify four target genes that are directly switched on by CO. Two of these were *FT* and *SOC1*. The authors were also able to demonstrate that FCA is required for full induction of *FT* and *SOC1* by CO meaning that the convergence of the autonomous/vernalisation and photoperiodic pathways is at the *CO* promoter itself. (It can also be concluded from this that *LFY* activation by CO is through an indirect route). Hepworth *et al.* (2002) subsequently carried out a deletion analysis of a *SOC1::GUS* reporter gene and identified a CArG-box (MADS-domain protein-binding element) within the *SOC1* promoter that was recognised by FLC, suggesting that FLC also binds directly to the *SOC1* promoter. The balance between regulation by the autonomous and photoperiodic pathways differs for *SOC1* and *FT*. Consistent with this, *soc1* mutants flower late under the influence of both long and short days, whereas *ft* mutants flower late in response to long days only (Figure 8.3).

The convergence of the gibberellin and photoperiodic pathways was investigated by Blazquez and Weigel (Blazquez *et al.*, 2002). Their approach involved carrying out deletion analysis of a *SOC1::GUS* reporter gene. They found that gibberellins activated the *LFY* promoter through *cis*-acting elements that differed from those inducing the day-length flowering response. An MYB transcription factor binding site was required for activation of *LFY* by the gibberellin-dependent pathway. When this MYB-deleted promoter was used to drive transcription of *LFY*, the construct was able to rescue an *lfy* mutant in long days but not in short days indicating that the photoperiodic pathway was still able to drive *LFY* expression through a distinct upstream sequence (Figure 8.3).

8.5   Conclusion

By being able to precisely regulate the timing of their flowering, plants gain a competitive advantage. The requirement for internal cues indicative of reaching a certain age or stage of development ensures that sufficient reserves are available. The requirement for external cues allows a plant avoid flowering in the unfavourable conditions of winter. Accelerated flowering in response to prolonged shade appears to supersede all of the above mechanisms of control. This may be regarded as a survival mechanism in an environment that may be so severe as to compromise the survival of the plant. Here energy is redirected to make reproduction a higher priority meaning that resilient structure is produced as a 'genetic lifeboat', ensuring that genes are passed on to the next generation when favourable conditions trigger germination. However, all of these mechanisms are essentially survival mechanisms when regarded from the point of view of the species. Survival of the fittest is not so much 'who survives' but 'how many descendants they get to have', and ensuring the optimum timing of flowering even under ideal conditions can increase this.

References

An, H., Roussot, C., Suarez-Lopez, P., Corbesier, L., *et al.* (2004) CONSTANS acts in the phloem to regulate a systemic signal that induces photoperiodic flowering of *Arabidopsis*. *Development* **131**, 3615–3626.

Ayre, B.G. and Turgeon, R. (2004) Graft transmission of a floral stimulant derived from CONSTANS. *Plant Physiol.* **135**, 2271–2278.

Bagnall, D.J. (1993) Light quality and vernalization interact in controlling late flowering in *Arabidopsis* ecotypes and mutants. *Ann. Bot.* **71**, 75–83.

Bernier, G. (1988) The control of floral evocation and morphogenesis. *Annu. Rev. Plant Physiol. Plant Mol. Biol.* **39**, 175–219.

Blazquez, M.A., Green, R., Nilsson, O., Sussman, M.R. and Weigel, D. (1998) Gibberellins promote flowering of *Arabidopsis* by activating the LEAFY promoter. *Plant Cell* **10**, 791–800.

Blazquez, M.A., Trenor, M. and Weigel, D. (2002) Independent control of gibberellin biosynthesis and flowering time by the circadian clock in *Arabidopsis*. *Plant Physiol.* **130**, 1770–1775.

Blazquez, M.A. and Weigel, D. (1999) Independent regulation of flowering by phytochrome B and gibberellins in *Arabidopsis*. *Plant Physiol.* **120**, 1025–1032.

Borner, R., Kampmann, G., Chandler, J., *et al.* (2000) A MADS domain gene involved in the transition to flowering in *Arabidopsis*. *Plant J.* **24**, 591–599.

Boxall, S.F., Foster, J.M., Bohnert, H.J., Cushman, J.C., Nimmo, H.G. and Hartwell, J. (2005) Conservation and divergence of circadian clock operation in a stress-inducible Crassulacean acid metabolism species reveals clock compensation against stress. *Plant Physiol.* **137**, 969–982.

Bünning, E. (1960) Biological clocks. Cold spring harbor symp. *Quant. Biol.* **15**, 1–9.

Carre, I.A. (1998) Genetic dissection of the photoperiod-sensing mechanism in the long day plant *Arabidopsis thaliana*. In: *Biological Rhythms and Photoperiodism in Plants* (eds Lumsden, P.J. and Millar, A.J.), pp. 257–270. BIOS Scientific, Oxford.

Cerdan, P.D. and Chory, J. (2003) Regulation of flowering time by light quality. *Nature* **423**, 881–885.

Chailakhyan, M.K. (1936) On the hormonal theory of plant development. *Dokl. Akad. Sci. SSSR* **12**, 443–447.

Coulter, M.W. and Hamner, K.C. (1964) Photoperiodic flowering response of biloxi soybean in 72 hour cycles. *Plant Physiol.* **39**, 848–856.

Covington, M.F., Panda, S., Liu, X.L., Strayer, C.A., Wagner, D.R. and Kay, S.A. (2001) ELF3 modulates resetting of the circadian clock in *Arabidopsis*. *Plant Cell* **13**, 1305–1315.

Darwin, C. (1859) *On the Origin of Species*. John Murray, London.

Deitzer, G.F. (1984) Photoperiodic induction in long-day plants. In: *Light and the Flowering Process* (eds Vince-Prue, D., Thomas, B. and Cockshull, K.E.), pp. 51–63. Academic Press, New York.

Devlin, P.F., Halliday, K.J., Harberd, N.P. and Whitelam, G.C. (1996) The rosette habit of *Arabidopsis thaliana* is dependent upon phytochrome action: novel phytochromes control internode elongation and flowering time. *Plant J.* **10**, 1127–1134.

Devlin, P.F. and Kay, S.A. (2000) Cryptochromes are required for phytochrome signaling to the circadian clock but not for rhythmicity. *Plant Cell* **12**, 2499–2510.

Devlin, P.F., Patel, S.R. and Whitelam, G.C. (1998) Phytochrome E influences internode elongation and flowering time in *Arabidopsis*. *Plant Cell* **10**, 1479–1487.

Devlin, P.F., Robson, P.R., Patel, S.R., Goosey, L., Sharrock, R.A. and Whitelam, G.C. (1999) Phytochrome D acts in the shade-avoidance syndrome in *Arabidopsis* by controlling elongation growth and flowering time. *Plant Physiol.* **119**, 909–915.

Devlin, P.F., Yanovsky, M.J. and Kay, S.A. (2003) A genomic analysis of the shade avoidance response in *Arabidopsis*. *Plant Physiol.* **133**, 1617–1629.

Dunlap, J.C. (1999) Molecular bases for circadian clocks. *Cell* **96**, 271–290.

Eichenberg, K., Baurle, I., Paulo, N., Sharrock, R.A., Rudiger, W. and Schafer, E. (2000) *Arabidopsis* phytochromes C and E have different spectral characteristics from those of phytochromes A and B. *FEBS Lett.* **470**, 107–112.

Farre, E.M., Harmer, S.L., Harmon, F.G., Yanovsky, M.J. and Kay, S.A. (2005) Overlapping and distinct roles of PRR7 and PRR9 in the *Arabidopsis* circadian clock. *Curr. Biol.* **15**, 47–54.

Fowler, S., Lee, K., Onouchi, H., *et al.* (1999) GIGANTEA: a circadian clock-controlled gene that regulates photoperiodic flowering in *Arabidopsis* and encodes a protein with several possible membrane-spanning domains. *EMBO J.* **18**, 4679–4688.

Green, R.M. and Tobin, E.M. (1999). Loss of the circadian clock-associated protein 1 in *Arabidopsis* results in altered clock-regulated gene expression. *Proc. Natl. Acad. Sci. USA* **96**, 4176–4179.

Guo, H.W., Yang, W.Y., Mockler, T.C. and Lin, C.T. (1998) Regulation of flowering time by *Arabidopsis* photoreceptors. *Science* **279**, 1360–1363.

Hall, A., Bastow, R.M., Davis, S.J., *et al.* (2003) The time for coffee (tic) gene maintains the amplitude and timing of *arabidopsis* circadian clocks. *Plant Cell* **15**, 2719–2729.

Halliday, K.J., Salter, M.G., Thingnaes, E. and Whitelam, G.C. (2003) Phytochrome control of flowering is temperature sensitive and correlates with expression of the floral integrator FT. *Plant J.* **33**, 875–885.

Halliday, K.J. and Whitelam, G.C. (2003) Changes in photoperiod or temperature alter the functional relationships between phytochromes and reveal roles for phyD and phyE. *Plant Physiol.* **131**, 1913–1920.

Hayama, R., Izawa, T. and Shimamoto, K. (2002) Isolation of rice genes possibly involved in the photoperiodic control of flowering by a fluorescent differential display method. *Plant Cell Physiol.* **43**, 494–504.

Hayama, R., Yokoi, S., Tamaki, S., Yano, M. and Shimamoto, K. (2003) Adaptation of photoperiodic control pathways produces short-day flowering in rice. *Nature* **422**, 719–722.

Hazen, S.P., Schultz, T.F., Pruneda-Paz, J.L., *et al.* (2005) LUX ARRHYTHMO encodes a Myb domain protein essential for circadian rhythms. *Proc. Natl. Acad. Sci. USA* **102**, 10387–10392.

Hecht, V., Foucher, F., Ferrandiz, C., *et al.* (2005) Conservation of *Arabidopsis* flowering genes in model legumes. *Plant Physiol.* **137**, 1420–1434.

Hempel, F.D., Weigel, D., Mandel, M.A., *et al.* (1997) Floral determination and expression of floral regulatory genes in *Arabidopsis*. *Development* **124**, 3845–3853.

Henderson, I.R. and Dean, C. (2004) Control of *Arabidopsis* flowering: the chill before the bloom. *Development* **131**, 3829–3838.

Hendricks, S.B. and Siegelman, H.W. (2006) Phytochrome and photoperiodism in plants. *Comp. Biochem.* **27**, 211–235.

Hepworth, S.R., Valverde, F., Ravenscroft, D., Mouradov, A. and Coupland, G. (2002) Antagonistic regulation of flowering-time gene SOC1 by CONSTANS and FLC via separate promoter motifs. *EMBO J.* **21**, 4327–4337.

Huang, T., Bohlenius, H., Eriksson, S., Parcy, F. and Nilsson, O. (2005) The mRNA of the *Arabidopsis* gene FT moves from leaf to shoot apex and induces flowering. *Science* **309**, 1694–1696.

Imaizumi, T., Tran, H.G., Swartz, T.E., Briggs, W.R. and Kay, S.A. (2003) FKF1 is essential for photoperiodic-specific light signalling in *Arabidopsis*. *Nature* **426**, 302–306.

Ito, S., Nakamichi, N., Matsushika, A., Fujimori, T., Yamashino, T. and Mizuno, T. (2005) Molecular dissection of the promoter of the light-induced and circadian-controlled APRR9 gene encoding a clock-associated component of *Arabidopsis thaliana*. *Biosci. Biotechnol. Biochem.* **69**, 382–390.

Izawa, T., Oikawa, T., Tokutomi, S., Okuno, K. and Shimamoto, K. (2000) Phytochromes confer the photoperiodic control of flowering in rice (a short-day plant). *Plant J.* **22**, 391–399.

Jacobsen, S.E. and Olszewski, N.E. (1993) Mutations at the SPINDLY locus of *Arabidopsis* alter gibberellin signal transduction. *Plant Cell* **5**, 887–896.

Johanson, U., West, J., Lister, C., Michaels, S., Amasino, R. and Dean, C. (2000) Molecular analysis of FRIGIDA, a major determinant of natural variation in *Arabidopsis* flowering time. *Science* **290**, 344–347.

Johnson, E., Bradley, M., Harberd, N.P. and Whitelam, G.C. (1994) Photoresponses of light-grown phyA mutants of *Arabidopsis*. Phytochrome A is required for the perception of daylength extensions. *Plant Physiol.* **105**, 141–149.

Kardailsky, I., Shukla, V.K., Ahn, J.H., *et al.* (1999) Activation tagging of the floral inducer FT. *Science* **286**, 1962–1965.

Kikis, E.A., Khanna, R. and Quail, P.H. (2005) ELF4 is a phytochrome-regulated component of a negative-feedback loop involving the central oscillator components CCA1 and LHY. *Plant J.* **44**, 300–313.

Kim, W.Y., Hicks, K.A. and Somers, D.E. (2005) Independent roles for early flowering 3 and zeitlupe in the control of circadian timing, hypocotyl length, and flowering time. *Plant Physiol.* **139**, 1557–1569.

Kim, J.Y., Song, H.R., Taylor, B.L. and Carre, I.A. (2003) Light-regulated translation mediates gated induction of the *Arabidopsis* clock protein LHY. *EMBO J.* **22**, 935–944.

Kojima, S., Takahashi, Y., Kobayashi, Y., *et al.* (2002) Hd3a, a rice ortholog of the *Arabidopsis* FT gene, promotes transition to flowering downstream of Hd1 under short-day conditions. *Plant Cell Physiol.* **43**, 1096–1105.

Koornneef, M., Hanhart, C.J. and Van Der Veen, J.H. (1991) A genetic and physiological analysis of late flowering mutants in *Arabidopsis thaliana. Mol. Gen. Genet.* **229**, 57–66.

Langridge, J. (1957) Effect of day-length and gibberellic acid on the flowering of *Arabidopsis. Nature* **180**, 36–37.

Lariguet, P. and Dunand, C. (2005) Plant photoreceptors: phylogenetic overview. *J. Mol. Evol.* **61**, 559–569.

Lee, H., Suh, S.S., Park, E., *et al.* (2000) The AGAMOUS-LIKE 20 MADS domain protein integrates floral inductive pathways in *Arabidopsis. Genes Dev.* **14**, 2366–2376.

Liu, J., Yu, J., McIntosh, L., Kende, H. and Zeevaart, J.A. (2001) Isolation of a CONSTANS ortholog from Pharbitis nil and its role in flowering. *Plant Physiol.* **125**, 1821–1830.

Locke, J.C., Southern, M.M., Kozma-Bognar, L., *et al.* (2005) Extension of a genetic network model by iterative experimentation and mathematical analysis. *Mol. Syst. Biol.* 28 June 2005 doi:10.1038/msb4100018 [online].

Macknight, R., Duroux, M., Laurie, R., Dijkwel, P., Simpson, G. and Dean, C. (2002) Functional significance of the alternative transcript processing of the *Arabidopsis* floral promoter FCA. *Plant Cell* **14**, 877–888.

Makino, S., Matsushika, A., Kojima, M., Oda, Y. and Mizuno, T. (2001) Light response of the circadian waves of the APRR1/TOC1 quintet: when does the quintet start singing rhythmically in *Arabidopsis? Plant Cell Physiol.* **42**, 334–339.

Makino, S., Matsushika, A., Kojima, M., Yamashino, T. and Mizuno, T. (2002) The APRR1/TOC1 quintet implicated in circadian rhythms of *Arabidopsis thaliana.* I. Characterization with APRR1-overexpressing plants. *Plant Cell Physiol.* **43**, 58–69.

Martinez-Garcia, J.F., Huq, E. and Quail, P.H. (2000) Direct targeting of light signals to a promoter element-bound transcription factor. *Science* **288**, 859–863.

Martinez-Garcia, J.F., Virgos-Soler, A. and Prat, S. (2002) Control of photoperiod-regulated tuberization in potato by the *Arabidopsis* flowering-time gene CONSTANS. *Proc. Natl. Acad. Sci. USA* **99**, 15211–15216.

Mas, P., Devlin, P.F., Panda, S. and Kay, S.A. (2000) Functional interaction of phytochrome B and cryptochrome 2. *Nature* **408**, 207–211.

Mas, P., Kim, W.Y., Somers, D.E. and Kay, S.A. (2003) Targeted degradation of TOC1 by ZTL modulates circadian function in *Arabidopsis thaliana. Nature* **426**, 567–570.

Matsushika, A., Makino, S., Kojima, M. and Mizuno, T. (2000) Circadian waves of expression of the APRR1/TOC1 family of pseudo-response regulators in *Arabidopsis thaliana*: insight into the plant circadian clock. *Plant Cell Physiol.* **41**, 1002–1012.

McDaniel, C.N., Singer, S.R. and Smith. S.M. (1992) Development states associated with the floral transition. *Dev. Biol.* **153**(1), 59–69.

McWatters, H.G., Bastow, R.M., Hall, A. and Millar, A.J. (2000) The ELF3 zeitnehmer regulates light signalling to the circadian clock. *Nature* **408**, 716–720.

Michaels, S.D. and Amasino, R.M. (1999) FLOWERING LOCUS C encodes a novel MADS domain protein that acts as a repressor of flowering. *Plant Cell* **11**, 949–956.

Millar, A.J. and Kay, S.A. (1996) Integration of circadian and phototransduction pathways in the network controlling *CAB* gene transcription in *Arabidopsis. Proc. Natl. Acad. Sci. USA* **93**, 15491–15496.

Millar, A.J., Straume, M., Chory, J., Chua, N.-H. and Kay, S.A. (1995) The regulation of circadian period by phototransduction pathways in *Arabidopsis. Science* **267**, 1163–1166.

Mizoguchi, T., Wheatley, K., Hanzawa, Y., *et al.* (2002) LHY and CCA1 are partially redundant genes required to maintain circadian rhythms in *Arabidopsis. Dev. Cell* **2**, 629–641.

Mizoguchi, T., Wright, L., Fujiwara, S., *et al.* (2005) Distinct roles of GIGANTEA in promoting flowering and regulating circadian rhythms in *Arabidopsis. Plant Cell* **17**, 2255 2270.

Mizuno, T. and Nakamichi, N. (2005) Pseudo-response regulators (PRRs) or true oscillator components (TOCs). *Plant Cell Physiol.* **46**, 677–685.

Mockler, T.C., Guo, H., Yang, H., Duong, H. and Lin, C. (2006) Antagonistic actions of *Arabidopsis* cryptochromes and phytochrome B in the regulation of floral induction. *Development* **126**, 2073–2082.

Moon, J., Suh, S.S., Lee, H., *et al.* (2003) The SOC1 MADS-box gene integrates vernalization and gibberellin signals for flowering in *Arabidopsis. Plant J.* **35**, 613–623.

Nagatani, A., Chory, J. and Furuya, M. (1991) Phytochrome B is not detectable in the *hy3* mutant of *Arabidopsis*, which is deficient in responding to end-of-day far-red light treatments. *Plant Cell Physiol.* **32**, 1119–1122.

Nakamichi, N., Kita, M., Ito, S., Yamashino, T. and Mizuno, T. (2005) PSEUDO-RESPONSE REGULATORS, PRR9, PRR7 and PRR5, together play essential roles close to the circadian clock of *Arabidopsis thaliana. Plant Cell Physiol.* **46**, 686–698.

Nemoto, Y., Kisaka, M., Fuse, T., Yano, M. and Ogihara, Y. (2003) Characterization and functional analysis of three wheat genes with homology to the CONSTANS flowering time gene in transgenic rice. *Plant J.* **36**, 82–93.

Onai, K., Katagiri, S., Akiyama, M. and Nakashima, H. (1998) Mutation of the gene for the second-largest subunit of RNA polymerase I prolongs the period length of the circadian conidiation rhythm in *Neurospora crassa. Mol. Gen. Genet.* **259**, 264–271.

Pittendrigh, C.S. and Minis, D.H. (1964) The entrainment of circadian oscillations by light and their role as photoperiodic clocks. *Amer. Nat.* **98**, 261–294.

Putterill, J., Robson, F., Lee, K., Simon, R. and Coupland, G. (1995) The *CONSTANS* gene of *Arabidopsis* promotes flowering and encodes a protein showing similarities to zinc finger transcription factors. *Cell* **80**, 847–857.

Quesada, V., Dean, C. and Simpson, G.G. (2005) Regulated RNA processing in the control of *Arabidopsis* flowering. *Int. J. Dev. Biol.* **49**, 773–780.

Quesada, V., Macknight, R., Dean, C. and Simpson, G.G. (2003) Autoregulation of FCA pre-mRNA processing controls *Arabidopsis* flowering time. *EMBO J.* **22**, 3142–3152.

Salisbury, F.B. (1963) Biological timing and hormone synthesis in flowering of Xanthium. *Planta* **49**, 518–524.

Salter, M.G., Franklin, K.A. and Whitelam, G.C. (2003) Gating of the rapid shade-avoidance response by the circadian clock in plants. *Nature* **426**, 680–683.

Samach, A., Onouchi, H., Gold, S.E., *et al.* (2000) Distinct roles of CONSTANS target genes in reproductive development of *Arabidopsis. Science* **288**, 1613–1616.

Schaffer, R., Ramsay, N., Samach, A., *et al.* (1998) The late elongated hypocotyl mutation of *Arabidopsis* disrupts circadian rhythms and the photoperiodic control of flowering. *Cell* **93**, 1219–1229.

Sheldon, C.C., Burn, J.E., Perez, P.P., *et al.* (1999) The FLF MADS box gene: a repressor of flowering in *Arabidopsis* regulated by vernalization and methylation. *Plant Cell* **11**, 445–458.

Simpson, G.G. (2003) Evolution of flowering in response to day length: flipping the CONSTANS switch. *Bioessays* **25**, 829–832.

Simpson, G.G. (2004) The autonomous pathway: epigenetic and post-transcriptional gene regulation in the control of *Arabidopsis* flowering time. *Curr. Opin. Plant Biol.* **7**, 570–574.

Somers, D.E., Devlin, P.F. and Kay, S.A. (1998a). Phytochromes and cryptochromes in the entrainment of the *Arabidopsis* circadian clock. *Science* **282**, 1488–1490.

Somers, D.E., Kim, W.Y. and Geng, R. (2004) The F-box protein ZEITLUPE confers dosage-dependent control on the circadian clock, photomorphogenesis, and flowering time. *Plant Cell* **16**, 769–782.

Somers, D.E., Schultz, T.F., Milnamow, M. and Kay, S.A. (2000) *ZEITLUPE*, a novel clock associated PAS protein from *Arabidopsis. Cell* **101**, 319–329.

Somers, D.E., Webb, A.A.R., Pearson, M. and Kay, S. (1998b). The short-period mutant, *toc1-1*, alters circadian clock regulation of multiple outputs throughout development in *Arabidopsis thaliana. Development* **125**, 485–494.

Strayer, C., Oyama, T., Schultz, T.F., *et al.* (2000) Cloning of the *Arabidopsis* clock gene TOC1, an autoregulatory response regulator homolog. *Science* **289**, 768–771.

Suarez-Lopez, P., Wheatley, K., Robson, F., Onouchi, H., Valverde, F. and Coupland, G. (2001) CONSTANS mediates between the circadian clock and the control of flowering in *Arabidopsis. Nature* **410**, 1116–1120.

Sun, T. and Kamiya, Y. (1994) The *Arabidopsis GA1* locus encodes the cyclase *ent*-kaurene synthetase A of gibberellin biosynthesis. *Plant Cell* **6**, 1509–1518.

Takada, S. and Goto, K. (2003) Terminal flower2, an *Arabidopsis* homolog of heterochromatin protein1, counteracts the activation of flowering locus T by constans in the vascular tissues of leaves to regulate flowering time. *Plant Cell* **15**, 2856–2865.

Thomas, B. and Vince-Prue, D. (1997) *Photoperiodism in Plants*, 2nd edn. Academic Press, London.

Valverde, F., Mouradov, A., Soppe, W., Ravenscroft, D., Samach, A. and Coupland, G. (2004) Photoreceptor regulation of CONSTANS protein in photoperiodic flowering. *Science* **303**, 1003–1006.

Vince-Prue, D. (1975) *Photoperiodism in Plants*. McGraw-Hill, London.

Weigel, D. and Nilsson, O. (1995) A developmental switch sufficient for flower initiation in diverse plants. *Nature* **377**, 495–500.

West, S.A., Lively, C.M. and Read, A.F. (1999) A pluralist approach to sex and recombination. *J. Evol. Biol.* **12**, 1003–1012.

Wilson, R.N., Heckman, J.W. and Somerville, C.R. (1992) Gibberellin is required for flowering in *Arabidopsis thaliana* under short days. *Plant Physiol.* **100**, 403–408.

Yano, M., Katayose, Y., Ashikari, M., *et al.* (2000) Hd1, a major photoperiod sensitivity quantitative trait locus in rice, is closely related to the *Arabidopsis* flowering time gene CONSTANS. *Plant Cell* **12**, 2473–2484.

Yanovsky, M.J., Casal, J.J. and Whitelam, G.C. (1995) Phytochrome A, phytochrome B and HY4 are involved in hypocotyl growth responses to natural radiation in *Arabidopsis*: weak de-etiolation of the *phyA* mutant under dense canopies. *Plant Cell Environ.* **18**, 788–794.

Yanovsky, M.J. and Kay, S.A. (2002) Molecular basis of seasonal time measurement in *Arabidopsis*. *Nature* **419**, 308–312.

Zeevaart, J.A.D. (1976) Physiology of flower formation. *Ann. Rev. Plant Physiol.* **27**, 321–348.

9 Red:far-red ratio perception and shade avoidance

Keara A. Franklin and Garry C. Whitelam

9.1 Introduction

Competition for light to drive photosynthesis is a key determinant regulating the growth of plants in crowded communities. In addition to providing energy, light signals convey important environmental information to plants, enabling both seasonal prediction and the determination of spatial orientation. Plants measure the quantity, quality and direction of incident light, using specialised photoreceptors, the red/far-red (R/FR) light-absorbing phytochromes and the UV-A/blue light-absorbing cryptochromes (Cashmore *et al.*, 1999) and phototropins (Briggs and Huala, 1999). The interaction of light signals with the endogenous circadian oscillator also provides plants with a means to monitor daylength (photoperiod). Taken together, this information can be used to direct developmental strategy, allowing the optimisation of morphological form and photosynthetic activity to the ambient surroundings. The effective perception, transduction and interpretation of light signals is therefore paramount to an individual's success in natural environments. In stands of mixed vegetation, competition for light requires developmental adaptation to either tolerate or avoid shading by neighbours. In plants displaying the latter strategy, alterations in both light quality and light quantity can invoke a suite of 'escape' responses, collectively termed shade avoidance.

9.2 Natural light environment

The natural light environment is variable with daily fluctuations in both quantity and spectral quality. Prior to reaching the earth's surface, solar radiation is attenuated within the atmosphere. Oxygen and water vapour strongly absorb longer wavelength radiation whereas shorter wavelength radiation is selectively attenuated by the ozone layer (Smith, 1975). Plants use the R/FR-reversible phytochrome family of photoreceptors to measure light quality. The proportion of R to FR wavelengths in a plant's ambient light environment can alter the balance of phytochrome molecules in their active and inactive forms (phytochrome photoequilibrium, see Section 9.4) and thereby determine phytochrome activity and physiological response. For this reason, a frequently used parameter to describe the spectral distribution of natural radiation is the ratio of spectral photon irradiance in the R region of the spectrum to that in the FR region. This is precisely defined as follows:

$$R:FR = \frac{\text{Photon irradiance between 660 and 670 nm}}{\text{Photon irradiance between 725 and 735 nm}}$$

The R:FR ratio of both direct sunlight and scattered daylight is typically around 1.15 and is not dramatically affected by weather conditions or season (Smith, 1982). However, as solar elevation decreases below $10°$, enhanced absorption, scattering and refraction of the solar beam in the atmosphere lead to an enhancement of longer wavelengths reaching the earth's surface. The onset of both dawn and dusk is therefore associated with a significant drop in R:FR ratio. On cloudless days, R:FR ratio at dusk has been recorded as low as 0.7 (Holmes and Smith, 1977a). An alteration in the R:FR ratio of daylight also occurs underwater. Water absorbs strongly in the FR and infrared regions of the spectrum, resulting in an increase in R:FR ratio with increasing depth (Smith, 1982).

Perhaps the greatest influencing factor affecting the spectral quality of light perceived by plants is the presence of neighbouring vegetation. The photosynthetic pigments, chlorophylls and carotenoids, absorb light over most of the visible spectrum. A proportion of green wavelengths is, however, reflected or transmitted, thereby making chlorophyllous vegetation appear green to the human eye. Radiation in the FR region of the spectrum is the most poorly absorbed waveband. Indeed, if it were not for the insensitivity of our visual systems to wavelengths beyond approximately 700 nm, leaves would appear FR in colour, not green! Light reflected from or transmitted through green tissues is depleted in R and enriched in FR wavelengths, resulting in a significantly reduced R:FR ratio when compared with daylight. The spectral energy distribution of daylight and reflected daylight is shown in Figure 9.1. These measurements were taken on a clear sunny day in Leicester, UK. Here, the R:FR ratio of daylight was recorded as 1.25 (Figure 9.1A). The spectral energy distribution of daylight reflected 10 mm from the edge of a stand of wheat seedlings is shown in Figure 9.1B. Here, the R:FR ratio decreased to 0.421. A similar alteration in light quality would be perceived by plants growing in close proximity to (but not directly shaded by) neighbouring vegetation. Reductions in R:FR ratio, perceived by the phytochromes, provide plants with an early and unambiguous warning that competitors are nearby. Furthermore, the extent of reduction in R:FR ratio is directly proportional to the density and proximity of neighbouring vegetation (Smith and Whitelam, 1997). Of course, when plants are subject to actual vegetational shading, light is both transmitted through and reflected within the canopy. This results not only in a reduced R:FR ratio, but also in a marked decrease in photosynthetically active radiation (PAR) reaching stems. Figure 9.2 shows the spectral energy distributions of daylight within the same stand of wheat seedlings. Readings were taken at the top, middle and bottom of the canopy and are represented in Figures 9.2A, 9.2B and 9.2C, respectively. The photon irradiance of PAR recorded at each position decreased dramatically with increasing canopy depth. At the top of the canopy, the PAR of transmitted daylight was reduced from 233 to 134 μmol m^{-2} s^{-1}. A lower PAR (16 μmol m^{-2} s^{-1}) was recorded in the middle

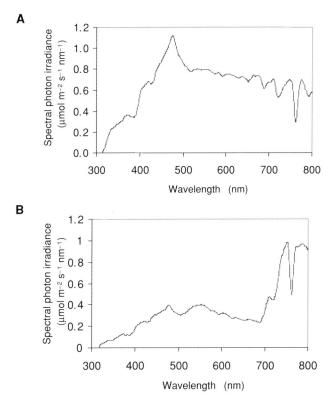

Figure 9.1   The spectral energy distribution of (A) daylight and (B) light reflected from a stand of wheat seedlings.

of the canopy with very low levels (0.1 μmol m^{-2} s^{-1}) being detected at the soil surface. The relative proportion of FR wavelengths behaved oppositely, resulting in a decrease in R:FR ratio with increasing canopy depth. Despite detecting very low amounts of visible radiation (Figures 9.2B and 9.2C), spectral measurements within the canopy revealed considerable levels of FR, lowering the R:FR ratio to below 0.1.

The detection of both light quality (R:FR ratio) and light quantity (in particular blue (400–500 nm) wavelengths) in different tissues provides plants with some mechanism to distinguish between the threat of shading (proximity perception) and actual shading (shade perception). The proximity of neighbouring vegetation is largely detected in stem tissue via the perception of horizontally reflected FR signals (Ballaré et al., 1990; Smith et al., 1990). During actual shading, a proportion of R and B wavelengths are absorbed by the shading canopy. Filtered, FR-enriched light is propagated downwards on to leaves before multiple scattering and reflection occur within the lower vegetational strata. In shade-tolerant species, energy-conserving slow growth rates are often accompanied by adaptations in photosynthetic structures to optimise efficiency at low light levels. Such adaptations

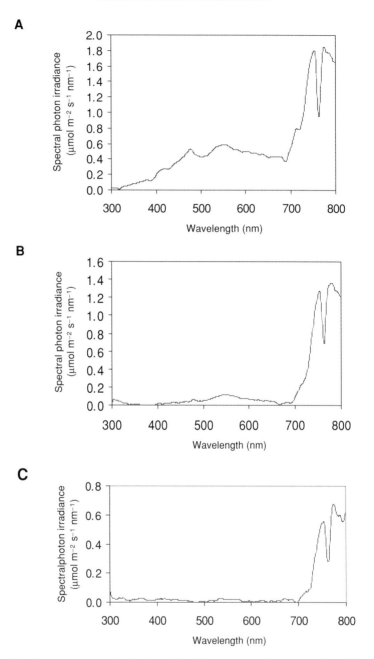

Figure 9.2   The spectral energy distribution of daylight recorded at the (A) top, (B) middle and (C) bottom of a stand of wheat seedlings.

include thinner leaves, higher chlorophyll content and lens-shaped epidermal cells to focus light within the mesophyll tissue (Boardman, 1977; Middleton, 2001). In shade-avoiding species, both proximity perception and shade perception can result in a plastic alteration in plant development, commonly termed the shade avoidance syndrome.

9.3   Shade avoidance syndrome

The term shade avoidance syndrome was first used, by Smith and colleagues in the 1970s, to describe a range of developmental traits observed in laboratory-grown plants subject to FR-enrichment of their light environment. The same group had previously established a link between R:FR ratio and the status of phytochrome photoequilibrium, using quantitative spectral measurements of natural radiation above and below vegetational canopies and spectrophotometric quantification of active phytochrome levels (Holmes and Smith, 1975, 1977b). Variations in R:FR ratio were related to estimated phytochrome photoequilibrium and were found to form a rectangular hyperbola. Reductions in R:FR ratio therefore correlated closely with decreases in the proportion of active phytochrome molecules. Variation in R:FR ratio was achieved in the laboratory by supplementing the output of white fluorescent tubes with FR light. In this way, PAR remained constant in both treatments and plant responses to reductions in R:FR ratio could be examined. In a series of physiological experiments by Morgan and Smith (1976, 1978, 1981), using *Sinapis alba* and *Chenopodium album*, a range of developmental responses were characterised which closely correlated with the responses of plants subject to actual shading in natural light environments. The principle traits observed in dicotyledonous species were an elongation of stems and petioles, increased apical dominance and early flowering – often at the expense of leaf and storage organ development. The elongation of stems is the most pronounced and easily observable phenotype of the shade avoidance syndrome and, as such, was used by Smith and colleagues in a pioneering study to examine the kinetics of R:FR ratio signalling. In this work, linear voltage displacement transducers were used to measure the real-time growth rate of *S. alba* seedlings in response to supplementary FR (Morgan *et al.*, 1980). Using fibre optic light guides to target supplementary FR to individual internodes, the authors reported an acceleration of growth rate after a lag phase of just 10 min. Moreover, an increase in growth rate of up to fivefold was recorded within 30 min of the addition of supplementary FR (Morgan *et al.*, 1980; Child and Smith, 1987). Such observations not only revealed the stem tissue of these plants to act as a site of R:FR ratio perception, but also provided an exciting insight into the remarkable rapidity of R:FR ratio signal transduction.

Many plants have also been shown to reorientate their leaves upwards in response to reductions in R:FR ratio (Whitelam and Johnson, 1982). This process, termed leaf hyponasty, presumably increases light capture in dense vegetational canopies. A reduction in leaf thickness and decrease in leaf chlorophyll content are also common phenotypes associated with shade avoidance (McLaren and Smith, 1978). In

Figure 9.3   The R:FR-mediated shade avoidance response. The appearance of (A) *Arabidopsis*, (B) *Vicia faba*, (C) sunflower and (D) radish plants grown in high and low R:FR ratio conditions. All plants were grown under white fluorescent light providing equal photosynthetically active radiation (400–700 nm). For each species, plants on the right received supplementary FR to reduce R:FR ratio.

monocotyledonous species, growth in low R:FR ratio leads to an elongation of leaves and increased apical dominance, manifested as a reduced number of tillers (Casal *et al.*, 1986; Barnes and Bugbee, 1991). Multiple species displaying shade avoidance phenotypes are shown in Figure 9.3. Petiole elongation and reduced leaf area are clearly visible in *Arabidopsis thaliana* plants grown in low R:FR ratio (Figure 9.3A). Pronounced stem elongation is the most characteristic phenotype displayed in low R:FR ratio-grown broad bean (*Vicia faba*) (Figure 9.3B) and sunflower (*Helianthus annuus*) (Figure 9.3C) plants. Both sunflower and radish (*Raphanus sativus*) (Figure 9.3D) plants grown in low R:FR ratio displayed reduced leaf area and chlorophyll content. Physiological adaptations to low R:FR ratio are accompanied by changes in the distribution of assimilates between leaves, stems and roots. Studies using radish reported leaves of plants grown in low R:FR ratio to contain more hexose sugar and less starch accumulation than high R:FR ratio-grown controls (Keiller and Smith, 1989). The shade avoidance syndrome is therefore often associated with a reduction in plant productivity as resources are reallocated towards the development of reproductive structures. Indeed, radish plants grown in low R:FR ratio light display a significant reduction in tuber size (Figure 9.3D). Decreases in the specific stem weight, leaf area and whole plant biomass of *Arabidopsis* and *Brassica rapa* have been recorded in laboratory-grown plants subject to reductions in R:FR ratio and end-of-day far-red (EOD-FR) treatments, respectively (Robson *et al.*, 1993; Devlin *et al.*, 1996, 1999). In plants grown in light/dark cycles, the latter mimic growth in low R:FR ratio by depleting active phytochrome levels prior to the onset of darkness. These studies also used mutants deficient in phytochrome B (phyB) grown in high R:FR ratio to mimic growth of wild-type plants in low R:FR (see Section 9.4). Surprisingly and in contrast to observations using wild-type plants grown in low R:FR ratio, Robson and colleagues reported phyB-deficient mutants of *Arabidopsis* grown in high R:FR to display an *increased* leaf area and therefore *increased* biomass than wild-type controls, yet offer no explanation for this apparent discrepancy (Robson *et al.*, 1993).

9.4   Phytochrome regulation of shade avoidance

Higher plants contain multiple phytochromes, the apoproteins of which are encoded by a family of divergent genes (Quail, 1994). Three major phytochrome types have been identified in angiosperms, phytochromes A, B and C (phyA, phyB and phyC), encoded by the *PHYA*, *PHYB* and *PHYC* genes, respectively. In the model plant species *Arabidopsis*, five genes (*PHYA–E*) have been sequenced and characterised (Sharrock and Quail, 1989; Clack *et al.*, 1994). The protein products of the *PHYB* and *PHYD* genes share the closest sequence similarity (~80%) and together are more related to *PHYE* (~55% identity) than to *PHYA* or *PHYC*. Phytochromes B, D and E are therefore considered to form a more recently diverged subgroup of the *Arabidopsis PHY* family (Goosey *et al.*, 1997). Phytochromes are synthesised in darkness in their inactive R light-absorbing (Pr) form. Upon transfer to light, photon absorption converts phytochrome molecules to their active FR light-absorbing (Pfr) form. This reaction occurs optimally at R wavelengths (~660 nm) and is reversed following photon absorption by Pfr molecules. The conversion of Pfr molecules back to the Pr form occurs optimally at wavelengths in the FR region of the spectrum (>700 nm). Phytochromes therefore exist in an equilibrium of Pr and Pfr forms in almost all natural irradiation conditions. Reductions in R:FR ratio favour the conversion of phytochrome molecules to their inactive Pr form. The 'shade avoidance syndrome' must therefore be suppressed under high R:FR ratio conditions. In this way, shade avoidance represents the relief of suppression rather than the induction of physiological responses.

In contrast to other family members, which display relative stability in the Pfr form, phyA undergoes rapid light-induced proteolysis and therefore accumulates to high levels only in etiolated seedlings (Quail *et al.*, 1973; Clough and Vierstra, 1997). Three distinct response modes of phytochrome action have been characterised and are determined by fluence requirements and R/FR reversibility. These are very low fluence response (VLFR), low fluence response (LFR) and high irradiance response (HIR) (reviewed in Schäfer and Bowler, 2002). VLFRs are mediated by phyA and are saturated by very low concentrations of Pfr, therefore preventing

R/FR reversibility. In contrast, LFRs display robust R/FR reversibility and are mediated by phytochromes stable in the Pfr form (phyB–phyE). HIRs require prolonged irradiation, do not display R/FR reversibility and are mediated by phyA. The reversibility of R:FR ratio perception therefore supports the involvement of light-stable phytochromes acting in the LFR mode.

9.4.1 The role of phytochromes B, D and E in R:FR ratio signalling

The isolation and characterisation of mutant plants, null for one or more phytochromes, has been paramount in elucidating the roles of individual family members in mediating shade avoidance responses. The suppression of shade avoidance by Pfr implies that mutants null for the relevant phytochromes should display constitutive shade avoidance phenotypes in high R:FR ratio conditions. Early analyses of the long hypocotyl (*lh*) mutant of cucumber revealed a number of aberrant responses to light (Adamse *et al.*, 1987). Seedlings grown in high R:FR ratio displayed increased stem elongation, reduced cotyledon expansion and decreased chlorophyll synthesis – phenotypes similar to those of wild-type plants grown in low R:FR ratio (Adamse *et al.*, 1987; López-Juez *et al.*, 1990; Ballaré *et al.*, 1991a). Immunochemical analysis of the mutant confirmed an absence of the phyB-like photoreceptor (Adamse *et al.*, 1988; López-Juez *et al.*, 1992). Mutants deficient in phyB have since been identified in a number of species including *Arabidopsis*, *B. rapa* and tomato and shown to display a range of phenotypes termed *constitutive shade avoidance*, including enhanced elongation of stems and petioles, decreased leaf and cotyledon expansion, reduced chlorophyll synthesis and early flowering when compared to wild-type controls (Somers *et al.*, 1991; Devlin *et al.*, 1992; Reed *et al.*, 1993). The developmental characteristics of *phyB* null mutants suggested phyB to be the predominant photoreceptor mediating R:FR ratio signalling in these plants. The participation of additional phytochromes was, however, implicated following observations of residual shade avoidance responses in *phyB* null mutants of multiple species subject to daytime reductions in R:FR ratio and EOD-FR treatments (Whitelam and Smith, 1991; Robson *et al.*, 1993; Halliday *et al.*, 1994; Devlin *et al.*, 1996). Further confirmation that phyB is not the sole (or even predominant) photoreceptor mediating shade avoidance responses in all species emerged following analysis of the *tri* mutant of tomato (Kendrick *et al.*, 1997). This mutant was shown to be deficient in a homologue of phyB (van Tuinen *et al.*, 1995; Kerckhoffs *et al.*, 1996), yet did not display the shade avoidance syndrome in high R:FR ratio light-grown plants. Furthermore, responses to reductions in R:FR ratio and EOD-FR treatments were similar to wild-type controls (Kerckhoffs *et al.*, 1992). The identities of the other phytochromes involved in mediating shade avoidance responses eventually emerged following the isolation of mutants, null for other family members.

Sequencing of *PHYD* in the Wassilewskija accession of *Arabidopsis* revealed a naturally occurring mutation in this gene, thus leading the role of this phytochrome in R:FR ratio signalling to be examined (Aukerman *et al.*, 1997). Hypocotyl elongation responses to EOD-FR treatments were similar in *phyD* mutants and controls containing an introgressed functional *PHYD* (Aukerman *et al.*, 1997). Light-grown double mutants deficient in phyB and phyD, however, displayed longer hypocotyls, longer petioles and earlier flowering than either monogenic mutant (Aukerman *et al.*, 1997; Devlin *et al.*, 1999). Such observations suggested redundancy of function between phyB and phyD in mediating the suppression of shade avoidance responses (Devlin *et al.*, 1999), a proposal supported by their sequence similarity and parallel patterns of gene expression (Goosey *et al.*, 1997; Mathews and Sharrock, 1997).

Further advancement in our understanding of shade avoidance occurred following the isolation of an *Arabidopsis* mutant deficient in the phyE photoreceptor. Double mutants deficient in phytochromes A and B have been shown to display internode extension between rosette leaves, following EOD-FR treatments (Devlin *et al.*, 1996). A phenotypic screen of mutagenised *phyAphyB* double mutants grown in high R:FR ratio revealed a plant displaying clearly visible internode growth between rosette leaves and early flowering. Molecular characterisation of the plant revealed a single base pair deletion at the *PHYE* locus (Devlin *et al.*, 1998). The phenotypic similarity between *phyAphyBphyE* triple mutants grown in high R:FR ratio and *phyAphyB* double mutants subject to EOD-FR treatments implicated a role for phyE in the regulation of these responses (Devlin *et al.*, 1998). Furthermore, deficiency of phyE resulted in an attenuation of elongation and flowering responses to EOD-FR treatment in *phyAphyB* plants (Devlin *et al.*, 1998). Similar to the *phyD* mutant, however, monogenic *phyE* plants displayed no obvious impairment of R:FR ratio signalling (Aukerman *et al.*, 1997; Devlin *et al.*, 1998). The isolation of the *phyE* mutant enabled the subsequent construction of an *Arabidopsis* triple mutant, deficient in phytochromes B, D and E, which displayed insensitivity to both EOD-FR and low R:FR ratio treatments (Franklin *et al.*, 2003a). In this work, leaf morphology (recorded as leaf length/width ratio) and flowering time were recorded in multiple phytochrome-deficient mutant combinations. Loss of either phyD or phyE, in the absence of phyB, resulted in elongated leaves in high R:FR ratio, an effect that was not further exacerbated by EOD-FR treatments (Franklin *et al.*, 2003a). In contrast, *phyBphyD* and *phyBphyE* double mutants displayed an earlier flowering response when subject to either EOD-FR or low R:FR ratio treatments. Triple mutants deficient in phytochromes B, D and E displayed no acceleration of flowering in response to these treatments. Such findings provided confirmation that in *Arabidopsis* at least, the suppression of shade avoidance responses in high R:FR ratio is mediated exclusively by phytochromes B, D and E in a functionally redundant manner.

The majority of published studies use ambient growth temperatures in excess of 20°C when elucidating the roles of different phytochromes in *Arabidopsis* development. Recent investigations have, however, revealed that the hierarchy of phytochrome function can be modified by growth temperature. When grown at 16°C, *phyB* mutants in the La-*er* background did not display the early flowering phenotype characteristic of growth at higher temperatures (Halliday *et al.*, 2003). An early flowering response to reductions in R:FR ratio was, however, still observed in wild-type plants grown at 16°C. These data inferred that at lower growth temperatures, phytochromes other than phyB perform a predominant role in the suppression of flowering. The subsequent analysis of multiple phytochrome-deficient mutant combinations revealed phyE and, to a lesser extent, phyD to adopt this function (Halliday and Whitelam, 2003) (see Section 9.6.3 and Chapter 10). Moreover, the elongated internodes observed in *Arabidopsis phyAphyBphyE* triple mutants grown in high R:FR ratio light at 22°C were not visible in plants grown at 16°C, suggesting maintenance of rosette habit to be regulated by phytochrome in a temperature-dependent manner (Halliday and Whitelam, 2003).

9.4.2   The role of phyA in R:FR ratio signalling

The unique degradation behaviour of phyA enables this photoreceptor to operate as an effective FR sensor in the HIR mode (Hennig *et al.*, 2000). Indeed, an inability of seedlings to de-etiolate in continuous FR has been successfully exploited as a screen for mutants, null at the *PHYA* locus (e.g. Nagatani *et al.*, 1993; Parks and Quail, 1993; Whitelam *et al.*, 1993). The capacity of phyA to operate as a FR sensor has implications in R:FR ratio signalling. When plants are subject to a FR-enrichment of their natural or artificial light environment, the action of phyA can 'antagonise' the physiological consequences of phyB, phyD and phyE conversion to the inactive Pr form. Indeed, 'enhanced' shade avoidance responses have been observed in *phyA* null mutants. When grown in continuous low R:FR ratio, *Arabidopsis phyA* seedlings displayed longer hypocotyls than wild-type controls (Johnson *et al.*, 1994). The same study also revealed *phyAphyB* double mutants to display longer hypocotyls than the *phyB* monogenic parent, providing further support for a role for phyA in the inhibition of hypocotyl growth. The importance of phyA in antagonising shade avoidance in the field was elegantly demonstrated by Yanovsky and colleagues (Yanovsky *et al.*, 1995) who observed extreme elongation responses in *Arabidopsis phyA* seedlings germinated under dense vegetational shade. A significant proportion of the elongated seedlings failed to become established and died prematurely. These data suggest that a major role of phyA in natural light environments is to limit excessive elongation, which could ultimately prove lethal.

The role of phyA in antagonising shade avoidance responses is not, however, restricted to young seedlings. The comparative analysis of *phyBphyDphyE* triple and *phyAphyBphyDphyE* quadruple mutants grown in high R:FR ratio revealed the latter to display considerably elongated rosette leaves when compared to *phyBphyDphyE* plants (Franklin *et al.*, 2003a). In addition, *phyAphyBphyDphyE* quadruple mutant plants grown in high R:FR ratio displayed clear internode growth between rosette leaves – phenotypes not visible in *phyBphyDphyE* triple mutants. Such data provide indisputable evidence for the role of phyA in modulating the suppression of internode growth and leaf elongation in light-grown plants (Franklin *et al.*, 2003a). Furthermore, populations of *Impatiens capensis* have been identified which display *less* petiole elongation in low R:FR ratio than high R:FR ratio controls, suggesting significant phyA action in light-grown plants (see Section 9.7). Disruption, or indeed 'reversal' of shade avoidance responses has previously been reported in transgenic tobacco plants constitutively expressing an oat *PHYA* gene (McCormac *et al.*, 1991,

1992; Rousseaux *et al.*, 1997). In these experiments, internode and petiole elongation was actually inhibited by supplementary FR light, suggesting significant phyA activity.

Despite the relatively close phylogenetic relationship between phyA and phyC (Mathews and Sharrock, 1997), no role for phyC in R:FR ratio signalling has ever been reported. Observations revealing *phyAphyBphyDphyE* quadruple mutants to display insensitivity to reductions in R:FR ratio and EOD-FR treatments have suggested no obvious role for phyC in mediating shade avoidance responses (Franklin *et al.*, 2003a). This conclusion is supported by analyses of *phyC* mutant combinations, which displayed no identifiable aberrations in R:FR ratio signalling (Franklin *et al.*, 2003b).

9.5   The roles of other signals in shade avoidance

The majority of published shade avoidance studies examine plant responses to reductions in R:FR ratio. Such signals are characteristic of light reflected from chlorophyllous tissue and provide plants with information concerning their proximity to neighbouring vegetation (Ballaré *et al.*, 1990). When subject to actual shading, however, plants are exposed to a number of environmental signals. In addition to alterations in light quality, plants experience a reduction in PAR (in particular blue light (B) signals) and elevated levels of ethylene.

9.5.1   PAR and B signals

Changes in B quantity are detected in higher plants by the UV-A/-B photoreceptors, cryptochromes and phototropins (Briggs and Huala, 1999; Cashmore *et al.*, 1999). In *Arabidopsis*, the B-mediated inhibition of hypocotyl elongation is regulated by two cryptochromes, cry1 and cry2 (Ahmad *et al.*, 1995; Lin *et al.*, 1998). These differ in light lability and fluence rate specificity. cry1 is light stable and acts at higher fluence rates (>10 μmol m^{-2} s^{-1}) of B, whereas cry2 behaves oppositely (Lin *et al.*, 1998). Increased stem elongation in response to reduced quantities of B has been reported in multiple species (Ballaré *et al.*, 1991a,b; Casal and Sánchez, 1994). In developing stands of monocultures, the quantity of light reaching stems is reduced well before leaves are shaded (Ballaré *et al.*, 1987, 1991b). Glasshouse studies of sunlight-grown *Datura ferox* and *S. alba* have shown light quality and quantity signals to regulate stem elongation (Ballaré *et al.*, 1991b). In these experiments, cuvettes containing solutions of organic dyes or inorganic salts were used in combination with coloured acetate to manipulate the spectral distribution of sunlight reaching individual internodes. Reducing PAR with a green light-absorbing filter had minimal effects on stem elongation, whereas the same reduction using a B-absorbing filter resulted in significant internode extension (Ballaré *et al.*, 1991b). Similar results were obtained in the hypocotyls of wild-type cucumber seedlings (Ballaré *et al.*, 1991a). More interestingly, the phyB-deficient *lh* mutant of cucumber did not respond to reductions in B, suggesting that for cucumber at least, the

B-mediated inhibition of hypocotyl elongation requires the presence of phyB Pfr (Ballaré et al., 1991a). B signals have also been reported to regulate leaf hyponasty in tobacco plants – a response that may be crucial in the early stages of competition (Pierik et al., 2004a). In these experiments, reductions in the photon fluence rate of B were shown to initiate leaf hyponasty, and therefore decrease leaf angle to the stem – a response similar to that observed upon exposure of plants to low R:FR ratio.

9.5.2 Hormone signals

Until recently, the role of plant hormones in transducing R:FR ratio signals has remained largely speculative. The promotory effect of auxin on cell division and cell elongation has lead to suggestions that this hormone may be involved in mediating elongation growth responses to reductions in R:FR ratio. Furthermore, the exogenous application of auxin to wild-type plants has been reported to result in elongated hypocotyls, longer petioles and increased apical dominance – phenotypes comparable to those of shade avoidance (Smalle et al., 1997; Chatfield et al., 2000; Sawa et al., 2002). Circumstantial evidence linking auxin to the shade avoidance syndrome includes observations that the application of an auxin transport inhibitor to wild-type Arabidopsis seedlings significantly reduced their elongation response to FR-enriched light (Steindler et al., 1999). The same study also showed that an auxin-response mutant axr1 did not elongate significantly following a similar treatment. In a separate investigation, the auxin-insensitive mutants axr1-3 and axr2 displayed attenuated petiole elongation in response to reductions in light intensity (Vandenbussche et al., 2003). However, given the pleiotropic morphological phenotypes of axr mutants, interpretation of their physiological responses requires circumspection. The identification of an Arabidopsis mutant displaying reduced shade avoidance phenotypes suggested a role for auxin transport in R:FR ratio signalling. The mutant, designated asa1, or attenuated shade avoidance 1, was isolated from a mutagenesis screen phyAphyB double mutants (Kanyuka et al., 2003). An individual failing to display the elongated, early flowering phenotype of parent plants was characterised and found to carry a mutation in the BIG (also known as DOC1, TIR3, UMB, GA6) gene. This gene encodes a large (560 kDa) protein involved in polar auxin transport and hormone signalling (Kanyuka et al., 2003). It is therefore possible that shade avoidance phenotypes result, in part, from the differential transport of auxin within tissues. Transcriptomic analyses of Arabidopsis plants subject to both low R:FR ratio and low (35 μmol m^{-2} s^{-1}) light intensity treatments have reported increased transcript levels of a number of auxin-related genes, giving support to this hypothesis (Devlin et al., 2003; Vandenbussche et al., 2003).

The potential role of ethylene as a shade avoidance signalling component was first convincingly proposed by Pierik and colleagues in a series of experiments using tobacco plants insensitive to the hormone. Investigations were initiated following observations that the exposure of young tobacco plants to low concentrations of ethylene resulted in increased stem elongation and leaf hyponasty – responses identical to those displayed in the shade avoidance syndrome (Pierik et al., 2003). These ethylene-mediated developmental adaptations are also displayed in the flooding response of *Rumex palustris* and enable submerged leaves to reach the water sur-face (Cox *et al.*, 2003). Subsequent investigations of ethylene insensitive transgenic (Tetr) tobacco revealed delayed stem elongation and leaf hyponasty in response to crowding (Pierik *et al.*, 2003). Interestingly, the delayed shade avoidance re-sponses observed were shown to result from an insensitivity of transgenic plants to reduced fluence rates of B (Pierik *et al.*, 2004a). The response of transgenic plants to reductions in R:FR ratio remained similar to wild-type controls, suggest-ing that B signals play an important role in mediating shade avoidance during actual shading and that ethylene is an important regulatory component of these responses (Pierik *et al.*, 2004a). This study also reported that ethylene levels within the canopy of densely planted tobacco reached concentrations that could induce shade avoid-ance responses in wild-type plants. Such observations are supported by studies in *Arabidopsis*, which recorded an increase in ethylene production in response to low light intensity (Vandenbussche *et al.*, 2003). Moreover, mutations in *PHYB* (Finlayson *et al.*, 1999; Vandenbussche *et al.*, 2003) and exposure of wild-type plants to low R:FR ratio (Finlayson *et al.*, 1998, 1999; Pierik *et al.*, 2004b) have been demonstrated to enhance ethylene production in *Arabidopsis*, tobacco and *Sorghum bicolor*. Taken together, this work provides compelling evidence that el-evations of atmospheric ethylene can signal to plants the presence of competing vegetation and initiate escape responses before canopy closure. Aberrant responses to artificial shading were observed in the ethylene-insensitive *Arabidopsis* mutants, *etr-1* and *ein2-1*, which displayed an increase in leaf area following reductions in light intensity (Vandenbussche *et al.*, 2003). This response was reversed in wild-type plants, which displayed a decrease in leaf surface area, thus resembling plants grown in low R:FR ratio (Smith and Whitelam, 1997). The same study also reported elevated levels of auxin-induced ethylene biosynthesis genes in *Arabidopsis phyB* mutants, suggesting a complex interaction of these two hormones in regulating plant physiological responses to shade.

Our understanding of shade avoidance regulation by plant hormones is further complicated by observations implicating the additional involvement of gibberel-lic acid (GA). It has been demonstrated that the application of GA biosynthesis inhibitors can attenuate shade avoidance responses in wild-type tobacco plants sub-ject to both low R:FR ratio treatment and ethylene application (Pierik *et al.*, 2004b). The involvement of GA in the phytochrome regulation of plant growth was previ-ously suggested following observations that a mutation in a GA biosynthesis gene abolished the characteristic long-hypocotyl phenotype of *Arabidopsis phyB* mutants (Peng and Harberd, 1997). Furthermore, EOD-FR treatments have been shown to increase active GA content in cowpea (*Vigna sinensis*) epicotyls (Martínez-García *et al.*, 2000) and transcript levels of GA biosynthesis gene, gibberellin oxidase 20 (*GA20-ox*), in *Arabidopsis* rosettes (Hisamatsu *et al.*, 2005). Such data are supported by studies using transgenic potato (*Solanum tuberosum*) plants with reduced levels of *PHYB* (Jackson *et al.*, 2000). Transgenic plants displayed enhanced transcript levels of *GA20-ox1*, suggesting gibberellin biosynthesis to be regulated, in part, by phyB. Work by Hisamatsu and colleagues has also reported transgenic *Arabidop-sis* lines with RNA silencing of *GA20-ox2* to display reduced petiole elongation in response to EOD-FR treatments. The authors therefore suggest a possible link between gibberellin biosynthesis and petiole growth responses in shade avoidance. Overall, it can be concluded that shade avoidance responses in higher plants involve the complex interplay of multiple phytohormones, the intricacies of which remain to be elucidated.

9.6   Signalling in shade avoidance

Observations showing stem elongation to occur within 10 min of R:FR ratio perception in *S. alba* (Morgan *et al.*, 1980; Child and Smith, 1987) brought into question whether rapid growth responses to low R:FR ratio involved changes in gene expression or were mediated by existing proteins within the plant. The answer to such a question ultimately required not only the isolation of R:FR ratio-regulated genes but also a quantitative analysis of their expression kinetics.

9.6.1   ATHB-2

The first genes reported to be reversibly regulated by changes in R:FR ratio were the transcription factors *ATHB-2* (also known as *HAT4*) and *ATHB-4* (Carabelli *et al.*, 1993, 1996). Both contain a homeodomain linked to a leucine zipper motif and have been shown to interact with the DNA sequence CAATNATTG, suggesting them to function as transcriptional regulators (Sessa *et al.*, 1993; Henriksson *et al.*, 2005). Transcript levels of *ATHB-2* were shown to be low in light-grown plants, but rapidly elevated in response to low R:FR ratio or EOD-FR treatments (Carabelli *et al.*, 1993, 1996). Analysis of *ATHB-2* gene expression in multiple phytochrome-deficient mutants revealed phyB and phyE to regulate transcript levels in a functionally redundant manner (Franklin *et al.*, 2003a). The involvement of *ATHB-2* in shade avoidance was suggested on the basis of the phenotypes of transgenic *Arabidopsis* expressing elevated and reduced levels of transcript (Schena and Davies, 1992; Steindler *et al.*, 1999). Seedlings with reduced levels of *ATHB-2* displayed short stature and large leaves whereas overexpressing lines behaved oppositely, thus resembling wild-type plants grown in low R:FR ratio (Steindler *et al.*, 1999). Plants expressing elevated levels of *ATHB-2* displayed enhanced cell expansion in the hypocotyl, reduced secondary growth of vascular tissues and decreased lateral root formation (Steindler *et al.*, 1999). These phenotypes are consistent with auxin-regulated processes, leading the authors to speculate that some aspects of the shade avoidance syndrome result from changes in auxin transport, mediated by R:FR ratio-dependent changes in *ATHB-2* expression (Steindler *et al.*, 1999; Morelli and Ruberti, 2002).

9.6.2   PIL1

Genomic analysis of shade avoidance in adult *Arabidopsis* plants further revealed two genes *PIL1* (*PIF3-Like 1*) and *PIL2* (*PIF3-Like 2*) displaying reversible regulation by R:FR ratio (Salter *et al.*, 2003). Both encode basic helix-loop-helix (bHLH) transcription factors with homology to the phytochrome-interacting protein PIF3 (Ni *et al.*, 1998). Rapid increases in *PIL1* transcript were observed within 15 min of transfer to low R:FR, with maximum levels detected at 30 min. The derepression of *PIL2* transcript by low R:FR ratio occurred at a slower rate and required a lag time of at least 3 h (Salter *et al.*, 2003). The PIL1 protein was initially identified as an interacting partner of the circadian clock component *TOC1* (Makino *et al.*, 2003). Despite its similarity to the phytochrome-interacting PIF family of bHLH proteins, PIL1 has been shown not to bind *PHYB in vitro* (Khanna *et al.*, 2004). Detailed expression studies of *PIL1* and *PIL2* revealed the derepression of both genes by low R:FR ratio to be gated by the circadian clock, with maximum increases at subjective dawn (Salter *et al.*, 2003).

The gating of *PIL1* and *PIL2* gene expression by the circadian clock suggested that physiological responses to low R:FR ratio may be regulated in a similar manner. Observations that a 2 h transient reduction in R:FR ratio could elicit a 30% increase in hypocotyl elongation within the following 24 h enabled the circadian control of this response to be investigated (Salter *et al.*, 2003). The derepression of hypocotyl inhibition was also shown to be gated by the circadian clock, with maximum increases at subjective dusk (Salter *et al.*, 2003). This coincides with the natural rhythm of elongation growth in *Arabidopsis* seedlings (Dowson-Day and Millar, 1999). Inhibitions of growth were observed following low R:FR ratio treatment at subjective dawn, yet were absent in *phyA* mutants, confirming the role of this phytochrome in antagonising shade avoidance. The attenuated elongation phenotype of *pil1* null mutants to transient, but not prolonged, reductions in R:FR ratio suggested a putative role for this protein in mediating rapid responses to shade (Salter *et al.*, 2003). The requirement of PIL1 for rapid hypocotyl elongation does, however, present a temporal discrepancy. The derepression of *PIL1* transcript by low R:FR ratio occurs at subjective dawn, whereas the physiological response of hypocotyls at subjective dusk. The identification of PIL1 signalling components and signal transduction pathways should provide some insight into the mechanism of PIL1 function and address this issue.

Subsequent microarray analyses have since revealed a related bHLH transcription factor, *HFR1*, to display significant increases in transcript level upon transfer of *Arabidopsis* seedlings to low R:FR light (Sessa *et al.*, 2005). This gene was previously identified as a component of phyA signalling on the basis of the long-hypocotyl phenotype of mutant seedlings grown in continuous FR (Fairchild *et al.*, 2000; Fankhauser and Chory, 2000; Soh *et al.*, 2000). The authors report elevated levels of transcript of a number of low R:FR ratio derepressed genes (e.g. *PIL1*, *PIL2*) following 24 h of low R:FR ratio treatment in *hfr1* null mutants compared to wild-type seedlings and suggest the existence of a HFR1-mediated negative regulatory feedback loop controlling the magnitude of shade avoidance responses (Sessa *et al.*, 2005). Given the role of HFR1 as a phyA signalling component and the established role of phyA in antagonising shade avoidance responses (e.g. Salter *et al.*, 2003), it is, however, possible that the elevated levels of transcripts observed in *hfr1* mutants represent a defect in phyA signalling rather than a specific regulatory role for HFR1 per se.

9.6.3 R:FR ratio and flowering

A characteristic component of the shade avoidance syndrome is a pronounced acceleration in the timing of transition to reproductive development – a response mediated in *Arabidopsis* by phytochromes B, D and E (Halliday *et al.*, 1994; Franklin *et al.*, 2003a). The regulation of flowering time in *Arabidopsis* is determined by the complex interplay of multiple environmental signals that act together to determine the fate of the shoot apical meristem through regulation of meristem identity genes such as *LFY* (for review, see Simpson and Dean, 2002). The use of indicators such as temperature and photoperiod can enable a degree of seasonal prediction, allowing floral initiation to be coordinated with conditions of favourable climate and/or competitive advantage. The expression of meristem identity genes has been shown to be controlled by a number of floral integrators such as *FT* and *SOC1*, which are themselves regulated by transcriptional activators such as *FLC* and *CO* (Simpson and Dean, 2002). Indeed, the photoperiodic promotion of flowering in *Arabidopsis* by long days is thought to result from the activation of *FT* expression, caused by the coincidence of a photoreceptor-derived signal with high levels of *CO* (Yanovsky and Kay, 2002). In this way, the circadian-regulation of *CO* levels provides plants with a molecular mechanism to discriminate between long and short days.

The promotion of flowering by reductions in R:FR ratio is thought to operate through *FT*, independently of *CO*, despite an earlier report by Blázquez and Weigel (1999) suggesting phyB-deficiency (and thereby low R:FR ratio) to operate independently of both transcriptional activators (Cerdán and Chory, 2003; Halliday *et al.*, 2003). Halliday *et al.* (2003) observed the temperature-conditional early flowering response of *Arabidopsis phyB* mutants to correlate with elevated levels of *FT* transcript (see Chapter 10). When grown at 16°C, the wild-type flowering response observed in these plants was paralleled by near wild-type levels of *FT* expression. Comparison of *phyAphyBphyD* triple and *phyAphyBphyDphyE* quadruple mutants grown at 16°C revealed phyE to perform a predominant role in suppressing *FT* expression and consequently flowering at this temperature (Halliday *et al.*, 2003). The involvement of *FT* in the regulation of flowering time by light quality was further supported by Cerdán and Chory (2003), who independently revealed a correlation between the early flowering response of *phyB* mutants and elevated *FT* transcript levels. In addition, this study revealed a possible signalling component in the pathway linking phyB action and *FT* expression. A recessive mutation *pft1* was identified from a screen of mutagenised *Arabidopsis* plants showing aberrant flowering behaviour. The late flowering phenotype of *pft1* plants in both long and short days suggested PFT1 to be an essential signalling component in the phyB-mediated regulation of flowering time. The cloning of *PFT1* revealed a nuclear-localised protein with similarity to some transcriptional activators.

Despite abolishing the early flowering response associated with phyB-deficiency, *phyBpft1* double mutants displayed petiole lengths similar to those of *phyB* controls. The mutation appeared not to significantly affect the regulation of flowering time by photoperiod, but considerably impaired flowering responses to EOD-FR treatments (Cerdán and Chory, 2003). The latter are mediated in wild-type plants by phytochromes B, D and E (Franklin *et al.*, 2003a). The authors therefore propose that PFT1 functions downstream of phytochromes B, D and E in a photoperiod-independent manner. A separation of the responses of the shade avoidance syndrome has previously been reported in a number of natural *Arabidopsis* accessions that elongate upon reduction of the R:FR ratio, but remain unresponsive with respect to flowering time (Botto and Smith, 2002). Of these, the *Bla-6* accession displayed an extreme phenotype, showing pronounced elongation, yet only a minor acceleration of flowering time, in response to low R:FR ratio. Such data provide compelling evidence of a branched signal transduction pathway in R:FR ratio signalling.

9.7 The adaptive value of shade avoidance

The shade avoidance syndrome is one of the more radical developmental strategies displayed by higher plants. The adaptive value of such plasticity has been assessed in multiple field studies, the majority of which suggest that rapid elongation in response to signals from neighbouring vegetation confers high relative fitness in dense stands (Schmitt *et al.*, 2003). This notion was convincingly demonstrated in an investigation of transgenic tobacco plants, constitutively expressing an oat *PHYA* cDNA. In these plants, elevated levels of phyA result in the persistent antagonism of shade avoidance responses in FR-rich light environments (McCormac *et al.*, 1991, 1992). When grown at high density, transgenic plants were unable to elongate in response to reductions in R:FR ratio and displayed decreased fitness, as measured by dry biomass accumulation (Schmitt *et al.*, 1995; Robson *et al.*, 1996). A similar finding was obtained using transgenic tobacco plants insensitive to the hormone ethylene. The delayed stem elongation and leaf hyponasty responses of transgenic plants reduced competitive advantage when grown in mixed populations with wild-type controls (Pierik *et al.*, 2003). When grown in dense monocultures, however, all transgenic plants displayed a similar biomass (Pierik *et al.*, 2003). Such observations infer that fitness costs in mixed populations arise from inequality in adaptive plasticity. Despite conferring selective advantage in dense stands, excessive elongation growth can prove severely disadvantageous in the absence of competition, resulting in decreased fitness and a risk of mechanical damage (Casal and Smith, 1989). Indeed, decreases in dry biomass and numbers of reproductive structures were recorded in the elongated *ein* mutant of *B. rapa* grown at low density, in addition to lodging and mechanical damage to stems (Schmitt *et al.*, 1995). An increase in stem damage was also recorded in elongated *lh* mutants of cucumber grown individually in the field (Casal *et al.*, 1994). Even in dense stands, the selective advantage of shade avoidance responses can be compromised by other environmental factors. When water levels are limiting, the reallocation of resources towards elongation growth at the expense of root development can result in reduced fitness (Huber *et al.*, 2004). It is therefore possible that selection of shade avoidance traits in different species relates, in part, to environmental habitat. Species in which ecotypes have evolved in contrasting selective conditions have been shown to display variation in response to R:FR ratio. Early studies by Morgan and Smith (1979) showed species from open habitats (e.g. *C. album, Chamaenerion angustifolium, Senecio ja-cobea*) to display greater stem elongation in response to low R:FR ratio than species from permanently shaded woodland habitats (e.g. *Mercurialis perennis, Teucreum scorodonia*). Such diversity in responsivity to low R:FR ratio is not restricted to different species. Indeed different ecotypes of the same species can respond differently to the low R:FR ratio signal. Ecotypic variation in response to low R:FR ratio has been reported in *Stellaria longipes*. Populations from densely vegetated prairies displayed stem elongation responses to low R:FR ratio, whereas populations from less competitive alpine environments remained unresponsive (Alokam *et al.*, 2002). More intriguingly, recent observations have reported woodland populations of *I. capensis* to display *less* elongation in low R:FR ratio than under ambient control conditions. In these experiments, populations from open woodland displayed characteristic shade avoidance responses (von Wettberg and Schmitt, 2005). The authors propose that this difference may result from a persistent FR-HIR in woodland populations, a response which may prevent unprofitable elongation in a permanently shaded predicament.

The evolutionary implications of plastic adaptation to low R:FR ratio signals have been investigated in trees by Smith and colleagues (Gilbert *et al.*, 2001). In these experiments, stands of both early and late successional species were grown at different spacings throughout several growing seasons and their heights and leaf areas were measured. A reverse relationship was recorded between responsivity to low R:FR ratio and the magnitude of signal generation. Early successional species generated small proximity signals, but responded most strongly to them, whereas late successional species behaved oppositely. Such data suggest that the adaptive benefit of shade avoidance is dependent upon not only environmental surroundings but also the evolutionary time of development.

In contrast to elongation responses, which can occur within minutes of signal perception (Morgan *et al.*, 1980), the acceleration of flowering in shade avoidance requires a prolonged exposure to low R:FR ratio. Temporary shading occurs frequently in natural environments and can often be overcome by a brief period of elongation growth. Under such circumstances, a rapid transition to flowering would not prove beneficial to the success of a shaded individual. When plants are subject to the unfavourable situation of prolonged shading, however, a precocious switch to reproductive development may prove the best strategy for optimising survival to the next generation (Dudley and Schmitt, 1995; Donohue *et al.*, 2001; Botto and Smith, 2002).

9.8 Conclusions

The shade avoidance syndrome encompasses a variety of physiological responses observed when plants are subject to reduced light intensity and/or a reduction in the R:FR ratio of their ambient light environment. The ability to elongate stems and precociously initiate reproductive development upon perception of neighbouring vegetation enables shade-avoiding higher plants take opportunistic advantage of gaps in the canopy and over-top competitors. The transcriptomic analysis of shade avoidance responses in the laboratory has identified a number of genes displaying R:FR ratio-regulation of transcript abundance. Kinetic studies of these genes and parallel investigations of null mutants have provided a small insight into the molecular mechanisms operating to confer this adaptive plasticity. The complexity of shade avoidance signalling has been highlighted through studies revealing both temporal specificity and crosstalk with temperature and multiple hormone signalling pathways. A significant future challenge therefore exists to dissect not only the components of R:FR ratio signal transduction, but also their points of crosstalk with other environmental cues. The identification and investigation of such signalling networks should ultimately facilitate a more holistic understanding of this important biological phenomenon.

References

Adamse, P., Jaspers, P.A.P.M., Bakker, J.A., Kendrick, R.E and Koornneef, M. (1988) Photophysiology and phytochrome content of long-hypocotyl and wild-type cucumber seedlings. *Plant Physiol.* **87**, 264–268.

Adamse, P., Jaspers, P.A.P.M., Kendrick, R.E. and Koornneef, M. (1987) Photomorphogenetic responses of long hypocotyl mutant of *Cucumis sativus* . *J. Plant Physiol.* **127**, 481–491.

Ahmad, M., Lin, C. and Cashmore, A.R. (1995) Mutations throughout an *Arabidopsis* blue-light photoreceptor impair blue-light-responsive anthocyanin accumulation and inhibition of hypocotyl elongation. *Plant J.* **8**, 653–658.

Alokam, S., Chinnappa, C.C. and Reid, D.M. (2002) Red/far-red light mediated stem elongation and anthocyanin accumulation in *Stellaria longipes*: differential responses of alpine and prairie ecotypes. *Can. J. Bot.* **80**, 72–81.

Aukerman, M.J., Hirschfeld, M., Wester, L., *et al.* (1997) A deletion in the *PHYD* gene of the *Arabidopsis* Wassilewskija ecotype defines a role for phytochrome D in red/far-red light sensing. *Plant Cell* **9**, 1317–1326.

Ballaré, C.L., Casal, J.J. and Kendrick, R.E. (1991a) Responses of light-grown wild-type and long-hypocotyl mutant cucumber seedlings to natural and stimulated shade light. *Photochem. Photobiol.* **54**, 819–826.

Ballaré, C.L., Sánchez, R.A., Scopel, A.L., Casal, J.J. and Ghersa, C.M. (1987) Early detection of neighbour plants by phytochrome perception of spectral changes in reflected sunlight. *Plant Cell Environ.* **10**, 551–557.

Ballaré, C.L., Scopel, A.L. and Sánchez, R.A. (1990) Far-red radiation reflected from adjacent leaves: an early signal of competition in plant canopies. *Science* **247**, 329–332.

Ballaré, C.L., Scopel, A.L. and Sánchez, R.A. (1991b) Photocontrol of stem elongation in plant neighbourhoods: effects of photon fluence rate under natural conditions of radiation. *Plant Cell Environ.* **14**, 57–65.

Barnes, C. and Bugbee, B. (1991) Morphological responses of wheat to changes in phytochrome photoequilibrium. *Plant Physiol.* **97**, 359.

Blázquez, M.A. and Weigel, D. (1999) Independent regulation of flowering by phytochrome B and gibberellins in *Arabidopsis*. *Plant Physiol.* **120**, 1025–1032.

Boardman, N.K. (1977) Comparative photosynthesis of sun and shade plants. *Annu. Rev. Plant Physiol.* **28**, 355–377.

Botto, J.F. and Smith, H. (2002) Differential genetic variation in adaptive strategies to a common environmental signal in *Arabidopsis* accessions: phytochrome-mediated shade avoidance. *Plant Cell Environ.* **25**, 53–63.

Briggs, W.R. and Huala, E. (1999) Blue-light photoreceptors in higher plants. *Annu. Rev. Cell Dev. Biol.* **15**, 33–62.

Carabelli, M., Morelli, G., Whitelam, G.C. and Ruberti, I. (1996) Twilight-zone and canopy shade induction of the *ATHB-2* homeobox gene in green plants. *Proc. Natl. Acad. Sci. USA* **93**, 3530–3535.

Carabelli, M., Sessa, G., Ruberti, I. and Morelli, G. (1993) The *Arabidopsis ATHB-2* and *-4* genes are strongly induced by far-red-rich light. *Plant J.* **4**, 469–479.

Casal, J.J., Ballaré, C.L., Tourn, M. and Sánchez, R.A. (1994) Anatomy, growth and survival of a long-hypocotyl mutant of *Cucumis sativus* deficient in phytochrome B. *Ann. Bot.* **73**, 569–575.

Casal, J.J. and Sánchez, R.A. (1994) Impaired stem growth response to blue light irradiance in light-grown transgenic tobacco seedlings overexpressing *Avena* phytochrome A. *Physiol. Plant.* **91**, 268–272.

Casal, J.J., Sánchez, R.A. and Deregibus, V.V. (1986) The effect of plant density on tillering: the involvement of R/FR ratio and the proportion of radiation intercepted per plant. *Environ. Exp. Bot.* **26**, 365–371.

Casal, J.J. and Smith, H. (1989) The function, action and adaptive significance of phytochrome in light-grown plants. *Plant Cell Environ.* **12**, 855–862.

Cashmore, A.R., Jarillo, J.A., Wu, Y.J. and Liu, D. (1999) Cryptochromes: blue light receptors for plants and animals. *Science* **284**, 760–765.

Cerdán, P.D. and Chory, J. (2003) Regulation of flowering time by light quality. *Nature* **423**, 881–885.

Chatfield, S.P., Stirnberg, P., Forde, B.G. and Leyser, O. (2000) The hormonal regulation of axillary bud growth in *Arabidopsis*. *Plant J.* **24**, 159–169.

Child, R. and Smith, H. (1987) Phytochrome action in light-grown mustard: kinetics, fluence-rate compensation and ecological significance. *Planta* **172**, 219–229.

Clack, T., Mathews, S. and Sharrock, R.A. (1994) The phytochrome apoprotein family in *Arabidopsis* is encoded by five genes: the sequences and expression of *PHYD* and *PHYE*. *Plant Mol. Biol.* **25**, 413–427.

Clough, R.C. and Vierstra, R.D. (1997) Phytochrome degradation. *Plant Cell Environ.* **20**, 713–721.

Cox, M.C.H., Millenaar, F.F., de Jong van Berkel, Y.E.M., Peeters, A.J.M. and Voesenek, L.A.C.J. (2003) Plant movement: submergence-induced petiole elongation in *Rumex palustris* depends on hyponastic growth. *Plant Physiol.* **132**, 282–291.

Devlin, P.F., Halliday, K.J., Harberd, N.P. and Whitelam, G.C. (1996) The rosette habit of *Arabidopsis thaliana* is dependent upon phytochrome action: novel phytochromes control internode elongation and flowering time. *Plant J.* **10**, 1127–1134.

Devlin, P.F., Patel, S.R. and Whitelam, G.C. (1998) Phytochrome E influences internode elongation and flowering time in *Arabidopsis*. *Plant Cell* **10**, 1479–1487.

Devlin, P.F., Robson, P.R.H., Patel, S.R., Goosey, L., Sharrock, R.A. and Whitelam, G.C. (1999) Phytochrome D acts in the shade-avoidance syndrome in *Arabidopsis* by controlling elongation and flowering time. *Plant Physiol.* **119**, 909–915.

Devlin, P.F., Rood, S.B., Somers, D.E., Quail, P.H. and Whitelam, G.C. (1992) Photophysiology of the *elongated internode* (*ein*) mutant of *Brassica rapa*: *ein* mutant lacks a detectable phytochrome B-like protein. *Plant Physiol.* **100**, 1442–1447.

Devlin, P.F., Yanovsky, M.J. and Kay, S.A. (2003) A genomic analysis of the shade avoidance response in *Arabidopsis*. *Plant Physiol.* **133**, 1617–1629.

Donohue, K., Pyle, E.H., Messiqua, D., Heschel, M.S. and Schmitt, J. (2001) Adaptive divergence in plasticity in natural populations of *Impatiens capensis* and its consequences for performance in novel habitats. *Evolution* **55**, 692–702.

Dowson-Day, M.J. and Millar, A.J. (1999) Circadian dysfunction causes aberrant hypocotyl elongation patterns in *Arabidopsis*. *Plant J.* **17**, 63–71.

Dudley, S.A. and Schmitt, J. (1995) Genetic differentiation in morphological responses to simulated foliage shade between populations of *Impatiens capensis* from open and woodland sites. *Funct. Ecol.* **9**, 655–666.

Fairchild, C.D., Schumaker, M.A. and Quail, P.H. (2000) HFR1 encodes a typical bHLH protein that acts in phytochrome A signal transduction. *Genes Dev.* **14**, 2377–2391.

Fankhauser, C. and Chory, J. (2000) *RSF1*, an *Arabidopsis* locus implicated in phytochrome A signalling. *Plant Physiol.* **124**, 39–45.

Finlayson, S.A., Jung, I.-J., Mullet, J.E. and Morgan, P.W. (1999) The mechanism of rhythmic ethylene production in Sorghum: the role of phytochrome B and simulated shading. *Plant Physiol.* **119**, 1083–1089.

Finlayson, S.A., Lee, I.-J. and Moran, P.W. (1998) Phytochrome B and the regulation of circadian ethylene production in sorghum. *Plant Physiol.* **116**, 17–25.

Franklin, K.A., Davis, S.J., Stoddart, W.M., Vierstra, R.D. and Whitelam, G.C. (2003b) Mutant analyses define multiple roles for phytochrome C in *Arabidopsis thaliana* photomorphogenesis. *Plant Cell* **15**, 1981–1989.

Franklin, K.A., Praekelt, U., Stoddart, W.M., Billingham, O.E., Halliday, K.J. and Whitelam, G.C. (2003a) Phytochromes B, D and E act redundantly to control multiple physiological responses in *Arabidopsis*. *Plant Physiol.* **131**, 1340–1346.

Gilbert, I.R., Jarvis, P.G. and Smith, H. (2001) Proximity signal and shade avoidance differences between early and late successional trees. *Nature* **411**, 792–795.

Goosey, L., Palecanda, L. and Sharrock, R.A. (1997) Differential patterns of expression of the *Arabidopsis PHYB, PHYD*, and *PHYE* phytochrome genes. *Plant Physiol.* **115**, 959–969.

Halliday, K.J., Koornneef, M. and Whitelam, G.C. (1994) Phytochrome B and at least one other phytochrome mediate the accelerated flowering response of *Arabidopsis thaliana L.* to low red/far-red ratio. *Plant Physiol.* **104**, 1311–1315.

Halliday, K.J., Salter, M.G., Thingnaes, E. and Whitelam, G.C. (2003) Phytochrome control of flowering is temperature sensitive and correlates with expression of the floral integrator FT. *Plant J.* **33**, 875–885.

Halliday, K.J. and Whitelam, G.C. (2003) Changes in photoperiod or temperature reveal roles for phyD and phyE. *Plant Physiol.* **131**, 1913–1920.

Hennig, L., Büche, C. and Schäfer, E. (2000) Degradation of phytochrome A and the high irradiance response in *Arabidopsis*: a kinetic analysis. *Plant Cell Environ.* **23**, 727–734.

Henriksson, E., Olsson, A.S.B., Johannesson, H., Hanson, J., Engström, P. and Söderman, E. (2005) Homeodomain leucine zipper class I genes in *Arabidopsis*. Expression patterns and phylogenetic relationships. *Plant Physiol.* **139**, 509–518.

Hisamatsu, T., King, R.W., Helliwell, C.A. and Koshioka, M. (2005) The involvement of gibberellin 20-oxidase genes in phytochrome-regulated petiole elongation of *Arabidopsis*. *Plant Physiol.* **138**, 1106–1116.

Holmes, M.G. and Smith, H. (1975) The function of phytochrome in plants growing in the natural environment. *Nature* **254**, 512–514.

Holmes, M.G. and Smith, H. (1977a) The function of phytochrome in the natural environment. I. Characterisation of daylight for studies in photomorphogenesis and photoperiodism. *Photochem. Photobiol.* **25**, 533–538.

Holmes, M.G. and Smith, H. (1977b) The function of phytochrome in the natural environment. II. The influence of vegetation canopies on the spectral energy distribution of natural daylight. *Photochem. Photobiol.* **25**, 539–545.

Huber, H., Kane, N.C., Heschel, M.S., *et al.* (2004) Frequency and microenvironmental pattern of selection on plastic shade-avoidance traits in a natural population of *Impatiens capensis*. *Am. Nat.* **1634**, 548–563.

Jackson, S.D., James, P.E., Carrera, E., Pratt, S. and Thomas, B. (2000) Regulation of transcript levels of a potato gibberellin 20-oxidase gene by light and phytochrome B. *Plant Physiol.* **124**, 423–430.

Johnson, E., Bradley, J.M., Harberd, N.P. and Whitelam, G.C. (1994) Photoresponses of light-grown *phyA* mutants of *Arabidopsis*: phytochrome A is required for the perception of daylength extensions. *Plant Physiol.* **105**, 141–149.

Kanyuka, K., Praekelt, U., Billingham, O., *et al.* (2003) Mutations in the huge *Arabidopsis* gene *BIG* affect a range of hormone and light responses. *Plant J.* 35, 57–70.

Keiller, D. and Smith, H (1989) Control of carbon partitioning by light quality mediated by phytochrome. *Plant Sci.* **63**, 25–29.

Kendrick, R.E., Kerckhoffs, L.H.J., van Tuinen, A. and Koornneef, M. (1997) Photomorphogenic mutants of tomato. *Plant Cell Environ.* **20**, 746–751.

Kerckhoffs, L.H.J., Kendrick, R.E., Whitelam, G.C. and Smith, H. (1992) Extension growth and anthocyanin responses of photomorphogenic tomato mutants to changes in the phytochrome photoequilibrium during the daily photoperiod. *Photochem. Photobiol.* **56**, 611–616.

Kerckhoffs, L.H.J., van Tuinen, A., Hauser, B.A., *et al.* (1996) Molecular analysis of *tri* mutant alleles in tomato indicates the *TRI* locus is the gene encoding the apoprotein of phytochrome B1. *Planta* **199**, 152–157.

Khanna, R., Huq, E., Kikis, E.A., Al-Sady, B., Lanzatella, C. and Quail, P.H. (2004) A novel molecular recognition motif necessary for targeting photoactivated phytochrome signalling to specific basic helix-loop-helix transcription factors. *Plant Cell* **16**, 3033–3044.

Lin, C., Yang, H., Guo, H., Mockler, T., Chen, J. and Cashmore, A.R. (1998) Enhancement of blue light sensitivity of *Arabidopsis* seedlings by a blue light receptor cryptochrome 2. *Proc. Natl. Acad. Sci. USA* **95**, 7686–7699.

López-Juez, E., Buurmeijer, W.F., Heeringa, G.H., Kendrick, R.E. and Wesselius, J.C. (1990) Response of light-grown wild-type and long hypocotyl mutant cucumber plants to end-of-day far-red light. *Photochem. Photobiol.* **52**, 143–149.

López-Juez, E., Nagatani, A., Tomizawa, K.-I., *et al.* (1992) The cucumber long hypocotyl mutant lacks a light-stable PHYB-like phytochrome. *Plant Cell* **4**, 241–251.

Makino, S., Matsushika, A., Kojima, M., Yamashino, T. and Mizuno, T. (2002) The APRR1/TOC1 quintet implicated in circadian rhythms of *Arabidopsis thaliana*: I. Characterization with APRR1-overexpressing plants. *Plant Cell Physiol.* **43**, 58–69.

Martínez-García, J.F., Santes, C.M. and García-Matínez, J.L. (2000) The end-of-day far-red irradiation increases gibberellin A$_1$ content in cowpea (*Vigna sinensis*) epicotyls by reducing inactivation. *Physiol. Plant.* **108**, 426–434.

Mathews, S. and Sharrock, R.A. (1997) Phytochrome gene diversity. *Plant Cell Environ.* **20**, 666–671.

McCormac, A.C., Cherry, J.R., Hershey, H.P., Vierstra, R.D. and Smith, H. (1991) Photoresponses of transgenic tobacco plants expressing an oat phytochrome gene. *Planta* **185**, 162–170.

McCormac, A.C., Whitelam, G.C. and Smith, H. (1992) Light grown plants of transgenic tobacco expressing an introduced oat phytochrome A gene under the control of a constitutive viral promoter exhibit persistent growth inhibition by far-red light. *Planta* **188**, 173–181.

McLaren, J.S. and Smith, H. (1978) The function of phytochrome in the natural environment. VI. Phytochrome control of the growth and development of *Rumex obtusifolius* under simulated canopy light environments. *Plant Cell Environ.* **1**, 61–67.

Middleton, L. (2001) Shade-tolerant flowering plants: adaptations and horticultural implications. *Acta Hort. (ISHS)* **552**, 95–102.

Morelli, G. and Ruberti, I. (2002) Light and shade in the photocontrol of *Arabidopsis* growth. *Trends Plant Sci.* **7**, 399–404.

Morgan, D.C., O'Brien, T., and Smith, H. (1980) Rapid photomodulation of stem extension in light-grown *Sinapis alba* L. Studies on kinetics, site of perception and photoreceptor. *Planta* **150**, 95–101.

Morgan, D.C. and Smith, H. (1981) Control of development in *Chenopodium album* l by shadelight – the effect of light quantity (total fluence rate) and light quality (red-far-red ratio). *New Phytol.* **88**, 239–248.

Morgan, D.C. and Smith, H. (1978) The function of phytochrome in the natural environment. VII. The relationship between phytochrome photo-equilibrium and development in light-grown *Chenopodium album* L. *Planta* **132**, 187–193.

Morgan, D.C. and Smith, H. (1976) Linear relationship between phytochrome photoequilibrium and growth in plants under simulated natural radiation. *Nature* **262**, 210–212.

Morgan, D.C. and Smith, H. (1979) A systematic relationship between phytochrome-controlled development and species habitat, for plants grown in simulated natural radiation. *Planta* **145**, 253–258.

Nagatani, A., Reed, J.W. and Chory, J. (1993) Isolation and initial characterisation of *Arabidopsis* mutants that are deficient in functional phytochrome A. *Plant Physiol.* **102**, 269–277.

Ni, M., Tepperman, J.M. and Quail, P.H. (1998) PIF3, a phytochrome-interacting factor necessary for normal photoinduced signal transduction, is a novel basic helix-loop-helix protein. *Cell* **95**, 657–667.

Parks, B.M. and Quail, P.H. (1993) hy8, a new class of *Arabidopsis* long hypocotyl mutants deficient in functional phytochrome A . *Plant Cell* **3**, 39–48.

Peng, J. and Harberd, N.P (1997) Gibberellin deficiency and response mutations suppress the stem elongation phenotype of phytochrome-deficient mutants of *Arabidopsis*. *Plant Physiol.* **113**, 1051–1058.

Pierik, R., Cuppens, M.L.C., Voesenek, L.A.C.J. and Visser, E.J.W. (2004b) Interactions between ethylene and gibberellins in phytochrome-mediated shade avoidance responses in tobacco. *Plant Physiol.* **136**, 2928–2936.

Pierik, R., Visser, E.J.W., de Kroon, H., andVoesenek, L.A.C.J. (2003) Ethylene is required in tobacco to successfully complete with proximate neighbours. *Plant Cell Environ.* **26**, 1229–1234.

Pierik, R., Whitelam, G.C., Voesenek, L.A.C.J., de Kroon, H. and Visser, E.J.W. (2004a) Canopy studies on ethylene-insensitive tobacco identify ethylene as a novel element in blue light and plant-plant signalling. *Plant J.* **38**, 310–319.

Quail, P.H. (1994) Phytochrome genes and their expression . In: *Photomorphogenesis in Plants*, 2nd edn (eds Kendrick, R.E. and Kronenberg, G.H.M.), pp. 71–104. Kluwer, Dordrecht, Netherlands.

Quail, P.H., Schäfer, E. and Marmé, D. (1973) Turnover of phytochrome in pumpkin cotyledons. *Plant Physiol.* **52**, 128–131.

Reed, J.W., Nagpal, P., Poole, D.S., Furuya, M. and Chory, J. (1993) Mutations in the gene for red/far-red light receptor phytochrome B alter cell elongation and physiological responses throughout *Arabidopsis* development. *Plant Cell* **5**, 147–157.

Robson, P.R.H., McCormac, A.C., Irvine, A.S. and Smith, H. (1996) Genetic engineering of harvest index in tobacco through overexpression of a phytochrome gene. *Nat. Biotechnol.* **14**, 995–998.

Robson, P.R.H., Whitelam, G.C. and Smith, H. (1993) Selected components of the shade-avoidance syndrome are displayed in a normal manner in mutants of *Arabidopsis thaliana* and *Brassica rapa* deficient in phytochrome B. *Plant Physiol.* **102**, 1179–1184.

Rousseaux, M.C., Ballaré, C.L., Jordan, E.T. and Vierstra, R.D. (1997) Directed overexpression of *PHYA* locally suppresses stem elongation and leaf senescence responses to far-red radiation. *Plant Cell Environ.* **20**, 1551–1558.

Salter, M.G., Franklin, K.A. and Whitelam, G.C. (2003) Gating of the rapid shade avoidance response by the circadian clock in plants. *Nature* **426**, 680–683.

Sawa, S., Ohgishi, M., Goda, H., *et al.* (2002) The HAT2 gene, a member of the HD-Zip gene family, isolated as an auxin inducible gene by DNA microarray screening, affects auxin response in *Arabidopsis*. *Plant J.* **32**, 1011–1022.

Sessa, G., Carabelli, M., Sassi, M., *et al.* (2005) A dynamic balance between gene activation and repression regulates the shade avoidance response in *Arabidopsis*. *Genes Dev.* **19**, 2811–2815.

Sessa, G., Morelli, G. and Rubert, I. (1993) The Athb-1 and-2 HD-Zip domains homodimerize forming complexes of different DNA binding specificities. *EMBO J.* **12**, 3507–3517.

Schafer, E. and Bowler, C. (2002) Phytochrome-mediated photoperception and signal transduction in higher plants. *EMBO Rep.* **3**, 1042–1048.

Schena, M. and Davis, R.W. (1992) HD-Zip proteins: members of an *Arabidopsis* homeodomain protein superfamily. *Proc. Natl. Acad. Sci. USA* **89**, 3894–3898.

Schmitt, J., McCormac, A.C. and Smith, H. (1995). A test of the adaptive plasticity hypothesis using transgenic and mutant plants disabled in phytochrome-mediated elongation responses to neighbours. *Am. Nat.* **146**, 937–953.

Schmitt, J., Stinchcombe, J.R., Heschel, M.S. and Huber, H. (2003) The adaptive evolution of plasticity: phytochrome-mediated shade avoidance responses. *Integrative Comparative Biol.* **43**, 459–469.

Sharrock, R.A. and Quail, P.H. (1989) Novel phytochrome sequences in *Arabidopsis thaliana*: structure, evolution, and differential expression of a plant regulatory photoreceptor family. *Genes Dev.* **3**, 1745–1757.

Simpson, G.G. and Dean, C. (2002) *Arabidopsis*, the Rosetta stone of flowering time? *Science* **296**, 285–289.

Smalle, J., Haegman, M., Kurepa, J., Van Montagu, M., and Van Der Straeten, D. (1997) Ethylene can stimulate *Arabidopsis* hypocotyl elongation in the light. *Proc. Natl. Acad. Sci. USA* **94**, 2756–2761.

Smith, H. (1982) Light quality, photoperception and plant strategy. *Annu. Rev. Plant Physiol.* **33**, 481–518.

Smith, H. (1975) *Phytochrome and Photomorphogenesis*. McGraw-Hill, UK.

Smith, H., Casal, J.J. and Jackson, G.M. (1990) Reflection signals and the perception by phytochrome of the proximity of neighbouring vegetation. *Plant Cell Environ.* **13**, 73–78.

Smith, H. and Whitelam, G.C. (1997) The shade avoidance syndrome: multiple responses mediated by multiple phytochromes. *Plant Cell Environ.* **20**, 840–844.

Soh, M.-S., Kim, Y.-M., Han, S.-J. and Song, P.-S. (2000) *REP1*, a basic helix-loop-helix protein is required for a branch of phytochrome A signalling in *Arabidopsis*. *Plant Cell* **12**, 2061–2073.

Somers, D.E., Sharrock, R.A., Tepperman, J.M. and Quail, P.H. (1991) The *hy3* long hypocotyl mutant of *Arabidopsis* is deficient in phytochrome B. *Plant Cell* **3**, 1263–1274.

Steindler, C., Matteucci, A., Sessa, G., *et al.* (1999) Shade avoidance responses are mediated by the ATHB-2 HD-zip protein, a negative regulator of gene expression. *Development* **126**, 4235–4245.

Vandenbussche, P., Vriezen, W.H., Small, J., Laarhoven, L.J.J., Harren, F.J.M. and Van Der Straeten D. (2003) Ethylene and auxin control the *Arabidopsis* response to decreased light intensity. *Plant Physiol.* **133**, 517–527.

van Tuinen, A., Kerckhoffs, L.H.J., Nagatani, A., Kendrick, R.E. and Koornneef, M. (1995) A temporarily red light-insensitive mutant of tomato lacks a light-stable, B-like phytochrome. *Plant Physiol.* **108**, 939–947.

von Wettberg, E.J. and Schmitt, J. (2005) Physiological mechanisms of population differentiation in shade-avoidance responses between woodland and clearing genotypes of *Impatiens capensis*. *Am. J. Bot.* **95**, 868–874.

Whitelam, G.C. and Johnson, C.B. (1982) Photomorphogenesis in *Impatiens parviflora* and other plant species under simulated natural canopy radiation. *New Phytol.* **90**, 611–618.

Whitelam, G.C., Johnson, E., Peng, J., *et al.* (1993) Phytochrome A null mutants of *Arabidopsis* display a wild-type phenotype in white light. *Plant Cell* **5**, 757–768.

Whitelam, G.C. and Smith, H. (1991) Retention of phytochrome-mediated shade avoidance responses in phytochrome-deficient mutants of *Arabidopsis*, cucumber and tomato. *J. Plant Physiol.* **39**, 119–125.

Yanovsky, M.J., Casal, J.J. and Whitelam, G.C. (1995) Phytochrome A, phytochrome B and HY4 are involved in hypocotyl growth responses to natural radiation in *Arabidopsis*: weak de-etiolation of the *phyA* mutant under dense canopies. *Plant Cell Environ.* **18**, 788–794.

Yanovsky, M.J. and Kay, S.A. (2002) Molecular basis of seasonal time measurement in *Arabidopsis*. *Nature* **419**, 308–312.

10 Photoreceptor interactions with other signals

Eve-Marie Josse and Karen J. Halliday

10.1 Introduction

Light is an incredibly powerful external signal that is responsible for driving and shaping a multitude of plant developmental processes. Light exerts its influence by tapping into endogenous pathways or signals generated by external cues. This type of signalling network, linking internally and externally derived signals, provides a means to integrate information from the environment with the plants intrinsic developmental programme. The extensive molecular interplay between internal and external signals ensures that plant development is highly plastic, and therefore responsive to frequent changes in habitat conditions. Although environmental cues influence signal transduction in animals, their development does not exhibit the extreme plasticity that is so essential for survival in immobile organisms. As a result, many plant and animal molecular signalling networks have evolved in different ways. By studying these pathways, one can establish the components and signalling motifs that are conserved and those that differ in plants and animals. This type of information provides crucial insights into how signalling mechanisms have evolved in these two kingdoms.

To enable constant adaptation to a changing environment, plants have developed highly sophisticated networks that are highly connected. These adaptive changes are mediated by manipulating signal transduction in multiple internal pathways. In the natural environment, plants often have to assimilate and interpret more than one external cue at any given time. This appears to be achieved by channelling these signals though integration points in the signalling network. In this way plant development is continuously engineered by the ambient surroundings. In this chapter, we will focus on how light connects with both internal and external pathways to control plant development. Light plays a principal role in ensuring that the plant's internal programme is synchronised with the daily light/dark cycle. Thus, we will explore the multiple ways in which light interacts with the internal circadian system. We will examine how light influences hormonal signalling and how light and temperature signals intercept.

10.2 Light–clock connections

10.2.1 The clock

In plants, the circadian system controls daily changes in gene expression, growth, photosynthetic activity and seasonal flowering (Dodd *et al.*, 2005; Schoning and

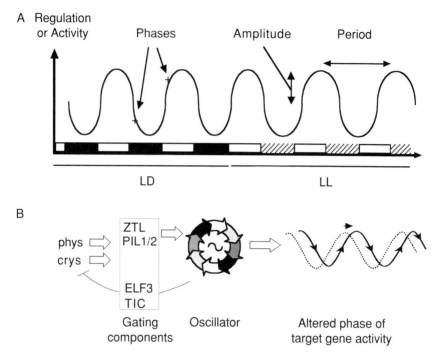

Figure 10.1 (A) Circadian rhythms. Clock-controlled gene regulation and/or activity oscillates through light/dark (LD) cycles (open and filled bars). Oscillation continues when transferred to constant conditions, shown here as constant light (LL) where the open and hatched bars represent alternate subjective days and nights. Phase is a specific point during the circadian cycle; amplitude is the maximum height of the waveform from peak to trough; period is the length of an entire cycle. (B) A simplified model of photoentrainment. Gating genes determine when, during the daily cycle, the clock components are receptive to light signals. These genes participate in re-phasing the clock and therefore clock-controlled genes to light signals.

Staiger, 2005). An overriding function of this rhythmic mechanism is to synchronise internal signalling processes with external light cues, which drive a vast array of metabolic and developmental responses. Biological clocks recognise and initiate responses to the changing daily light/dark cycle and as a result oscillator function is strongly influenced by photoperiod. However, a key feature of biological clocks is that they continue to run in constant light (LL) or dark (DD) conditions (Figure 10.1). Under such conditions, oscillation continues through the 'subjective' day and night. Using *Arabidopsis*, we are gradually piecing together the central components of the plant clock. This model is under constant review as we gather more experimental information that adds to the picture. Central clock genes include the MYB-domain transcription factors, *LATE ELONGATED HYPOCOTYL (LHY), CIRCADIAN CLOCK ASSOCIATED 1 (CCA1), LUX ARRHYTHMO (LUX)*, and the pseudo response regulator, *TIMING OF CAB 1 (TOC1)* (Millar, 2003; Salome and McClung, 2004; Hazen *et al.*, 2005). Levels and activity of *LHY/CCA1* and *TOC1/LUX* exhibit reciprocal patterns of oscillation in a 24-h autoregulatory circadian feedback loop. *LHY* and *CCA1* are induced by light but are expressed rhythmically with peaks of expression around dawn. Their cognate proteins have been shown to bind a motif called the evening element in the promoters of *TOC1* and *LUX* to block their transcription (Alabadi *et al.*, 2001; Hazen *et al.*, 2005). As LHY and CCA1 levels fall, *TOC1* and *LUX* mRNA levels rise to a peak at the end of the day (Strayer *et al.*, 2000; Hazen *et al.*, 2005). To complete the autoregulatory feedback loop, these morning genes activate *LHY* and *CCA1* transcription, though we do not yet know how this occurs. One thing we do know is that this central clock model is incomplete. The clock-associated genes *EARLY FLOWERING 3* and *4 (ELF3/4), TIME FOR COFFEE (TIC), GIGANTEA (GI), TEJ, SENSITIVITY TO RED LIGHT REDUCED (SRR1)* and the *PSEUDO-RESPONSE REGULATORS 5, 7* and *9 (PRR5/7/9)* all have roles in maintaining clock function in LL and DD (McWatters *et al.*, 2000; Covington *et al.*, 2001; Hicks *et al.*, 2001; Liu *et al.*, 2001; Doyle *et al.*, 2002; Panda *et al.*, 2002; Hall *et al.*, 2003; Staiger *et al.*, 2003; Kikis *et al.*, 2005; Mizuno and Nakamichi, 2005; Nakamichi *et al.*, 2005). Thus, some or all of these genes may function as principal oscillator components. This mechanism then imposes circadian regulation on multiple molecular, cellular processes.

10.2.2   Photoentrainment

Clock components and the photoreceptors have an intimate relationship. Light signals transduced by the phytochromes and cryptochromes ensure the clock is in tune with the daily light/dark cycles. This process, known as entrainment, is achieved by adjusting the phase and the period of the oscillator relative to the prevailing photoperiod (Figure 10.1). We have learned that several photoreceptors are involved in this process by studying the impact of photoreceptor action on clock-controlled genes. Several laboratories have used the circadian-regulated *CHLOROPHYLL A/B BINDING PROTEIN 2::LUCIFERASE (CAB2::LUC)* promoter::reporter construct to study this rhythmicity in vivo. In wild-type seedlings, period of *CAB2::LUC* is shortened by light and continues to shorten as light intensity increases (Somers *et al.*, 1998; Devlin and Kay, 2000). This adjustment of circadian period is achieved, at least in part, by phytochrome (phy) and cryptochrome (cry) photoreceptor action. Analysis of *CAB2::LUC* expression in phy null mutants indicates phyA, phyB, phyD and phyE have roles in this process in response to red light, whilst phyA, cry1 and cry2 fulfil this role under blue light (Somers *et al.*, 1998; Devlin and Kay, 2000). Interestingly, the *cry1* and *cry2* mutations also impair red light control of circadian period length. This suggests that cry1 and cry2 also act as phy-regulated signalling components, placing them in a central position in this response.

In the following sections we will examine some of the connections between the light receptors and central oscillator-associated components. To date, studies have only provided a partial view of how light and clock functions are integrated. However, genetic and molecular analysis has identified some important links, offering insights into how light signals connect with the circadian clockwork. A number of studies have provided evidence that, in *Arabidopsis*, ELF3 and TIC are key connections between light and the clock (Zagotta *et al.*, 1996; McWatters *et al.*, 2000; Reed *et al.*, 2000; Covington *et al.*, 2001; Hicks *et al.*, 2001; Liu *et al.*, 2001; Hall *et al.*, 2003). These genes participate in the differential regulation of day and night time sensitivity to light, a mechanism known as circadian gating. This ensures correct entrainment of the clock to changing dawn and dusk signals. In wild-type plants, changes in photoperiod adjust the circadian period by phase shifting the oscillator at dawn and dusk. *ELF3* and *TIC*, two clock controlled genes, have been shown to be important regulators of this process. In *elf3* and *tic* mutants, light induces high levels of *CAB::LUC* expression during the dark period, a time when this response is suppressed in the wild type (McWatters *et al.*, 2000; Hall *et al.*, 2003). Thus, *elf3* and *tic* do not display normal circadian gating of *CAB::LUC* expression observed in wild-type plants. Further evidence of this role for ELF3 comes from analysing the expression of *COLD CIRCADIAN REGULATED2::LUC (CCR2::LUC)* in response to light pulses that adjust the circadian phase (Covington *et al.*, 2001). In wild-type plants, red or blue light pulses provided at intervals through the subjective night are very effective in re-setting the phase of the clock. Light pulses given at the intervals from the subjective dusk cause phase delays, which increase in magnitude to a 'break point' in the subjective light. Thereafter, light pulses trigger phase advances, which decrease in magnitude towards subjective dawn. *ELF3* overexpression dampens this response, whilst *elf3-1* null mutants either exhibit larger phase shifts than wild type or become arrhythmic (Covington *et al.*, 2001). These experiments demonstrate that ELF3 has a prominent role in controlling phase setting by the phytochromes and cryptochromes. Other experiments, where the timing of clock arrest has been demonstrated for *elf3* and *tic*, suggest ELF3 and TIC work at different times of the day (McWatters *et al.*, 2000; Hall *et al.*, 2003). *ELF3* starts to operate at dusk, whilst *TIC* functions in the mid to late night. As *ELF3* expression peaks at subjective dusk, this supports a role for ELF3 at this time of day (Covington *et al.*, 2001; Hicks *et al.*, 2001; Liu *et al.*, 2001). In the case of ELF3, moderation of the phy signal may be direct as ELF3 is localised to the nucleus, the site of phyB action, and it has been shown to interact with phyB in vitro (Liu *et al.*, 2001).

Photoentrainment is also controlled by ZEITLUPE (ZTL), a member of the *ZTL/LKP3/FKF1* gene family, which encodes proteins that contain an LOV domain, and F-box and a kelch repeat. The LOV domains in *ZTL/LKP3/FKF1* are highly homologous to those in the blue-light photoreceptors PHOTOTROPIN 1 and 2 (PHOT1/2) where they act as light-sensing modules (see Chapter 3). This provides the possibility that ZTL/LKP3/FKF1 define a new class of light receptors (Imaizumi *et al.*, 2003). Like other F-box kelch proteins, the ZTL/LKP3/FKF1 family participates in the Skp/Cullin/F-box (SCF) E3 complex, recruiting specific substrates for ubiquitination and subsequent proteolysis by the 26S proteasome (Cope and Deshaies, 2003; Vierstra, 2003). ZTL has been shown to confer tight control of TOC1 protein levels via this mechanism (Mas *et al.*, 2003; Han *et al.*, 2004). *TOC1* mRNA levels increase during the day as transcriptional repression is relieved by falling CCA1/LHY levels. However, the time during which the protein is available, and therefore active, is regulated by ZTL. Analysis of TOC1 protein levels in *ztl* mutants suggests that ZTL plays a major role in degrading TOC1 during the dark period (Mas *et al.*, 2003). Like TOC1, the ZTL protein is itself subject to degradation by the proteasome. Levels peak at subjective dusk and trough at subjective dawn. These changes are under the control of the circadian clock and the daily light/dark cycle (Kim *et al.*, 2003). Indeed, ZTL oscillations are severely dampened in both LL and DD suggesting that the daily light/dark cycles are required to maintain ZTL protein rhythm. This may be a mechanism of gating the light input to the oscillator through its action on TOC1.

Another type of circadian gating is revealed through studying the mechanisms that underpin phy-mediated control of hypocotyl elongation. In seedling development, hypocotyl extension is known to be under circadian control, with daily arrests in growth occurring at dawn followed by periods of rapid elongation at dusk (Dowson-Day and Millar, 1999). These daily dawn and dusk rhythms are controlled photoreceptor action that is gated by the circadian clock. This response is acutely sensitive to phy status since depletion in active phy levels, induced by low red:far-red ratio light, relieves growth inhibition, and hypocotyl cells elongate as a consequence (Devlin *et al.*, 2003; Salter *et al.*, 2003). Low red:far-red ratio light simulates natural habitats where changes in light quality result from the selective absorption by green vegetation. The consequent alteration in red:far-red ratio signals the presence of neighbouring plants and lowers the proportion of active phy (Franklin and Whitelam, 2005). This triggers a striking series of 'shade-avoidance' responses, which appear to be an important survival strategy under unfavourable shade conditions (Donohue *et al.*, 2001; Botto and Smith, 2002). One component of the shade-avoidance response is enhanced hypocotyl elongation, and the basic helix-loop-helix gene *PHYTOCHROME INTERACTING FACTOR 3 (PIF3)-LIKE 1* (*PIL*) plays an important role in this process (Salter *et al.*, 2003). RT-PCR analysis has shown that PIL1 and PIL2, a close homologue, have increased expression in low red:far-red ratio light. However, when assayed over a 24-h period, *PIL1/PIL2* transcripts exhibited obvious circadian-gated expression patterns to transient reductions in low red:far-red ratio light. Increases in *PIL1* mRNA were detectable after only 8 min of low red:far-red light, whereas *PIL2* transcript levels rose more slowly. These expression patterns are consistent with PIL1 providing a rapid and PIL2 providing a more sustained response to low red:far-red ratio light. Analysis of the *pil1* mutant showed that PIL1 is required for the normal elongation response to low red:far-red ratio light. When compared to wild type seedlings, *pil1* exhibited reduced elongation responses that were phase-shifted, suggesting that PIL1 may operate by moderating oscillator function. It is possible that the observed effects in *pil1* are mediated through TOC1, as experiments using *in vitro* binding assays have demonstrated a PIL1-TOC1 interaction (Makino *et al.*, 2002). Furthermore, the *toc1-2* mutant was unable to mount an elongation response to low red:far-red ratio light, lending support to this notion (Salter *et al.*, 2003). Future work will reveal the precise nature of the molecular mechanism that controls this crucial response to neighbouring plants.

10.2.3   Light control of flowering time

Photoperiod (seasonal day length) and light-quality cues from neighbouring plants are potent regulators of flowering time. These environmental light signals are perceived and transduced by multiple photoreceptors, and it is their collective action that determines when the plant makes the switch from vegetative to reproductive development (Ni, 2005). The ability to assimilate and respond to these external signals is highly developed in plants, and this ensures that important developmental events like flowering occur under environmental conditions that are favourable for seed set and dispersal.

Arabidopsis is a facultative long-day plant which means that flowering is promoted as photoperiods lengthen. Early work based on physiological studies developed the external coincidence model to explain photoperiodic time measurement. This is described in detail in Chapter 8 and is discussed briefly here. It is light interaction with the transcriptional regulator CONSTANS (CO) that ensures that flowering occurs as the days lengthen (Putterill *et al.*, 2004; Searle and Coupland, 2004). *CO* mRNA is tightly regulated by the circadian oscillator. Under short-day conditions the peak of *CO* expression occurs during the night, whilst under long-day conditions the peak occurs during the day. Under these conditions, when the photoperiod coincides with elevated *CO* mRNA levels, photoreceptors enhance the levels and activity of CO protein (Valverde *et al.*, 2004). CO then triggers flowering by activating transcription of floral integrators such as *FLOWERING LOCUS T (FT)*.

10.3 Light–hormone connections

Phytohormones influence the whole of development, from germination through seedling establishment to reproductive development and senescence. To achieve optimal growth and development in a changing environment, internal cues, driven by hormone signalling, need to be coordinated with external cues. Signals that are generated by light quality, quantity or photoperiod provide accurate information on the immediate environment and the changing seasons. Many of these signals feed through to the hormonal pathways to manipulate their activity and the physiological processes they control. Indeed, light signalling has been shown to be associated with the biosynthesis and/or signalling of multiple phytohormones including auxin, gibberellins (GA), cytokinins, ethylene and brassinosteroids. Largely through genetic analysis, we have learned that the light-hormone pathways are integrated at many levels (Halliday and Fankhauser, 2003). Indeed, the auxin and GA pathways, in particular, appear to be strongly coupled to light signalling. In Section 10.4, the points at which light interfaces with auxin and GA signal transduction are examined.

10.4 Light and auxin signal integration

The phytohormone auxin (indole-3-acetic acid, IAA) regulates many different aspects of plant development, including cell division, elongation, differentiation and patterning. It is synthesised primarily in the shoot apex and young developing leaves, then transported downward to the root tip through the vasculature. Auxin also moves through tissues via polar transport (Blakeslee *et al.*, 2005; Leyser, 2005). This fine tunes its tissue distribution, a characteristic that is important for its mode of action. Several reports have demonstrated that light and auxin signalling are intimately connected. Light regulates phototropism and gravitropism, at least partly, through the asymmetrical distribution of auxin (see Section 10.5). End-of-day far-red light treatments that deplete phy levels at the end of the photoperiod trigger hypocotyl elongation and the expression of auxin-inducible genes (Tanaka *et al.*, 2002). Furthermore, stabilisation of SHY2/IAA3 leads to a constitutive photomorphogenic phenotype (Kim *et al.*, 1998; Tian and Reed, 1999). Thus, light and auxin regulatory pathways appear to intercept at multiple levels to control growth and development. In the following section we will examine the points of signal integration focusing on the molecular mechanisms through which light and auxin connect.

10.4.1   Light regulation of auxin biosynthesis and transport

Light signals have quite a grip on auxin signalling and this appears to start with the control of auxin biosynthesis. The *red1* mutant provides insight into how phy regulates auxin production. *red1* was originally identified in a screen for phy signalling components. It exhibits a long-hypocotyl phenotype that is specific to red light, suggesting that RED1 acts downstream of phyB to control this response (Wagner *et al.*, 1997). However, *RED1* was subsequently shown to be allelic to *ATR4/SUR2*, which encodes the cytochrome P450 monooxygenase, CYP83B1. This enzyme catalyses N-hydroxylation of the IAA precursor, indole-3-acetaldoxime (IAOx). The inhibition of IAOx hydroxylation leads to an accumulation of auxin since more IAOx is available for IAA synthesis. These studies suggest that RED1 provides a means for phy to control auxin homeostasis (Hoecker *et al.*, 2004).

Several studies have shown that light has a role in manipulating auxin transport through plant tissues. The auxin transport inhibitor naphthylphthalamic acid (NPA) reduces hypocotyl elongation in light-grown seedlings, but is completely ineffective when they are grown in darkness (Jensen *et al.*, 1998). This response was shown to be severely attenuated in *phyA*, *phyB* or *cry1* mutants when grown under far-red, red or blue light, respectively. Thus, photoreceptor-controlled inhibition of hypocotyl elongation, which is important for seedling establishment, appears to be mediated, at least partly, by regulating auxin transport (Jensen *et al.*, 1998).

Another connection between light and auxin transport was revealed through analysis of the homeodomain-leucine zipper transcription factor *ATHB-2*. Transcription of this gene is tightly regulated by phy. *ATHB-2* mRNA levels rise rapidly following seedling exposure to low red:far-red ratio light, which in the natural environment signals the presence of neighbouring plants (Carabelli *et al.*, 1996; Steindler *et al.*, 1999). This molecular shade-avoidance response is mainly under the control of phyB and phyE (Franklin *et al.*, 2003). ATHB-2 antisense seedlings exhibit an enhanced de-etiolation with shorter hypocotyls and enlarged cotyledons when compared to the wild type. In contrast, ATHB-2 overexpression (OX) lines resemble the elongated phy loss-of-function mutants, supporting a role for *ATHB-2* in the shade-avoidance response (Schena *et al.*, 1993). Like *phyB* mutants, ATHB-2 OX seedlings also produce fewer lateral roots as compared to the wild-type seedlings. As lateral root growth is known to be promoted by auxin derived from the shoot, the ATHB-2 OX phenotype was postulated to result from a decrease in auxin flow from the shoot to the root (Morelli and Ruberti, 2000; Bhalerao et al., 2002). Indeed, the ATHB-2 OX root phenotype can be rescued by exogenous auxin application lending support to this hypothesis and suggesting a role for phy in regulating shoot to root auxin transport.

Further evidence for connections between light and auxin transport comes from studies centred on the *tir3/doc1/asa1/umb3* mutant gene that codes for a large calossin-like protein aptly named BIG. The directional flow of auxin through cells is dependent on polarly localised (PIN-FORMED) PIN auxin efflux regulators, and the positioning of PINs at the membrane is controlled by auxin itself. Auxin achieves this by regulating PIN cycling between the plasma membrane and endosomes, which consequently alters the distribution of PINs at the membrane (Paciorek et al., 2005). BIG appears to participate in this auxin-regulated response and is therefore intimately involved in polar auxin transport. Indeed, mutant alleles of *big* have decreased polar auxin transport; however, they also exhibit altered photomorphogenic traits (Li et al., 1994; Gil et al., 2001; Kanyuka et al., 2003). One feature of the *big* mutant alleles is that they do not display the normal elongated hypocotyl phenotype in darkness, and as a consequence they are much shorter than wild-type seedlings. Microarray analysis of the *doc1* allele revealed that several genes that are normally light regulated were switched on in dark-grown *doc1* seedlings. However, their expression could be suppressed by elevated auxin levels. This suggests that auxin may be important in the CONSTITUTIVE PHOTOMORPHOGENESIS 1 (COP1) switch that maintains etiolated development by repressing the transcription of light-regulated genes (Gil et al., 2001). Analysis of BIG illustrates that normal auxin transport is necessary for etiolated seedling growth and disruption of this process interferes with the dark–light developmental switch.

10.4.2   Signalling components shared by light and auxin

Auxin signalling is mediated through the transcriptional regulation of at least three gene families: the *GH3*-related genes, the Aux/IAA genes and the SAURs (small auxin-up RNAs). Light has also been shown to control transcription and/or influence the activity of auxin-regulated genes (Abel et al., 1995; Tepperman et al., 2001; Devlin et al., 2003). In this section we will be examining how light controls development through the manipulation of GH3s and Aux/IAAs.

10.4.2.1   Light regulation of the GH3 gene family

In *Arabidopsis*, there are 20 genes belonging to the *GH3* family (Hagen and Guilfoyle, 2002). At least six members of this family, including *YDK/GH3-2, AtGH3a/GH3-5* and *DFL1/GH3-6*, act as IAA-amido synthetases catalysing the conjugation of amino acids to IAA which inactivates the molecule (Staswick et al., 2005). Thus, GH3s are important for the regulation of active, free auxin. Several *Arabidopsis* mutants in this family have altered photomorphogenic features suggesting

GH3-mediated IAA conjugation as another means by which light can regulate auxin levels. Analysis of the *dfl1-D/gh3-6* gain-of-function mutant showed that DFL1 is involved in the light-specific inhibition of hypocotyl cell elongation as well as lateral root production (Nakazawa *et al.*, 2001). The *dfl1-D/gh3-6* short-hypocotyl phenotype was observed under red, blue or far-red light indicating that DFL1 may work downstream of the phytochromes and cryptochromes. In a similar fashion to *dfl1*, overexpression of DFL2/GH3-10 also enhanced light-regulated inhibition of hypocotyl elongation (Takase *et al.*, 2003). In contrast, DFL2 antisense plants exhibited an elongated hypocotyl phenotype under red light, suggesting a role for DFL2 in light-stable phy signalling. Expression analysis of these genes provides some insights into how light and auxin moderate *GH3* action. *DFL1* transcription is regulated by auxin, and not by light, whilst *DFL2* transcript levels appear to be light, but not auxin, regulated. This indicates that light may regulate GH3s through both transcriptional and post-transcriptional mechanisms.

Elevated levels of expression of a third gene *YDK1/GH3-2* in this family alter the phenotype of light- and dark-grown seedlings. In this instance, *YDK1/GH3-2* gene expression was shown to be positively regulated by auxin and negatively regulated by blue and far-red light (Takase *et al.*, 2004). It is possible that light regulates *YDK1/GH3-2* through auxin; alternatively, it may be under dual control. Another auxin-regulated gene *AtGH3a/GH3-5* is also controlled by phyB. *AtGH3a/GH3-5* mRNA is elevated by end-of-day far-red treatments or in the *phyB* null mutant, suggesting that phyB negatively regulates *AtGH3a/GH3-5* transcription. However, phyB control of *AtGH3a/GH3-5* is not maintained in the gain-of-function *axr2-1/iaa7* mutant, suggesting that normal auxin signalling is required for this phyB-regulated response (Tanaka *et al.*, 2002). Taken together, these analyses demonstrate several modes of connection between auxin and light signalling through the regulation of GH3 family members. By regulating GH3 activity, in addition to auxin biosynthesis and transport (see Section 10.41), light appears to exert tight control on auxin homeostasis.

10.4.2.2 Role of Aux/IAAs and proteolysis in light and auxin signalling

In *Arabidopsis*, the Aux/IAAs are a family of 28 nuclear proteins, most of which are induced by auxin with varying response kinetics. Aux/IAAs operate by binding to AUXIN RESPONSE FACTOR (ARF) transcription factors to negatively regulate their action. This provides a mechanism through which auxin can modulate the expression of target genes (Liscum and Reed, 2002; Woodward and Bartel, 2005). Auxin has been shown to control Aux/IAA levels by stimulating the ubiquitin-mediated proteolysis of Aux/IAA proteins via the ubiquitin ligase SCF^{TIR1}, a process that feeds back to regulate *Aux/IAA* transcription. The F-box protein TIR1, a component of the SCF complex, targets Aux/IAAs for ubiquitination and subsequent degradation by the proteasome. Auxin controls this process by promoting the interaction between TIR1 and Aux/IAAs (Kepinski and Leyser, 2004; Dharmasiri *et al.*, 2005; Kepinski and Leyser, 2005). As ARFs target *Aux/IAA* genes, this auxin-induced negative feedback loop allows a high turnover, with newly synthesised Aux/IAA proteins quickly restoring ARF repression of auxin signalling.

This highly dynamic system is extremely responsive to alterations in input signals, and thus manipulation by light.

Aux/IAA turnover appears to be important for aspects of light-regulated development. Mutations that stabilise IAA3/SHY2 were isolated as suppressors of *hy2* and *phyB* phenotypes (Kim *et al.*, 1998; Tian and Reed, 1999). The *iaa3/shy2* gain-of-function mutants have short hypocotyl and expanded cotyledons in the dark, characteristics shared by *iaa7/axr2* and *iaa17/axr3*, also gain-of-function mutants (Kim *et al.*, 1996; Kim *et al.*, 1998; Reed *et al.*, 1998; Tian and Reed, 1999; Nagpal *et al.*, 2000). Furthermore, *iaa3/shy2-2* mutant seedlings have elevated levels of *CAB* mRNA, a gene that is repressed in dark-grown wild-type seedlings (Kim *et al.*, 1998; Tian and Reed, 1999). This suggests that, as for auxin transport (see Section 10.41), normal turnover of Aux/IAAs is important to repress photomorphogenesis in dark-grown seedlings. It is not entirely clear how light controls Aux/IAA activity in the switch to light-regulated development; however, the literature provides some insights. The transcription of several *Aux/IAA* genes, including *SHY/IAA3*, is regulated by phyB and phyA (Devlin *et al.*, 2003). Interestingly, *in vitro* studies have demonstrated that Aux/IAAs can be phosphorylated by oat phyA (Colon-Carmona *et al.*, 2000), and that IAA3/SHY2 can interact with *Arabidopsis* phyB (Tian *et al.*, 2003). Thus, it appears that the phytochromes may be able to influence *Aux/IAA* gene expression and post-translational activity, highlighting the strong links between light and Aux/IAA-mediated auxin signalling.

It is unclear how phy-mediated phosphorylation influences the activity of Aux/IAAs. In animals, SCF-substrate recognition requires phosphorylation (Moon *et al.*, 2004). However, studies have shown that phosphorylation was probably not involved in the SCF^{TIR1}-Aux/IAA interaction (Dharmasiri *et al.*, 2003; Kepinski and Leyser, 2004). Moreover, SCF^{TIR1}-substrate recognition is promoted by the binding of auxin to TIR1, the SCF component involved in target recognition (Dharmasiri *et al.*, 2005; Kepinski and Leyser, 2005).

One means by which light could regulate Aux/IAA levels is by targeting SCF^{TIR1} E3 ligase itself, and there is some support for this control mechanism. The activity of SCF^{TIR1} appears to be regulated by the COP9 signalosome (CSN) (Schwechheimer, 2004). More specifically, CSN5, a central component of the CSN complex, modifies SCF^{TIR1} activity by deneddylation: the removal of ubiquitin-like proteins NEDD8/RUB1. Like the *aux/iaa* gain-of-function mutants, *csn5* null mutants display constitutive photomorphogenic phenotypes in the dark. This suggests that CSN5 modification of SCF^{TIR1} activity is important for regulating Aux/IAA turnover in dark-grown seedlings. Therefore, this may represent a mechanism whereby light can control Aux/IAA degradation, a process that feeds back to regulate transcription (see above).

10.5 The tropisms

Plant tropic responses are characterised by the curvature of a plant organ towards or away from a directional stimulation (Esmon *et al.*, 2005). This reorientation is achieved by differential cellular elongation across the organ, a process that is triggered following the establishment of an auxin hormone gradient. In this section we will examine the molecular events involved in phototropism and gravitropism. Both these responses are controlled by the combined effects of light and hormonal signalling.

10.5.1 Light and auxin control of shoot phototropism

Positive phototropism of plant stems has been known to be induced by blue light for over a century (Briggs and Christie, 2002). We now know that the photoreceptors that control this response in *Arabidopsis* are the phototropins phot1 and phot2 (Briggs and Christie, 2002). However, we are only just beginning to understand how the phototropin signal is transduced. Screens for phototropism-deficient mutants have isolated two early signalling pathway components, *NON-PHOTOTROPIC HYPOCOTYL 3 (NPH3)* (Liscum and Briggs, 1995, 1996; Motchoulski and Liscum, 1999) and the related *ROOT PHOTOTROPISM 2 (RPT2)* (Okada and Shimura, 1992; Sakai *et al.*, 2000). NPH3 acts downstream of both phototropins in the regulation of phototropic curvature, whilst RPT2 action appears to be specific to phot1 (Motchoulski and Liscum, 1999; Inada *et al.*, 2004). Both NPH3 and RPT2 are able to interact with phot1, and the formation of a phot1–NPH3 complex is necessary for early phototropic signalling (Motchoulski and Liscum, 1999; Inada *et al.*, 2004). It has been proposed that NPH3 could act as a scaffold or an adaptor protein, allowing the assembly of a signalling complex containing phot1 at the plasma membrane (Motchoulski and Liscum, 1999). Interestingly, co-immunoprecipitation and yeast two-hybrid approaches have demonstrated that NPH3 and RPT2 interact; suggesting that RPT2 may be another component of the phot1–NPH3 complex (Inada *et al.*, 2004). Current thinking speculates that this plasma-membrane-associated complex, which may be regulated via changes in phosphorylation status, could be directly coupled to changes in auxin transport (Esmon *et al.*, 2005).

Early physiological analysis studying the role of the shoot apex in the tropic response led to the formulation of the Cholodny–Went hypothesis and the isolation of the phytohormone, auxin. This states that the bending of a phototropically stimulated shoot towards the light results from an increase in auxin concentration in the shaded flank of the stem, which leads to auxin-induced differential growth (Cholodny, 1927; Went and Thimann, 1937). Auxin gradients created by the movement of auxin through vasculature and polar transport through cells is now known to be central to many auxin-controlled responses (Friml, 2003). As phototropism is tightly coupled to the cellular distribution of auxin, it follows that auxin efflux regulation must be a key control point in this response. One suggested role for a plasma-membrane-associated phot1/NPH3/RPT2 complex is to influence auxin transport, possibly by modifying the cellular location of auxin transporter localisation (Esmon *et al.*, 2005). Regulating this process are members of the auxin efflux facilitator family of PIN proteins (Friml, 2003; Blakeslee *et al.*, 2005). From this family, two members appear to have predominant roles in phototropic responses: PIN1 (Geldner *et al.*, 2001) and PIN3 (Friml *et al.*, 2003). PIN1 delocalises from the basal wall of the cell upon blue stimulation in the mid-hypocotyl region, i.e. where phototropic bending occurs, and this delocalisation is impaired in a *phot1*-deficient mutant (Blakeslee *et al.*, 2004). Similarly, the asymmetric auxin distribution associated to the phototropic response requires laterally localised PIN3 (Friml *et al.*, 2003). Thus, the phot1/NPH3/RPT2 complex may operate, at least partly, by altering PIN1 and PIN3 localisation.

The isolation of the *nph4* mutant has provided insights into how phototropins moderate auxin signalling. The *NPH4* gene encodes the transcriptional activator *ARF7* (see Section 10.43), which appears to regulate localised cell elongation in response to an auxin-generated signal (Harper *et al.*, 2000; Liscum and Reed, 2002). The *nph4* mutation not only disrupts hypocotyl phototropism, but also other auxin-related phenotypes, suggesting that NPH4/ARF7 action is not confined to the phototropic response (Liscum and Briggs, 1995; Watahiki and Yamamoto, 1997; Stowe-Evans *et al.*, 1998). It is possible that other ARFs also participate in the phototropic response. ARF5 is a candidate as it has been shown to have overlapping functions and to interact with ARF7 *in planta*, suggesting that in some situations they may act as heterodimers (Hardtke *et al.*, 2004). Other candidates are ARF14 and ARF19, which have been shown to interact with NPH4/ARF7 *in vitro*. These observations combined with gene expression and genetic studies, using the *msg2/iaa19* and *slr/iaa14* gain-of-function mutants, suggest that ARF14 and ARF19 may act by repressing ARF7 activity in a range of physiological responses (Fukaki *et al.*, 2002; Tatematsu *et al.*, 2004; Fukaki *et al.*, 2005; Okushima *et al.*, 2005). Future work will determine whether these ARF transcription factors also play a role in phototropism.

10.5.2   *Phytochrome and cryptochrome modification of shoot phototropism*

Genetic studies have also provided insights into the interplay between photoreceptors in the modulation of phototropism. The phototropins and cryptochromes have been shown to act coordinately in the regulation of phototropism. Under moderate blue light fluence rates of 100 μmol m^{-2} s^{-1}, coaction of these photoreceptors attenuates phototropism. In contrast, their joint action enhances phototropism under low fluence rates (<1.0 μmol m^{-2} s^{-1}). phyA has also been shown to regulate phototropic curvature in response to blue light (Lariguet and Fankhauser, 2004). These experiments showed that phyA enhances phototropic curvature by suppressing gravitropism. Indeed, phyA does not appear to act through ARF7 (see above) as the *nph4/arf7* mutant retains its phyA-mediated modulation of the phototropic response (Liscum and Briggs, 1996; Stowe-Evans *et al.*, 2001). This lends support to phyA action via a separate pathway to regulate the phototropic response.

The amplitude of the blue-light-mediated phototropic curvature can also be enhanced by a prior exposure of seedlings to red light. Analysis of the *phyA*, *phyB* and *phyA phyB* deficient mutants have revealed that this modulation of phototropism is mediated by phyA and, to a lesser extent, phyB (Parks *et al.*, 1996; Janoudi *et al.*, 1997; Stowe-Evans *et al.*, 2001). It is not yet known whether this moderation of the phototropic response is through direct interaction with phot1 or the regulation of common signalling components.

10.5.3 Root phototropism

Like its shoots, *Arabidopsis* roots exhibit blue-light-mediated phototropism, but unlike shoots, roots curve away from, not towards, the light source. In roots, positive phototropic responses are generated by phyA and phyB in response to red light (Ruppel *et al.*, 2001; Kiss *et al.*, 2003). These phototropic responses are very weak when compared with gravitropism (Ruppel *et al.*, 2001; Kiss *et al.*, 2003). Indeed, *Arabidopsis* root phototropism is only visible in agravitropic mutants or when plants are grown under conditions that remove influence of the gravitational pull (Ruppel *et al.*, 2001; Kiss *et al.*, 2003). In nature, these phototropic responses may be working alongside gravitropic responses to control orientation of primary and lateral roots that are positioned close to the soil surface (Ruppel *et al.*, 2001). Precisely how these light-generated signals tie in with auxin signalling is not known; however, root phototropic responses are postulated to use similar mechanisms that operate in the shoot.

10.5.4 Gravitropism

In a similar fashion to phototropic responses, gravitropism is generated by an asymmetric distribution of auxin (Rashotte *et al.*, 2001; Boonsirichai *et al.*, 2003; Friml *et al.*, 2003; Ottenschlager *et al.*, 2003). However, the trigger for this response is different. The current hypothesis postulates that plants perceive gravity through the sedimentation of starch-filled plastids (statoliths) within specialised cells (statocytes) such as columella root cells or shoot endodermal cells (Boonsirichai *et al.*, 2002; Blancaflor and Masson, 2003; Morita and Tasaka, 2004). Several lines of evidence provide support for this proposition. Laser ablation of the central root columella cells produces an inhibitory effect on root curvature in response to gravity stimulation (Blancaflor *et al.*, 1998, 1999). Furthermore, both starchless and plastid-deficient mutants exhibit altered gravitropic response in roots and shoots (Kiss *et al.*, 1996; Weise and Kiss, 1999; Boonsirichai *et al.*, 2002; Yamamoto *et al.*, 2002). Recent work has provided some insights into the molecular and cellular events that trigger gravitropism. Gravity-induced statolith movement appears to be an important initiator of the gravitropic response. Statolith sedimentation seems to activate the actin-dependent relocalisation of PIN3 (Friml *et al.*, 2002a). As PINs facilitate the directional movement of auxin they are poised to rapidly relocate to a different membrane position. To achieve this, PIN proteins cycle between the plasma membrane and endosomes, a process that is cytoskeleton dependent. As stathiloths are enmeshed in actin, their movement is thought to reorganise the cytoskeleton. As a result PIN3 relocalises to the sides of the cells permitting the lateral transport of auxin across the organ (Friml *et al.*, 2002a). However, it is not only PIN3 that is involved in the gravity response. Other PINs also relocate in response to the gravity stimulus, and this movement is regulated by auxin itself (Paciorek *et al.*, 2005).

Although shoots exhibit positive and roots exhibit negative responses to gravity, the mechanisms involved in shoot and root gravitropism are postulated to be similar (Morita and Tasaka, 2004). However, a variety of shoot agravitropic mutants (*sgr*) have been isolated (Morita and Tasaka, 2004), indicating that at least some facets of the molecular gravitropic response are shoot specific (Morita *et al.*, 2002; Yano *et al.*, 2003).

10.5.5   Light regulation of gravitropism

Although the gravity signal is continuously present, once illuminated, the final growth orientation of a plant is defined by the combination of light and gravity signals. This balance is crucial for establishing the correct orientation of the developing shoot. Dark-grown hypocotyls of *Arabidopsis* seedlings exhibit a strong negative gravitropism. However, red and far-red light grown seedlings display randomly orientated growth, indicating that these wavelengths of light negatively regulate gravitropism in young seedlings (Liscum and Hangarter, 1993c; Hangarter, 1997). Analysis of phytochrome-deficient mutants has established that phyA and phyB participate in this modulation of the gravitropic response (Liscum and Hangarter, 1993c; Poppe *et al.*, 1996; Robson and Smith, 1996; Fankhauser and Casal, 2004). However, phyA also regulates gravitropism in response to blue light (see Section 10.52), where it has been shown to enhance phototropic curvature under these conditions (Lariguet and Fankhauser, 2004).

Light most likely regulates gravitropism by manipulating many aspects of the pathway. One point of regulation could be at the level of PIN3. The expression of *PIN3* appears to be negatively regulated by phyB, and this action is antagonised by phyA (Devlin *et al.*, 2003). Moreover, the *pin3* mutant displays a light-specific short hypocotyl phenotype suggesting the involvement of PIN3 in phyB signalling (Friml *et al.*, 2002b). The *MDR*-like genes *MDR1* and *PGP1* have been shown to be intimately involved in gravitropism and auxin transport (Noh *et al.*, 2003). Like *PIN3*, these genes also appear to be light controlled. Overexpression of sense or antisense *PGP1* constructs leads to light-specific elongation or shortening of the hypocotyl, respectively. Furthermore, *MDR1* expression is decreased by light, suggesting a tight link between light input and auxin transport facilitators.

For a variety of plant species, root gravitropism is also controlled by light, (Kiss *et al.*, 1996; Correll and Kiss, 2005). Gravity responses of primary and lateral roots are attenuated with increasing severity in *phyB* and *phyA phyB* double mutants (Correll and Kiss, 2005). This suggests that the orientation of roots is regulated by the redundant actions of phyA and phyB (Mullen and Hangarter, 2003). Interestingly, the *hy5* mutant root phenotype is similar to that of the *phyA phyB* mutant (Oyama *et al.*, 1997). HY5, a bZIP transcription factor, has an important role in light signalling, for it is regulated by multiple photoreceptors, including pyhA, phyB and cry1. In dark-grown seedlings, the COP1 E3 ligase targets HY5 for degradation (Casal and Yanovsky, 2005). In the switch to photomorphogenic development, light signals inactivate COP1, which relieves the repression of HY5-regulated gene transcription. The lateral root agravitropic traits of the *hy5* mutant result, at least partly, from altered auxin signalling (Cluis *et al.*, 2004). Several *Aux/IAA* genes, repressors of auxin-mediated signal transduction, were shown to be down-regulated in *hy5*. This work also provided evidence that HY5 could directly regulate the transcription of *AXR2/IAA7*. Thus, HY5 provides a means to integrate multiple light signals with auxin signalling.

10.6 Light and GA signal integration

The hormone gibberellic acid (GA) controls multiple aspects of plant development. These include germination, elongation and flowering, responses that are regulated by light (Halliday and Fankhauser, 2003). So, like auxin, the GA pathways provide a path via which environmental light signals can shape development. We shall be examining some of the mechanisms through which light can impose its influence on GA.

10.6.1 *Phytochrome regulation of GA biosynthesis and homeostasis*

In many species GA is essential for germination. It appears to have dual roles in this response: GA is needed to stimulate embryo growth potential, and it also promotes hydrolysis, which weakens the coat surrounding the embryo (Yamaguchi and Kamiya, 2002). The importance of GA in germination is illustrated well with the *Arabidopsis ga1* mutant. In *ga1*, where GA production is blocked at an early step in biosynthesis pathway, germination is completely inhibited. In the natural environment, light and temperature signals are important regulators of germination, and their effects are mediated, at least in part, by altering GA levels. In *Arabidopsis*, light-regulated germination is largely mediated through the phytochromes, with phyA, phyB and phyE playing prominent roles (Shinomura *et al.*, 1994; Poppe and Schafer, 1997; Shinomura *et al.*, 1998; Hennig *et al.*, 2002). During germination, phytochromes have been shown to regulate *GA3ox1* and *GA3ox2*, genes that encode GA biosynthesis enzymes (Yamaguchi *et al.*, 1998; Yamaguchi *et al.*, 2001). Transcription of these genes is controlled in a red:far-red reversible manner; however, red light control of *GA3ox2,* but not *GA3ox1*, is lost in a *phyB* mutant. This tells us that *GA3ox2* is regulated by *phyB* and *GA3ox1* by other light-stable phytochromes (Yamaguchi *et al.*, 1998).

More recent work has provided a mechanism through which phytochrome could regulate *GA3ox* levels. Central to this mechanism are the bHLH transcription factors *SPATULA (SPT)* and *PIL5*, which are related to *PIF3*. Genes in the bHLH family act as homo or heterodimers to regulate transcription and a subset of these interact with (mainly) phyB (Ni *et al.*, 1999; Martinez-Garcia *et al.*, 2000; Huq and Quail, 2002). Thus, PIF3-like genes appear to act early in phytochrome signal transduction. The *spt* mutant has elevated, and SPT OX lines have reduced, red-light-induced germination. PIL5 OX lines have reduced germination under red light, whilst under dark conditions, where the wild type is dormant, *pil5* germinates (Oh *et al.*, 2004; Penfield *et al.*, 2005). This illustrates that both SPT and PIL5 act to repress phyB-regulated germination. Furthermore, levels of *GA3ox1* and *GA3ox2* transcripts are elevated in *spt* and *pil5* mutants under conditions where germination is enhanced. Thus, SPT and PIL5 appear to antagonise phyB-regulated germination through the repression of *GA3ox1* and *GA3ox2* transcription. Interestingly, germination responses to cold are also altered in these mutants (see Section 10.7). Thus, SPT and PIL5 are involved in the integration of environmental light and cold signals in the regulation of germination.

GA is an important regulator of elongation growth. Indeed, GA levels have been shown to tightly correlate growth rate (Symons and Reid, 2003). Whilst the application of exogenous GA leads to an enhanced elongation, GA deficiency is associated with a dwarfed phenotype (Koornneef and Van der Veen, 1980). This action of GA on elongation growth is under the control of light. In cowpea seedlings, a far-red light treatment, which reduces active light-stable phytochrome levels, controls epicotyl elongation by increasing amount of bioactive GA (Martínez-García *et al.*, 2000). Furthermore, phyA has been shown to reduce bioactive GA in pea seedlings by the simultaneous up-regulation of catalytic gene expression and down-regulation of biosynthetic genes (Ait-Ali *et al.*, 1999; Reid *et al.*, 2002).

In *Arabidopsis*, GA homeostasis during elongation growth seems to be mainly controlled by regulating *GA20ox* isoform transcript abundance. This contrasts with germination where regulation of the *GA3ox* isoform appears to be more important (see Section 10.61). Overexpression of any of the *GA20ox* genes leads to a rise in bioactive GA levels and physiological changes that include elongation of hypocotyls and stems, and early flowering (Coles *et al.*, 1999). Interestingly, low red:far-red light-induced depletion in active phytochrome has a similar impact on plant growth and it elevates *GA20ox2* and *GA20ox3* transcript levels (Devlin *et al.*, 2003; Hisamatsu *et al.*, 2005). Thus, phytochrome-regulated cell elongation, and possibly flowering, is mediated, at least partly, by the manipulation of active GA levels.

10.6.2 *Light regulation of GA signalling*

Several genetic studies suggest that phytochrome action is not restricted to GA biosynthesis, it also appears to be involved in regulating aspects of GA signal transduction. This is illustrated well in studies using the *phyB ga1*, which has a longer hypocotyl when compared to the *ga1* parental line; furthermore, it exhibits an enhanced elongation response to applied GA (Reed *et al.*, 1996). As *ga1* severely impairs GA biosynthesis, this suggests that depleting phyB levels enhances GA signalling. One means via which phyB could regulate sensitivity of the GA-mediated response is by regulating DELLA action. DELLA proteins are a subgroup of the GRAS family of putative transcriptional regulators. In *Arabidopsis*, the DELLA family, which act as growth repressors, comprises *GIBBERELLIC ACID INSENSTIVE (GAI), REPRESSOR OF ga1-3 (RGA), RGA-LIKE 1-3 (RGL1, RGL2 and RGL3)* (Itoh *et al.*, 2003; Sun and Gubler, 2004; Alvey and Harberd, 2005). GA operates by repressing DELLA protein activity and this is achieved by promoting SCF[SLY1] E3 ligase targeting of DELLA proteins for subsequent destruction by the 26S proteasome.

Recent work has provided insights into the role of DELLAs in the control of germination. These studies were conducted in a (GA-deficient) *ga1* background, which renders the seed unable to germinate. Removal of DELLA genes restored germination in *ga1* to different extents (Lee *et al.*, 2002; Tyler *et al.*, 2004; Cao *et al.*, 2005). The degree of restoration depended on the individual or specific combinations of *della* mutant alleles. This work demonstrated that RGA, GAI, RGL1 and RGL2 act to enhance seed dormancy, with RGL2 playing the most prominent role in this response. Furthermore, particular combinations of *della* alleles were reported to confer either light- or dark-specific germination. These experiments suggest that light may control germination by inactivating specific combinations of DELLAs. However, additional work will be required to confirm and elucidate the mechanism of photoreceptor action in this context.

GA and therefore DELLA action extends to many aspects of plant growth and development. The DELLA genes, *RGA* and *GAI*, have been shown to have important roles in skotomorphogenesis and stem growth (King *et al.*, 2001; Alabadi *et al.*, 2004). As these processes are dramatically affected by light, it is possible that DELLAs are integral components of photoreceptor signalling. In this context, it is interesting that DELLAs are also regulated by auxin and ethylene, so they appear to be focal points for several pathways (Achard *et al.*, 2003).

Although the SPINDLY (SPY) protein is unrelated to the DELLA family, *spy* is another suppressor of the *ga1* mutation. SPY is an *O*-linked ß-*N*-acetylglucosamine transferase that represses GA signalling, possibly by regulating the phosphorylation status of target proteins (Sun and Gubler, 2004). SPY was shown to physically interact with the phyB- and clock-associated protein GIGANTEA (GI), and the *spy-4* allele was epistatic to *gi-2* for hypocotyl and flowering phenotypes (Tseng *et al.*, 2004). The current model proposes that the light acts through GI to inactivate SPY; therefore, this may represent another means via which the light and GA pathways intercept.

10.7   The thermosensory pathways

The effects of photoperiod and sustained periods of cold, experienced during winter months in temperate climates, have been well documented (Hayama and Coupland, 2003; Henderson and Dean, 2004). These conditions are a prerequisite for flowering in many plants that overwinter in a vegetative state. However, recent work has revealed that flowering time is also influenced by relatively small changes in ambient temperature (Blazquez *et al.*, 2003; Halliday *et al.*, 2003; Halliday and Whitelam, 2003). In this context, photoreceptor action appears to buffer the effects of environmental temperature fluctuations. cry1, cry2 and phyA are all positive regulators of flowering time triggered by long days (see above) (Johnson *et al.*, 1994; Mockler *et al.*, 1999; Mazzella *et al.*, 2001). However, analysis of mutants and different ambient temperatures provides a more accurate picture of how they act in the natural environment. When grown under long days, at around 23°C, *cry1* mutants flower at about the same time, whilst *cry2* mutants flower late relative to wild-type plants (Blazquez *et al.*, 2003). A small drop in ambient temperature to 16°C induces a minor flowering delay in wild-type plants, but it has a dramatic effect on flowering time

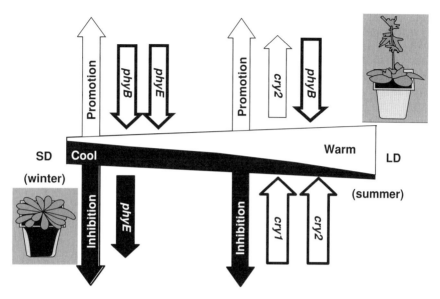

Figure 10.2   Photoreceptors buffer the impact of temperature on flowering time. During the short days (SD) of winter, phyB and phyE inhibit flowering counterbalancing the effects of warm spells, which promote flowering. phyE also inhibits flowering in cool ambient temperatures. The collective action of phyB and phyE ensures that flowering is not triggered during the less favourable winter months. During the long days (LD) of summer, cry1 and cry2, which promote flowering, buffer the inhibitory effects of cold periods. cry2 also acts as a potent regulator of flowering in warm temperatures. Thus, the collective action of cry1 and cry2 is to promote flowering over a range of ambient temperatures in inductive LD photoperiods. phyB antagonises cry2 action as temperatures rise. This may guard against precocious flowering in hot spells.

in the *cry* mutants. Growth at 16°C induces late flowering in *cry1* and further delays flowering in the already late flowering *cry2* mutant. Thus, cooler temperatures delay flowering and the cryptochromes appear to antagonise this action (Figure 10.2).

Several studies using *phyB* mutants indicate that phyB is a potent suppressor of flowering time. *phyB* mutants generally flower much earlier than wild-type plants under both long-day and short-day conditions (Halliday *et al.*, 1994; Halliday *et al.*, 2003). These studies suggest that unlike cry1, cry2 and phyA, phyB operates in-dependently of photoperiod. However, in-depth genetic and molecular analysis has revealed a more complex picture. Contrary to expectations, phyB has a role in the CO-photoperiod pathway antagonising cry1/cry2/phyA action (see Section 10.23), but its activity is not confined to this pathway (Valverde *et al.*, 2004). phyB exerts strong control on flowering by acting on key integration points of the flowering network. In addition to controlling CO, phyB also regulates flowering through the terminal floral integrator *FT* (Halliday *et al.*, 2003). By targeting *FT*, phyB can to some extent override signals through other pathways that control *FT* levels. *PHY-TOCHROME AND FLOWERING TIME 1 (PFT1)* appears to act downstream of phyB in this response (Cerdan and Chory, 2003). When compared to wild-type plants, the *pft1* mutant suppresses *FT* expression and flowers late as a consequence. Double mutant analysis showed that *pft1* is completely epistatic to *phyB* for the regulation of *FT* mRNA and flowering, in both long and short days, and this occurs independently of CO levels. This suggests that whilst phyB can control *FT* through regulation of CO, it also targets *FT* through a separate pathway that requires PFT1. These studies reveal that phyB has such a powerful impact on flowering because it can regulate the terminal floral integrator *FT* through more than one mechanism.

Interestingly, the effects of the *phyB* monogenic mutation are only seen at warmer ambient temperatures: *phyB* mutants grown at 16°C do not flower early (Halliday *et al.*, 2003). So, as for *cry1* and *cry2*, the *phyB* phenotype flowering is temperature sensitive. Thus, temperature interactions may be a common feature in light-mediated flowering responses. However, the nature of these temperature interactions is a little complex. In a similar fashion to cry1, the phyB phenotype is temperature conditional. However, the *cry1* late-flowering phenotype is only observed at cooler temperatures, whilst the *phyB* early flowering phenotype only occurs at warmer ambient temperatures (Blazquez *et al.*, 2003; Halliday *et al.*, 2003) (Figure 10.2). *phyE* mutants, which flower early under short-day photoperiods, do so at both warm and cool ambient conditions (Halliday *et al.*, 2003). In this respect, phyE has a similar operating range to cry2, as like *phyE*, the *cry2* mutant phenotype is observed over the 16°C–23°C temperature range (Figure 10.2). Thus, phyB and phyE appear to be important for antagonising the promotory effects of increased temperature on flowering. In contrast, cry1 and cry2 antagonise the inhibitory effects of decreased temperature on flowering. So, collective action of the photoreceptors appears to buffer the effects of temperature on flowering. In the natural environment, this might be a way of safeguarding the flowering response from day-to-day fluctuations in temperature. In seasonal climates, cry1 and cry2 would only be active as day length increases towards the summer, phyE would be active in short winter days and phyB would act all year round. In this scenario, cry1 and cry2 would buffer the effects of cool spells, whilst phyB would buffer the effects of warm spells during the summer months. The balance of cry1 and cry2 versus phyB action would ensure a robust flowering response regardless of small changes in temperature. In winter, phyB and phyE buffer against warm periods, ensuring that flowering remains repressed under these otherwise non-inductive conditions (Figure 10.2).

The mechanisms for thermosensory interactions in the flowering response are not yet understood. However, there is evidence that the flowering pathway genes *FCA* and *FVE* play a role (Blazquez *et al.*, 2003). The mutant alleles for these flowering time genes were originally identified in one of the early *Arabidopsis* genetic screens (Koornneef *et al.*, 1991). Fairly extensive analysis has placed them in the autonomous (non-photoperiodic) flowering pathway where their respective gene products act as positive regulators (Koornneef *et al.*, 1991, 1998; Michaels and Amasino, 2001; Rouse *et al.*, 2002). Similar to other genes in this pathway they act principally by moderating the MADS box transcription factor *FLOWERING LOCUS C (FLC)*, a key integrator of the autonomous pathway (Michaels and Amasino, 2001; Simpson and Dean, 2002; Putterill *et al.*, 2004). Studies have shown that raised *FLC* mRNA levels strongly correlate with delayed flowering, implicating FLC as a powerful negative regulator of flowering. Thus, in *fca* and *fve* mutants, *FLC* transcript levels are very high and as a result these mutants are late flowering (Rouse *et al.*, 2002). Removal of functional FLC effectively restores a wild-type flowering response to both *fca* and *fve* (Michaels and Amasino, 2001).

Reducing ambient growth conditions from 23°C to 16°C evokes a modest delay in flowering in wild-type plants. The late-flowering *fca* and *fve* mutants are completely insensitive to this temperature change, flowering at the same time in 23°C and 16°C (Blazquez *et al.*, 2003). In wild-type plants, *FLC* transcript levels are moderately higher in plants grown at the cooler temperatures, and this may account for the moderate delays in flowering observed under these conditions. The retention of a response to temperature in the *flc* mutant argues against this role for FLC. However, it is equally possible that, in *flc*, one or more genes can substitute for lack of FLC action. The latter scenario may be more probable as the *flc* mutation can restore temperature sensitivity to mutants lacking *FVE*. This requirement of functional FLC for the *fve* phenotype implicates FLC as a downstream component in the FVE-dependent temperature pathway. Support for a role for FVE in temperature responses comes from the finding that cold responsive *(COR)* genes containing the cold responsive C/DRE element are up-regulated in *fve/acg1* mutant alleles (Ausin *et al.*, 2004). Furthermore, the *COR15a promoter::GUS* gene fusions were more highly expressed in *fve/acg1* at 3°C than 23°C, suggesting a role for FVE in the temperature-regulated *COR15a* gene expression. This study also showed that the delayed flowering observed in wild-type plants subject to intermittent cold treatments was absent in *fve/acg1* mutants. These conditions also lead to enhanced levels of *FLC* in *fve/acg1*. Other studies have provided evidence that FVE acts to repress *FLC* transcription by modifying the *FLC* chromatin structure by histone deacetylation (He *et al.*, 2003; Ausin *et al.*, 2004). Thus, FVE appears to be able to integrate temperature and flowering signals at least partly through chromatin remodelling at the *FLC* locus. FCA, an RNA processing protein, negatively regulates *FLC* mRNA; however, a direct link between FCA control of *FLC* and temperature sensing has not been established (Simpson *et al.*, 2004).

Thermosensory control of development is not confined to the flowering pathways. Many vegetative processes are also subject to dual control by temperature and light. Such an interaction has been documented for *Abutilon theophrasti* (velvetleaf) an annual weed widely distributed throughout the United States and Canada (Weinig, 2000). This study demonstrated that temperature has a major impact on phytochrome-controlled elongation responses to low red:far-red ratio light. Such changes in light quality signal the presence of neighbouring plants, and trigger elongation and flowering responses (see above). This change in growth strategy enhances survival chances in a competitive environment. In *A. theophrasti,* the enhanced hypocotyl elongation observed in seedlings exposed to low red:far-red ratio light is greatly exaggerated when seedlings are grown under warm ambient temperatures (Weinig, 2000). Thus, temperature appears to be an important moderator of the response to low red:far-red ratio light in *A. theophrasti*. This contrasts with the situation in *Arabidopsis* where only minor temperature effects are observed for hypocotyl elongation in *phyA, phyB and cry1* mutants or plants carrying combinations of these mutations (Mazzella *et al.*, 2000). However, light has been shown to be important in maintaining the *Arabidopsis* rosette habit when ambient temperature increases (Mazzella *et al.*, 2000; Halliday and Whitelam, 2003). In species that form a compact rosette, internode elongation is almost entirely arrested during normal development. However, analysis of mutants null for one, two or more photoreceptors has revealed roles for both the phytochromes and cryptochromes in this response. When kept at 20°C or at alternate 20°C/30°C (15-h day/9-h night), wild-type plants grow with compact rosettes. In contrast, the *phyB, phyA phyB, phyB cry1, phyA phyB cry1* mutants display increasing degrees of internode elongation (Mazzella *et al.*, 2000). A similar situation has been observed for the *phyA phyB phyE* mutant, which has a pronounced internode phenotype when grown at 22°C (Halliday and Whitelam, 2003). These two studies suggest a hierarchy of photoreceptor action in the suppression of internode elongation, with phyB playing the most prominent role. However, the internode elongation observed in the photoreceptor mutants was temperature conditional. When grown at cooler temperatures, even the most severe photoreceptor mutants (*phyA phyB phyE and phyA phyB cry1*) showed no signs of internode elongation. So, for this response it appears that the phytochromes and cryptochromes both play a role in suppressing elongation induced by elevated ambient temperature. In this instance, the light receptor action appears to be important for maintaining the rosette habit in the natural environment, which is subject to changes in ambient temperature.

A response that is acutely sensitive to light and cold is germination. Either light or cold stratification can break dormancy in newly harvested seed. However, the combined actions of light and cold have a synergistic effect, providing a potent germination signal. Recent work has provided insights into how light and cold signals are integrated to regulate this response. The PIF3-like bHLH transcription factors *SPT* and *PIL5/PIF1* appear to be central to this mechanism (see Section 10.61). *spt* and *pil5* mutants have altered phytochrome and temperature-controlled germination responses, which result, at least partly, from enhanced GA biosynthesis (Oh *et al.*, 2004; Penfield *et al.*, 2005). In these mutants, elevated germination in *spt* and *pil5* correlates with rises in GA biosynthesis gene *GA3ox1* and *GA3ox2* transcript abundance. Thus, SPT and PIL5/PIF1 appear to act as integration points for light and cold in the regulation of germination. As SPT and PIL5/PIF1 are putative transcription factors they may act by regulating *GA3ox* transcription directly; however, this has not yet been tested and it remains possible that *GA3oxs* are not their only target. Future work will reveal whether their action is confined to the regulation of GA biosynthesis or if it extends to additional moderators of this response.

10.8   Summary

Coordinated development requires a fully integrated signalling network that is responsive to a range of external signals. Light is an incredibly influential environmental cue providing spatial and temporal information that shapes plant growth and development. We have known for some time that the light signals through the circadian system and the hormone pathways. However, it is not until recently that we have begun to understand the complexity of the network and how other environmental cues impact on shared key network connections. Recent studies demonstrating light and ambient temperature interactions in the control of development have provided preliminary insights into how information from external signals is assimilated. It is clear at this stage that several mechanisms are likely to integrate environmental signals. Understanding how the molecular network accommodates and transduces these signals represents a new intellectual challenge. To help us meet this challenge we will need to take a more holistic approach that incorporates alternative methods such as mathematical modelling. In this way, large data sets, such as transcriptome regulation through development, can be processed to provide a view of how the network is connected and how this changes through developmental time. This type of information can be combined with complementary metabolomics and protein function data to establish more precisely how the network operates. Other types of analyses can provide a means to test components or predict outcomes that can feed back to inform experimental design. It will be interesting to see how these more integrative approaches will inform the way we view signal transduction.

References

Abel, S., Nguyen, M.D. and Theologis, A. (1995) The PS-IAA4/5-like family of early auxin-inducible mRNAs in *Arabidopsis thaliana*. *J. Mol. Biol.* **251**, 533–549.

Achard, P., Vriezen, W.H., Van Der Straeten, D. and Harberd, N.P. (2003) Ethylene regulates *Arabidopsis* development via the modulation of DELLA protein growth repressor function. *Plant Cell* **15**, 2816–2825.

Ait-Ali, T., Frances, S., Weller, J.L., Reid, J.B., Kendrick, R.E. and Kamiya, Y. (1999) Regulation of gibberellin 20-oxidase and gibberellin 3 beta-hydroxylase transcript accumulation during de-etiolation of pea seedlings. *Plant Physiol.* **121**, 783–791.

Alabadi, D., Gil, J., Blazquez, M.A. and Garcia-Martinez, J.L. (2004) Gibberellins repress photomorphogenesis in darkness. *Plant Physiol.* **134**, 1050–1057.

Alabadi, D., Oyama, T., Yanovsky, M.J., Harmon, F.G., Mas, P. and Kay, S.A. (2001) Reciprocal regulation between TOC1 and LHY/CCA1 within the *Arabidopsis* circadian clock. *Science* **293**, 880–883.

Alvey, L. and Harberd, N.P. (2005) DELLA proteins: integrators of multiple plant growth regulatory inputs? *Physiol. Plant* **123**, 153–160.

Ausin, I., Alonso-Blanco, C., Jarillo, J.A., Ruiz-Garcia, L. and Martinez-Zapater, J.M. (2004) Regulation of flowering time by FVE, a retinoblastoma-associated protein. *Nat. Genet.* **36**, 162–166.

Bhalerao, R.P., Eklof, J., Ljung, K., Marchant, A., Bennett, M. and Sandberg, G. (2002) Shoot-derived auxin is essential for early lateral root emergence in *Arabidopsis* seedlings. *Plant J.* **29**, 325–332.

Blakeslee, J.J., Bandyopadhyay, A., Peer, W.A., Makam, S.N. and Murphy, A.S. (2004) Relocalization of the PIN1 auxin efflux facilitator plays a role in phototropic responses. *Plant Physiol.* **134**, 28–31.

Blakeslee, J.J., Peer, W.A. and Murphy, A.S. (2005) Auxin transport. *Curr. Opin. Plant Biol.* **8**, 494.

Blancaflor, E.B., Fasano, J.M. and Gilroy, S. (1999) Laser ablation of root cap cells: implications for models of graviperception. *Adv. Space Res.* **24**, 731–738.

Blancaflor, E.B., Fasano, J.M. and Gilroy, S. (1998) Mapping the functional roles of cap cells in the response of *Arabidopsis* primary roots to gravity. *Plant Physiol.* **116**, 213–222.

Blancaflor, E.B. and Masson, P.H. (2003) Plant gravitropism. Unraveling the ups and downs of a complex process. *Plant Physiol.* **133**, 1677–1690.

Blazquez, M.A., Ahn, J.H. and Weigel, D.(2003) A thermosensory pathway controlling flowering time in *Arabidopsis thaliana*. *Nat. Genet.* **33**, 168–171.

Boonsirichai, K., Guan, C., Chen, R. and Masson, P.H. (2002) Root gravitropism: an experimental tool to investigate basic cellular and molecular processes underlying mechanosensing and signal transmission in plants. *Annu. Rev. Plant Biol.* **53**, 421–447.

Boonsirichai, K., Sedbrook, J.C., Chen, R., Gilroy, S. and Masson, P.H. (2003) ALTERED RESPONSE TO GRAVITY is a peripheral membrane protein that modulates gravity-induced cytoplasmic alkalinization and lateral auxin transport in plant statocytes. *Plant Cell* **15**, 2612–2625.

Botto, J.F. and Smith, H. (2002) Differential genetic variation in adaptive strategies to a common environmental signal in *Arabidopsis* accessions: phytochrome-mediated shade avoidance. *Plant Cell Environ.* **25**, 53–63.

Briggs, W.R. and Christie, J.M. (2002) Phototropins 1 and 2: versatile plant blue-light receptors. *Trends Plant Sci.* **7**, 204–210.

Cao, D., Hussain, A., Cheng, H. and Peng, J. (2005) Loss of function of four DELLA genes leads to light- and gibberellin-independent seed germination in *Arabidopsis*. *Planta* **223**, 105–113.

Carabelli, M., Morelli, G., Whitelam, G. and Ruberti, I. (1996) Twilight-zone and canopy shade induction of the Athb-2 homeobox gene in green plants. *Proc. Natl. Acad. Sci. USA* **93**, 3530–3535.

Casal, J.J. and Yanovsky, M.J. (2005) Regulation of gene expression by light. *Int. J. Dev. Biol.* **49**, 501–511.

Cerdan, P.D. and Chory, J. (2003) Regulation of flowering time by light quality. *Nature* **423**, 881–885.

Cholodny, N. (1927) Wuchshormone und tropismen bei den pflanzen. *Biol. Zentralbl.* **47**, 604–626.

Cluis, C.P., Mouchel, C.F. and Hardtke, C.S.(2004) The *Arabidopsis* transcription factor HY5 integrates light and hormone signaling pathways. *Plant J.* **38**, 332–347.

Coles, J.P., Phillips, A.L., Croker, S.J., Garcia-Lepe, R., Lewis, M.J. and Hedden, P. (1999) Modification of gibberellin production and plant development in *Arabidopsis* by sense and antisense expression of gibberellin 20-oxidase genes. *Plant J.* **17**, 547–556.

Colon-Carmona, A., Chen, D.L., Yeh, K.C. and Abel, S. (2000) Aux/IAA proteins are phosphorylated by phytochrome in vitro. *Plant Physiol.* **124**, 1728–1738.

Cope, G.A. and Deshaies, R.J. (2003) COP9 signalosome: a multifunctional regulator of SCF and other cullin-based ubiquitin ligases. *Cell* **114**, 663–671.

Correll, M.J. and Kiss, J.Z. (2005) The roles of phytochromes in elongation and gravitropism of roots. *Plant Cell Physiol.* **46**, 317–323.

Covington, M.F., Panda, S., Liu, X.L., Strayer, C.A., Wagner, D.R. and Kay, S.A.(2001) ELF3 modulates resetting of the circadian clock in *Arabidopsis*. *Plant Cell* **13**, 1305–1315.

Devlin, P.F. and Kay, S.A. (2000) Cryptochromes are required for phytochrome signaling to the circadian clock but not for rhythmicity. *Plant Cell* **12**, 2499–2510.

Devlin, P.F., Yanovsky, M.J. and Kay, S.A. (2003) A genomic analysis of the shade avoidance response in *Arabidopsis*. *Plant Physiol.* **133**, 1617–1629.

Dharmasiri, N., Dharmasiri, S. and Estelle, M. (2005) The F-box protein TIR1 is an auxin receptor. *Nature* **435**, 441–445.

Dharmasiri, S., Dharmasiri, N., Hellmann, H. and Estelle, M. (2003) The RUB/Nedd8 conjugation pathway is required for early development in *Arabidopsis*. *EMBO J.* **22**, 1762–1770.

Dodd, A.N., Salathia, N., Hall, A., *et al.* (2005) Plant circadian clocks increase photosynthesis, growth, survival, and competitive advantage. *Science* **309**, 630–633.

Donohue, K., Pyle, E.H., Messiqua, D., Heschel, M.S. and Schmitt, J.(2001) Adaptive divergence in plasticity in natural populations of *Impatiens capensis* and its consequences for performance in novel habitats. *Evolution* **55**, 692–702.

Dowson-Day, M.J. and Millar, A.J. (1999) Circadian dysfunction causes aberrant hypocotyl elongation patterns in *Arabidopsis*. *Plant J.* **17**, 63–71.

Doyle, M.R., Davis, S.J., Bastow, R.M., *et al.* (2002) The ELF4 gene controls circadian rhythms and flowering time in *Arabidopsis thaliana*. *Nature* **419**, 74–77.

Esmon, C.A., Pedmale, U.V. and Liscum, E. (2005) Plant tropisms: providing the power of movement to a sessile organism. *Int. J. Dev. Biol.* **49**, 665–674.

Fankhauser, C. and Casal, J.J. (2004) Phenotypic characterization of a photomorphogenic mutant. *Plant J.* **39**, 747–760.

Franklin, K.A., Praekelt, U., Stoddart, W.M., Billingham, O.E., Halliday, K.J. and Whitelam, G.C. (2003) Phytochromes B, D, and E act redundantly to control multiple physiological responses in *Arabidopsis. Plant Physiol.* **131**, 1340–1346.

Franklin, K.A. and Whitelam, G.C. (2005) Phytochromes and shade-avoidance responses in plants. *Ann. Bot. (Lond.)* **96**, 169–175.

Friml, J. (2003) Auxin transport – shaping the plant. *Curr. Opin. Plant Biol.* **6**, 7–12.

Friml, J., Benkova, E., Blilou, I., *et al.* (2002b) AtPIN4 mediates sink-driven auxin gradients and root patterning in *Arabidopsis. Cell* **108**, 661–673.

Friml, J., Vieten, A., Sauer, M., *et al.* (2003) Efflux-dependent auxin gradients establish the apical-basal axis of *Arabidopsis. Nature* **426**, 147–153.

Friml, J., Wisniewska, J., Benkova, E., Mendgen, K. and Palme, K.(2002a) Lateral relocation of auxin efflux regulator PIN3 mediates tropism in *Arabidopsis. Nature* **415**, 806–809.

Fukaki, H., Nakao, Y., Okushima, Y., Theologis, A. and Tasaka, M. (2005) Tissue-specific expression of stabilized SOLITARY-ROOT/IAA14 alters lateral root development in *Arabidopsis. Plant J.* **44**, 382–395.

Fukaki, H., Tameda, S., Masuda, H. and Tasaka, M. (2002) Lateral root formation is blocked by a gain-of-function mutation in the SOLITARY-ROOT/IAA14 gene of *Arabidopsis. Plant J.* **29**, 153–168.

Geldner, N., Friml, J., Stierhof, Y.D., Jurgens, G. and Palme, K. (2001) Auxin transport inhibitors block PIN1 cycling and vesicle trafficking. *Nature* **413**, 425–428.

Gil, P., Dewey, E., Friml, J., *et al.* (2001) BIG: a calossin-like protein required for polar auxin transport in *Arabidopsis. Genes Dev.* **15**, 1985–1997.

Hagen, G. and Guilfoyle, T. (2002) Auxin-responsive gene expression: genes, promoters and regulatory factors. *Plant Mol. Biol.* **49**, 373–385.

Hall, A., Bastow, R.M., Davis, S.J., *et al.* (2003) The TIME FOR COFFEE gene maintains the amplitude and timing of *Arabidopsis* circadian clocks. *Plant Cell* **15**, 2719–2729.

Halliday, K.J. and Fankhauser, C. (2003) Phytochrome-hormonal signaling networks. *New Phytol.* **157**, 449–463.

Halliday, K.J., Koornneef, M. and Whitelam, G.C. (1994) Phytochrome B and at least one other phytochrome mediate the accelerated flowering response of *Arabidopsis thaliana* l. to low red/far-red ratio. *Plant Physiol.* **104**, 1311–1315.

Halliday, K.J., Salter, M.G., Thingnaes, E. and Whitelam, G.C. (2003) Phytochrome control of flowering is temperature sensitive and correlates with expression of the floral integrator FT. *Plant J.* **33**, 875–885.

Halliday, K.J. and Whitelam, G.C. (2003) Changes in photoperiod or temperature alter the functional relationships between phytochromes and reveal roles for phyD and phyE. *Plant Physiol.* **131**, 1913–1920.

Han, L., Mason, M., Risseeuw, E.P., Crosby, W.L. and Somers, D.E. (2004) Formation of an SCF(ZTL) complex is required for proper regulation of circadian timing. *Plant J.* **40**, 291–301.

Hangarter, R.P. (1997) Gravity, light and plant form. *Plant Cell Environ.* **20**, 796–800.

Hardtke, C.S., Ckurshumova, W., Vidaurre, D.P., *et al.* (2004) Overlapping and non-redundant functions of the *Arabidopsis* auxin response factors MONOPTEROS and NONPHOTOTROPIC HYPOCOTYL 4. *Development* **131**, 1089–1100.

Harper, R.M., Stowe-Evans, E.L., Luesse, D.R., *et al.* (2000) The NPH4 locus encodes the auxin response factor ARF7, a conditional regulator of differential growth in aerial *Arabidopsis* tissue. *Plant Cell* **12**, 757–770.

Hayama, R. and Coupland, G. (2003) Shedding light on the circadian clock and the photoperiodic control of flowering. *Curr. Opin. Plant Biol.* **6**, 13–19.

Hazen, S.P., Schultz, T.F., Pruneda-Paz, J.L., Borevitz, J.O., Ecker, J.R. and Kay, S.A. (2005) LUX ARRHYTHMO encodes a Myb domain protein essential for circadian rhythms. *Proc. Natl. Acad. Sci. USA* **102**, 10387–10392.

He, Y., Michaels, S.D. and Amasino, R.M. (2003) Regulation of flowering time by histone acetylation in *Arabidopsis*. *Science* **302**, 1751–1754.

Henderson, I.R. and Dean, C. (2004) Control of *Arabidopsis* flowering: the chill before the bloom. *Development* **131**, 3829–3838.

Hennig, L., Stoddart, W.M., Dieterle, M., Whitelam, G.C. and Schafer, E. (2002) Phytochrome E controls light-induced germination of *Arabidopsis*. *Plant Physiol.* **128**, 194–200.

Hicks, K.A., Albertson, T.M. and Wagner, D.R. (2001) EARLY FLOWERING3 encodes a novel protein that regulates circadian clock function and flowering in *Arabidopsis*. *Plant Cell* **13**, 1281–1292.

Hisamatsu, T., King, R.W., Helliwell, C.A. and Koshioka, M. (2005) The involvement of gibberellin 20-oxidase genes in phytochrome-regulated petiole elongation of *Arabidopsis*. *Plant Physiol.* **138**, 1106–1116.

Hoecker, U., Toledo-Ortiz, G., Bender, J. and Quail, P.H. (2004) The photomorphogenesis-related mutant red1 is defective in CYP83B1, a red light-induced gene encoding a cytochrome P450 required for normal auxin homeostasis. *Planta* **219**, 195–200.

Huq, E. and Quail, P.H. (2002) PIF4, a phytochrome-interacting bHLH factor, functions as a negative regulator of phytochrome B signaling in *Arabidopsis*. *EMBO J.* **21**, 2441–2450.

Imaizumi, T., Tran, H.G., Swartz, T.E., Briggs, W.R. and Kay, S.A. (2003) FKF1 is essential for photoperiodic-specific light signalling in *Arabidopsis*. *Nature* **426**, 302–306.

Inada, S., Ohgishi, M., Mayama, T., Okada, K. and Sakai, T. (2004) RPT2 is a signal transducer involved in phototropic response and stomatal opening by association with phototropin 1 in *Arabidopsis thaliana*. *Plant Cell* **16**, 887–896.

Itoh, H., Matsuoka, M. and Steber, C.M. (2003) A role for the ubiquitin-26S-proteasome pathway in gibberellin signaling. *Trends Plant Sci.* **8**, 492–497.

Janoudi, A.K., Gordon, W.R., Wagner, D., Quail, P. and Poff, K.L. (1997) Multiple phytochromes are involved in red-light-induced enhancement of first-positive phototropism in *Arabidopsis thaliana*. *Plant Physiol.* **113**, 975–979.

Jensen, P.J., Hangarter, R.P. and Estelle, M. (1998) Auxin transport is required for hypocotyl elongation in light-grown but not dark-grown *Arabidopsis*. *Plant Physiol.* **116**, 455–462.

Johnson, E., Bradley, M., Harberd, N.P. and Whitelam, G.C. (1994) Photoresponses of light-grown phya mutants of *Arabidopsis* (phytochrome a is required for the perception of daylength extensions). *Plant Physiol.* **105**, 141–149.

Kanyuka, K., Praekelt, U., Franklin, K.A., *et al.* (2003) Mutations in the huge *Arabidopsis* gene BIG affect a range of hormone and light responses. *Plant J.* **35**, 57–70.

Kepinski, S. and Leyser, O. (2004) Auxin-induced SCFTIR1-Aux/IAA interaction involves stable modification of the SCFTIR1 complex. *Proc. Natl. Acad. Sci. USA* **101**, 12381–12386.

Kepinski, S. and Leyser, O. (2005) The *Arabidopsis* F-box protein TIR1 is an auxin receptor. *Nature* **435**, 446–451.

Kikis, E.A., Khanna, R. and Quail, P.H. (2005) ELF4 is a phytochrome-regulated component of a negative-feedback loop involving the central oscillator components CCA1 and LHY. *Plant J.* **44**, 300–313.

Kim, B.C., Soh, M.S., Hong, S.H., Furuya, M. and Nam, H.G. (1998) Photomorphogenic development of the *Arabidopsis* shy2-1D mutation and its interaction with phytochromes in darkness. *Plant J.* **15**, 61–68.

Kim, B.C., Soh, M.C., Kang, B.J., Furuya, M. and Nam, H.G. (1996) Two dominant photomorphogenic mutations of *Arabidopsis thaliana* identified as suppressor mutations of hy2. *Plant J.* **9**, 441–456.

Kim, W.Y., Geng, R. and Somers, D.E. (2003) Circadian phase-specific degradation of the F-box protein ZTL is mediated by the proteasome. *Proc. Natl. Acad. Sci. USA* **100**, 4933–4938.

King, K.E., Moritz, T. and Harberd, N.P. (2001) Gibberellins are not required for normal stem growth in *Arabidopsis thaliana* in the absence of GAI and RGA. *Genetics* **159**, 767–776.

Kiss, J.Z., Mullen, J.L., Correll, M.J. and Hangarter, R.P. (2003) Phytochromes A and B mediate red-light-induced positive phototropism in roots. *Plant Physiol.* **131**, 1411–1417.

Kiss, J.Z., Wright, J.B. and Caspar, T. (1996) Gravitropism in roots of intermediate-starch mutants of *Arabidopsis*. *Physiol. Plant* **97**, 237–244.

Koornneef, M., Alonso-Blanco, C., Blankestijn-de Vries, H., Hanhart, C.J. and Peeters, A.J. (1998) Genetic interactions among late-flowering mutants of *Arabidopsis*. *Genetics* **148**, 885–892.

Koornneef, M., Hanhart, C.J. and Van Der Veen, J.H. (1991) A genetic and physiological analysis of late flowering mutants in *Arabidopsis thaliana*. *Mol. Gen. Genet.* **229**, 57–66.

Koornneef, M. and Van Der Veen, J.H. (1980) Induction and analysis of gibberellin-insensitive mutants in *Arabidopsis thaliana* (L.) *Heynh. Theor. Appl. Genet.* **58**, 257–263.

Lariguet, P. and Fankhauser, C. (2004) Hypocotyl growth orientation in blue light is determined by phytochrome A inhibition of gravitropism and phototropin promotion of phototropism. *Plant J.* **40**, 826–834.

Lee, S., Cheng, H., King, K.E., *et al.* (2002) Gibberellin regulates *Arabidopsis* seed germination via RGL2, a GAI/RGA-like gene whose expression is up-regulated following imbibition. *Genes Dev.* **16**, 646–658.

Leyser, O. (2005) Auxin distribution and plant pattern formation: how many angels can dance on the point of PIN? *Cell* **121**, 819–822.

Li, H.M., Altschmied, L. and Chory, J. (1994) *Arabidopsis* mutants define downstream branches in the phototransduction pathway. *Genes Dev.* **8**, 339–349.

Liscum, E. and Briggs, W.R. (1995) Mutations in the NPH1 locus of *Arabidopsis* disrupt the perception of phototropic stimuli. *Plant Cell* **7**, 473–485.

Liscum, E. and Briggs, W.R. (1996) Mutations of *Arabidopsis* in potential transduction and response components of the phototropic signaling pathway. *Plant Physiol.* **112**, 291–296.

Liscum, E. and Hangarter, R.P. (1993c) Genetic evidence that the red-absorbing form of phytochrome B mediates gravitropism in *Arabodipsis thaliana*. *Plant Physiol.* **103**, 15–19.

Liscum, E. and Reed, J.W. (2002) Genetics of Aux/IAA and ARF action in plant growth and development. *Plant Mol. Biol.* **49**, 387–400.

Liu, X.L., Covington, M.F., Fankhauser, C., Chory, J. and Wagner, D.R. (2001) ELF3 encodes a circadian clock-regulated nuclear protein that functions in an *Arabidopsis* PHYB signal transduction pathway. *Plant Cell* **13**, 1293–1304.

Makino, S., Matsushika, A., Kojima, M., Yamashino, T. and Mizuno, T. (2002) The APRR1/TOC1 quintet implicated in circadian rhythms of *Arabidopsis thaliana*. I: Characterization with APRR1-overexpressing plants. *Plant Cell Physiol.* **43**, 58–69.

Martinez-Garcia, J.F., Huq, E. and Quail, P.H. (2000) Direct targeting of light signals to a promoter element-bound transcription factor. *Science* **288**, 859–863.

Martínez-García, J.F., Santes, C.M. and García-Martínez, J.L. (2000) The end-of-day far-red irradiation increases gibberellin A1 content in cowpea (Vigna sinensis) epicotyls by reducing its inactivation. *Physiol. Plant* **108**, 426–434.

Mas, P., Kim, W.Y., Somers, D.E. and Kay, S.A. (2003) Targeted degradation of TOC1 by ZTL modulates circadian function in *Arabidopsis thaliana*. *Nature* **426**, 567–570.

Mazzella, M.A., Bertero, D. and Casal, J.J. (2000) Temperature-dependent internode elongation in vegetative plants of *Arabidopsis thaliana* lacking phytochrome B and cryptochrome 1. *Planta* **210**, 497–501.

Mazzella, M.A., Cerdan, P.D., Staneloni, R.J. and Casal, J.J. (2001) Hierarchical coupling of phytochromes and cryptochromes reconciles stability and light modulation of *Arabidopsis* development. *Development* **128**, 2291–2299.

McWatters, H.G., Bastow, R.M., Hall, A. and Millar, A.J. (2000) The ELF3 zeitnehmer regulates light signalling to the circadian clock. *Nature* **408**, 716–720.

Michaels, S.D. and Amasino, R.M. (2001) Loss of FLOWERING LOCUS C activity eliminates the late-flowering phenotype of FRIGIDA and autonomous pathway mutations but not responsiveness to vernalization. *Plant Cell* **13**, 935–941.

Millar, A.J. (2003) A suite of photoreceptors entrains the plant circadian clock. *J. Biol. Rhythms* **18**, 217–226.

Mizuno, T. and Nakamichi, N. (2005) Pseudo-response regulators (PRRs) or true oscillator components (TOCs). *Plant Cell Physiol.* **46**, 677–685.

Mockler, T.C., Guo, H., Yang, H., Duong, H. and Lin, C. (1999) Antagonistic actions of *Arabidopsis* cryptochromes and phytochrome B in the regulation of floral induction. *Development* **126**, 2073–2082.

Moon, J., Parry, G. and Estelle, M. (2004) The ubiquitin-proteasome pathway and plant development. *Plant Cell* **16**, 3181–3195.

Morelli, G. and Ruberti, I. (2000) Shade avoidance responses. Driving auxin along lateral routes. *Plant Physiol.* **122**, 621–626.

Morita, M.T., Kato, T., Nagafusa, K., *et al.* (2002) Involvement of the vacuoles of the endodermis in the early process of shoot gravitropism in *Arabidopsis*. *Plant Cell* **14**, 47–56.

Morita, M.T. and Tasaka, M. (2004) Gravity sensing and signaling. *Curr. Opin. Plant Biol.* **7**, 712–718.

Motchoulski, A. and Liscum, E. (1999) *Arabidopsis* NPH3: a NPH1 photoreceptor-interacting protein essential for phototropism. *Science* **286**, 961–964.

Mullen, J.L. and Hangarter, R.P. (2003) Genetic analysis of the gravitropic set-point angle in lateral roots of *Arabidopsis*. *Adv. Space Res.* **31**, 2229–2236.

Nagpal, P., Walker, L.M., Young, J.C., *et al.* (2000) AXR2 encodes a member of the Aux/IAA protein family. *Plant Physiol.* **123**, 563–574.

Nakamichi, N., Kita, M., Ito, S., Sato, E., Yamashino, T. and Mizuno, T. (2005) The *Arabidopsis* pseudo-response regulators, PRR5 and PRR7, coordinately play essential roles for circadian clock function. *Plant Cell Physiol.* **46**, 609–619.

Nakazawa, M., Yabe, N., Ichikawa, T., *et al.* (2001) DFL1, an auxin-responsive GH3 gene homologue, negatively regulates shoot cell elongation and lateral root formation, and positively regulates the light response of hypocotyl length. *Plant J.* **25**, 213–221.

Ni, M. (2005) Integration of light signaling with photoperiodic flowering and circadian rhythm. *Cell Res.* **15**, 559–566.

Ni, M., Tepperman, J.M. and Quail, P.H. (1999) Binding of phytochrome B to its nuclear signalling partner PIF3 is reversibly induced by light. *Nature* **400**, 781–784.

Noh, B., Bandyopadhyay, A., Peer, W.A., Spalding, E.P. and Murphy, A.S. (2003) Enhanced gravi- and phototropism in plant mdr mutants mislocalizing the auxin efflux protein PIN1. *Nature* **423**, 999–1002.

Oh, E., Kim, J., Park, E., Kim, J.I., Kang, C. and Choi, G. (2004) PIL5, a phytochrome-interacting basic helix-loop-helix protein, is a key negative regulator of seed germination in *Arabidopsis thaliana*. *Plant Cell* **16**, 3045–3058.

Okada, K. and Shimura, Y. (1992) Mutational analysis of root gravitropism and phototropism of *Arabidopsis thaliana* seedlings. *Aust. J. Plant Physiol.* **19**, 439–448.

Okushima, Y., Overvoorde, P.J., Arima, K., *et al.* (2005) Functional genomic analysis of the AUXIN RESPONSE FACTOR gene family members in *Arabidopsis thaliana*: unique and overlapping functions of ARF7 and ARF19. *Plant Cell* **17**, 444–463.

Ottenschlager, I., Wolff, P., Wolverton, C., *et al.* (2003) Gravity-regulated differential auxin transport from columella to lateral root cap cells. *Proc. Natl. Acad. Sci. USA* **100**, 2987–2991.

Oyama, T., Shimura, Y. and Okada, K. (1997) The *Arabidopsis* HY5 gene encodes a bZIP protein that regulates stimulus-induced development of root and hypocotyl. *Genes Dev.* **11**, 2983–2995.

Paciorek, T., Zazimalova, E., Ruthardt, N., *et al.* (2005) Auxin inhibits endocytosis and promotes its own efflux from cells. *Nature* **435**, 1251–1256.

Panda, S., Poirier, G.G. and Kay, S.A. (2002) tej defines a role for poly(ADP-ribosyl)ation in establishing period length of the *Arabidopsis* circadian oscillator. *Dev. Cell* **3**, 51–61.

Parks, B.M., Quail, P.H. and Hangarter, R.P. (1996) Phytochrome A regulates red-light induction of phototropic enhancement in *Arabidopsis*. *Plant Physiol.* **110**, 155–162.

Penfield, S., Josse, E.M., Kannangara, R., Gilday, A.D., Halliday, K.J. and Graham, I.A. (2005) Cold and light control seed germination through the bHLH transcription factor SPATULA. *Curr. Biol.* **15**, 1998–2006.

Poppe, C., Hangarter, R.P., Sharrock, R.A., Nagy, F. and Schafer, E. (1996) The light-induced reduction of the gravitropic growth-orientation of seedlings of *Arabidopsis thaliana* (L.) Heynh is a photomorphogenic response mediated synergistically by the far-red-absorbing forms of phytochromes A and B. *Planta* **199**, 511–514.

Poppe, C. and Schafer, E. (1997) Seed germination of *Arabidopsis thaliana* phyA/phyB double mutants is under phytochrome control. *Plant Physiol.* **114**, 1487–1492.

Putterill, J., Laurie, R. and Macknight, R. (2004) It's time to flower: the genetic control of flowering time. *Bioessays* **26**, 363–373.

Rashotte, A.M., DeLong, A. and Muday, G.K. (2001) Genetic and chemical reductions in protein phosphatase activity alter auxin transport, gravity response, and lateral root growth. *Plant Cell* **13**, 1683–1697.

Reed, J.W., Elumalai, R.P. and Chory, J. (1998) Suppressors of an *Arabidopsis thaliana* phyB mutation identify genes that control light signaling and hypocotyl elongation. *Genetics* **148**, 1295–1310.

Reed, J.W., Foster, K.R., Morgan, P.W. and Chory, J. (1996) Phytochrome B affects responsiveness to gibberellins in *Arabidopsis*. *Plant Physiol.* **112**, 337–342.

Reed, J.W., Nagpal, P., Bastow, R.M., *et al.* (2000) Independent action of ELF3 and phyB to control hypocotyl elongation and flowering time. *Plant Physiol.* **122**, 1149–1160.

Reid, J.B., Botwright, N.A., Smith, J.J., O'Neill, D.P. and Kerckhoffs, L.H. (2002) Control of gibberellin levels and gene expression during de-etiolation in pea. *Plant Physiol.* **128**, 734–741.

Robson, P.R. and Smith, H. (1996) Genetic and transgenic evidence that phytochromes A and B act to modulate the gravitropic orientation of *Arabidopsis thaliana* hypocotyls. *Plant Physiol.* **110**, 211–216.

Rouse, D.T., Sheldon, C.C., Bagnall, D.J., Peacock, W.J. and Dennis, E.S. (2002) FLC, a repressor of flowering, is regulated by genes in different inductive pathways. *Plant J.* **29**, 183–191.

Ruppel, N.J., Hangarter, R.P. and Kiss, J.Z. (2001) Red-light-induced positive phototropism in *Arabidopsis* roots. *Planta* **212**, 424–430.

Sakai, T., Wada, T., Ishiguro, S. and Okada, K. (2000) RPT2. A signal transducer of the phototropic response in *Arabidopsis*. *Plant Cell* **12**, 225–236.

Salome, P.A. and McClung, C.R. (2004) The *Arabidopsis thaliana* clock. *J. Biol. Rhythms* **19**, 425–435.

Salter, M.G., Franklin, K.A. and Whitelam, G.C. (2003) Gating of the rapid shade-avoidance response by the circadian clock in plants. *Nature* **426**, 680–683.

Schena, M., Lloyd, A.M. and Davis, R.W. (1993) The HAT4 gene of *Arabidopsis* encodes a developmental regulator. *Genes Dev.* **7**, 367–379.

Schoning, J.C. and Staiger, D. (2005) At the pulse of time: protein interactions determine the pace of circadian clocks. *FEBS Lett.* **579**, 3246–3252.

Schwechheimer, C. (2004) The COP9 signalosome (CSN): an evolutionary conserved proteolysis regulator in eukaryotic development. *Biochim. Biophys. Acta.* **1695**, 45–54.

Searle, I. and Coupland, G. (2004) Induction of flowering by seasonal changes in photoperiod. *EMBO J.* **23**, 1217–1222.

Shinomura, T., Hanzawa, H., Schafer, E. and Furuya, M. (1998) Mode of phytochrome B action in the photoregulation of seed germination in *Arabidopsis thaliana*. *Plant J.* **13**, 583–590.

Shinomura, T., Nagatani, A., Chory, J. and Furuya, M. (1994) The induction of seed germination in *Arabidopsis thaliana* is regulated principally by phytochrome B and secondarily by phytochrome A. *Plant Physiol.* **104**, 363–371.

Simpson, G.G. and Dean, C. (2002) *Arabidopsis*, the Rosetta stone of flowering time? *Science* **296**, 285–289.

Simpson, G.G., Quesada, V., Henderson, I.R., Dijkwel, P.P., Macknight, R. and Dean, C. (2004) RNA processing and *Arabidopsis* flowering time control. *Biochem. Soc. Trans.* **32**, 565–566.

Somers, D.E., Devlin, P.F. and Kay, S.A. (1998) Phytochromes and cryptochromes in the entrainment of the *Arabidopsis* circadian clock. *Science* **282**, 1488–1490.

Staiger, D., Allenbach, L., Salathia, N., *et al.* (2003) The *Arabidopsis* SRR1 gene mediates phyB signaling and is required for normal circadian clock function. *Genes Dev.* **17**, 256–268.

Staswick, P.E., Serban, B., Rowe, M., *et al.* (2005) Characterization of an *Arabidopsis* enzyme family that conjugates amino acids to indole-3-acetic acid. *Plant Cell* **17**, 616–627.

Steindler, C., Matteucci, A., Sessa, G., *et al.* (1999) Shade avoidance responses are mediated by the ATHB-2 HD-zip protein, a negative regulator of gene expression. *Development* **126**, 4235–4245.

Stowe-Evans, E.L., Harper, R.M., Motchoulski, A.V. and Liscum, E. (1998) NPH4, a conditional modulator of auxin-dependent differential growth responses in *Arabidopsis*. *Plant Physiol.* **118**, 1265–1275.

Stowe-Evans, E.L., Luesse, D.R. and Liscum, E. (2001) The enhancement of phototropin-induced phototropic curvature in *Arabidopsis* occurs via a photoreversible phytochrome A-dependent modulation of auxin responsiveness. *Plant Physiol.* **126**, 826–834.

Strayer, C., Oyama, T., Schultz, T.F., *et al.* (2000) Cloning of the *Arabidopsis* clock gene TOC1, an autoregulatory response regulator homolog. *Science* **289**, 768–771.

Sun, T.P. and Gubler, F. (2004) Molecular mechanism of gibberellin signaling in plants. *Annu. Rev. Plant Biol.* **55**, 197–223.

Symons, G.M. and Reid, J.B. (2003) Hormone levels and response during de-etiolation in pea. *Planta* **216**, 422–431.

Takase, T., Nakazawa, M., Ishikawa, A., Manabe, K. and Matsui, M. (2003) DFL2, a new member of the *Arabidopsis* GH3 gene family, is involved in red light-specific hypocotyl elongation. *Plant Cell Physiol.* **44**, 1071–1080.

Takase, T., Nakazawa, M., Ishikawa, A., *et al.* (2004) ydk1-D, an auxin-responsive GH3 mutant that is involved in hypocotyl and root elongation. *Plant J.* **37**, 471–483.

Tanaka, S., Mochizuki, N. and Nagatani, A. (2002) Expression of the AtGH3a gene, an *Arabidopsis* homologue of the soybean GH3 gene, is regulated by phytochrome B. *Plant Cell Physiol.* **43**, 281–289.

Tatematsu, K., Kumagai, S., Muto, H., *et al.* (2004) MASSUGU2 encodes Aux/IAA19, an auxin-regulated protein that functions together with the transcriptional activator NPH4/ARF7 to regulate differential growth responses of hypocotyl and formation of lateral roots in *Arabidopsis thaliana*. *Plant Cell* **16**, 379–393.

Tepperman, J.M., Zhu, T., Chang, H.S., Wang, X. and Quail, P.H. (2001) Multiple transcription-factor genes are early targets of phytochrome A signaling. *Proc. Natl. Acad. Sci. USA* **98**, 9437–9442.

Tian, Q., Nagpal, P. and Reed, J.W. (2003) Regulation of *Arabidopsis* SHY2/IAA3 protein turnover. *Plant J.* **36**, 643–651.

Tian, Q. and Reed, J.W. (1999) Control of auxin-regulated root development by the *Arabidopsis thaliana* SHY2/IAA3 gene. *Development* **126**, 711–721.

Tseng, T.S., Salome, P.A., McClung, C.R. and Olszewski, N.E. (2004) SPINDLY and GIGANTEA interact and act in *Arabidopsis thaliana* pathways involved in light responses, flowering, and rhythms in cotyledon movements. *Plant Cell* **16**, 1550–1563.

Tyler, L., Thomas, S.G., Hu, J., *et al.* (2004) Della proteins and gibberellin-regulated seed germination and floral development in *Arabidopsis*. *Plant Physiol.* **135**, 1008–1019.

Valverde, F., Mouradov, A., Soppe, W., Ravenscroft, D., Samach, A. and Coupland, G. (2004) Photoreceptor regulation of CONSTANS protein in photoperiodic flowering. *Science* **303**, 1003–1006.

Vierstra, R.D. (2003) The ubiquitin/26S proteasome pathway, the complex last chapter in the life of many plant proteins. *Trends Plant Sci.* **8**, 135–142.

Wagner, D., Hoecker, U. and Quail, P.H. (1997) RED1 is necessary for phytochrome B-mediated red light-specific signal transduction in *Arabidopsis*. *Plant Cell* **9**, 731–743.

Watahiki, M.K. and Yamamoto, K.T. (1997) The massugu1 mutation of *Arabidopsis* identified with failure of auxin-induced growth curvature of hypocotyl confers auxin insensitivity to hypocotyl and leaf. *Plant Physiol.* **115**, 419–426.

Weinig, C. (2000) Differing selection in alternative competitive environments: shade-avoidance responses and germination timing. *Evolution Int. J. Org. Evolution* **54**, 124–136.

Weise, S.E. and Kiss, J.Z. (1999) Gravitropism of inflorescence stems in starch-deficient mutants of *Arabidopsis*. *Int. J. Plant Sci.* **160**, 521–527.

Went, F.W. and Thimann, K.V. (1937) *Phytohormones*. Macmillan, New York.

Woodward, A.W. and Bartel, B. (2005) Auxin: regulation, action, and interaction. *Ann. Bot. (Lond.)* **95**, 707–735.

Yamaguchi, S. and Kamiya, Y. (2002) Gibberellins and light-stimulated seed germination. *J. Plant Growth Regul.* **20**, 369–376.

Yamaguchi, S., Kamiya, Y. and Sun, T. (2001) Distinct cell-specific expression patterns of early and late gibberellin biosynthetic genes during *Arabidopsis* seed germination. *Plant J.* **28**, 443–453.

Yamaguchi, S., Smith, M.W., Brown, R.G., Kamiya, Y. and Sun, T. (1998) Phytochrome regulation and differential expression of gibberellin 3beta-hydroxylase genes in germinating *Arabidopsis* seeds. *Plant Cell* **10**, 2115–2126.

Yamamoto, K., Pyke, K.A. and Kiss, J.Z. (2002) Reduced gravitropism in inflorescence stems and hypocotyls, but not roots, of *Arabidopsis* mutants with large plastids. *Physiol. Plant* **114**, 627–636.

Yano, D., Sato, M., Saito, C., Sato, M.H., Morita, M.T. and Tasaka, M. (2003) A SNARE complex containing SGR3/AtVAM3 and ZIG/VTI11 in gravity-sensing cells is important for *Arabidopsis* shoot gravitropism. *Proc. Natl. Acad. Sci. USA* **100**, 8589–8594.

Zagotta, M.T., Hicks, K.A., Jacobs, C.I., Young, J.C., Hangarter, R.P. and Meeks-Wagner, D.R. (1996) The *Arabidopsis* ELF3 gene regulates vegetative photomorphogenesis and the photoperiodic induction of flowering. *Plant J.* **10**, 691–702.

Part IV Applied aspects of photomorphogenesis

11 Photoreceptor biotechnology

Matthew Hudson

11.1 Introduction and background

Plant photoreceptors influence or control almost all aspects of plant metabolism, growth and development. Only the extent and timing of this control are variable (see other chapters in this volume). Many agriculturally relevant traits are either heavily influenced or completely controlled by photoreceptors. These include seed germination, circadian timing, seedling architecture, bud dormancy, leaf shape and size, stem length and curvature, photosynthetic resource allocation, chloroplast development, chloroplast positioning, flowering time, grain filling and dormancy. In terms of metabolism, the enzymes and other protein components that mediate most of the reactions of photosynthesis are light regulated. Processes such as nitrogen fixation and gas exchange are also regulated by light and influenced by the circadian clock (which is itself under direct photoreceptor control; see Chapter 8). The expression of many other genes and processes is controlled by, or interacts with, photoreceptor pathways (see Chapter 10). One agronomically relevant example of this is that defense pathways are strongly influenced by photoreceptor signals (Genoud *et al.*, 2002). Since many photoreceptor-controlled processes are important in determining the yield and suitability of crops, there has been significant interest for many years in using or modifying photoperception for crop improvement.

The control exerted by photoreceptors over so many aspects of plant biology makes them an appealing target for biotechnology approaches. Engineering or smart breeding of photoreceptor genes or their signal transduction components could be used to modify many aspects of plant development and metabolism. There are three major families of plant photoreceptors, the red (R) and far-red (FR) light sensing phytochromes (see Schafer, this volume), the blue (B)/ultraviolet A (UVA) sensing cryptochromes (see Batschauer *et al.*, this volume) and the B/UVA sensing phototropins (see Chapter 3). The strategies used to influence desirable traits using these receptors are discussed in Section 11.2.

Of the three families, the phytochromes have so far attracted the most interest for biotechnology applications. This is in part because the phytochromes have been known for longer than other photoreceptors, and constructs to overexpress the genes have been available for some time. Phytochrome overexpressors have been shown to produce strong phenotypes, many of which are desirable from the perspective of yield, harvest time or plant architecture. The B/UVA sensing cryptochromes influence most of the same processes as the phytochromes. Cryptochrome overexpression thus can also be used to confer desirable phenotypes by overexpression. The phototropins control blue-light-induced phototropic responses, chloroplast positioning, leaf expansion and stomatal opening (Kagawa, 2003). While there is potential to modify these responses by altering expression of the phototropins or their signaling partners, there is currently minimal published work on biotechnology applications of the phototropins. Photoreceptor overexpression as a biotechnology tool is discussed in Section 11.3.

Further evidence of the ability of phytochrome photoreceptors to control many aspects of plant growth and development is given by the severe pleiotropic phenotypes of mutants in multiple photoreceptors, or mutants in the synthesis of the phytochrome chromophore (Hudson, 2000). The techniques and challenges of exploiting photoreceptor mutations and natural genetic diversity are discussed in Section 11.4.

Applications of plant photoreceptors are not limited to the engineering of plant development and metabolism. The unique biochemical properties of plant photoreceptors, particularly the phytochromes, make them attractive candidates for molecular biotechnology with wide-ranging applications. The phytochrome holoprotein is photoconverted between two states by pulses of R and FR. This bistable property of phytochromes could potentially be exploited in a number of ways as a molecular switch. Examples include using phytochrome as a light-controlled switch to regulate gene expression, or as a highly fluorescent molecular marker to monitor other biological processes. Details of these *ex planta* applications are described in Section 11.5 of this chapter.

11.2 Approaches to modification of photomorphogenic responses in crop plants

11.2.1 Dwarfing plants using photoreceptors

The first application of photoreceptor biotechnology became apparent when the first photoreceptor overexpressing plants showed a dwarfed, dark-green phenotype (Boylan and Quail, 1989; Kay *et al.*, 1989; Keller *et al.*, 1989). Dwarfing is a widely utilized method of increasing yield or other desirable characteristics by reducing the resources allocated to structural growth. In the case of a dwarf cereal such as wheat or rice, yield is increased by partitioning more photosynthate to the grain at the expense of the structural components of the plant (Salamini, 2003). Dwarfing also renders crops more resistant to mechanical flattening by wind or rain. Dwarf wheat and rice varieties have thus become the choice of most growers.

Dwarf crop varieties in wide use generally carry mutant alleles that affect gibberellin pathways (Peng *et al.*, 1999). However, creation of such mutants in crop species or varieties where they do not yet exist, particularly those with duplicated genomes, is by no means straightforward. Modification of growth regulator pathways using transgenic techniques is a powerful tool, but can be difficult to control. It can lead to pleiotropic dwarfing effects that substantially alter growth and development, making leaves much smaller and hence reducing photosynthetic capacity and potential yield (Curtis *et al.*, 2000). Photoreceptor overexpression is a more controllable tool, with the ability to create dwarfed plants that are not compromised in any aspect of their development that reduces photosynthetic capacity. Such a tool has obvious potential for agronomic application. A recent example of such an approach is the dwarfing of aromatic rice varieties by overexpression of *Arabidopsis* phytochrome A (phyA; Garg *et al.*, 2006). Details of this and other applications of photoreceptor-mediated dwarfing are given in Section 11.3.

11.2.2   The shade-avoidance response

The vast majority of plants have strong competitive morphological and physiological responses to crowding. Vegetation shade is an indicator of the presence of other plants. The responses of plants to vegetation shade are mediated by the perception of light spectral quality, and are collectively termed the 'shade-avoidance syndrome' (Chapter 9; Smith, 1995). The shade-avoidance syndrome strongly influences both resource partitioning and growth patterns in almost all plant species investigated, including *Arabidopsis*, maize, tobacco and some tree species (Smith, 1981, 1983, 1995; Robson *et al.*, 1996; Gilbert *et al.*, 2001). The syndrome displays common elements in all the species in which it has been identified. Shade-avoiding plants display rapid elongation growth and accelerated reproduction at the expense of leaf expansion and photosynthetic pigment production. The number of embryos that develop on each plant is usually reduced, leading to reduced yield of grain or seed. Shade-avoiding plants also allocate more photosynthate to stem elongation, and less to storage organs such as tubers. Consequently, although shade avoidance is an adaptive response in wild populations, in a densely grown crop it can lead to yield loss, poor harvest timing and undesirable morphology. Although most modern crops achieve optimal yields when grown at high planting densities, few are bred for responses to light spectral quality. Modification of shade avoidance thus has substantial potential for crop improvement.

Plants distinguish variations in light quality resulting from absorbance of solar irradiation by chlorophyll, even when the total photosynthetically active radiation is high. This is possible because the ratio of red to far-red light (R:FR), and hence the equilibrium between the active (Pfr) and inactive (Pr) forms of phytochrome, is strongly proportional to the density of vegetation in the immediate vicinity (Smith, 1995; Smith and Whitelam, 1997). This proportionality is caused by the depletion of R, with respect to FR, in the light transmitted through or reflected from the leaf canopy. Plants are therefore sensitive to crowding to a large extent because they respond to the spectral quality of vegetation shade via the phytochrome family of photoreceptors (Chapter 9; Casal *et al.*, 1997).

Importantly, mutants in the genes encoding light-stable phytochromes, particularly phytochrome B (phyB), have a pleiotropic phenotype that includes increased elongation, increased apical dominance, reduced chlorophyll levels per unit leaf area and early flowering. This simulates the effect of vegetation shade, probably because both leaf area and flowering time are controlled by Pfr levels in light-grown plants. Although phyB plays a dominant role in shade avoidance (Quail, 1994), other phytochromes also contribute (Smith and Whitelam, 1997). Just as *phyB* mutants can simulate an extreme shade-avoidance response, plants that overexpress phytochromes can be used to reduce the extent to which the shade-avoidance syndrome influences plant morphology and development. PhyB overexpression has thus been used successfully to modify shade-avoidance characteristics, and achieve increased yield in field crops (see Section 11.3).

The phyA photoreceptor has a unique ability to respond more strongly to FR than to R light, and phyA overexpression (see Section 11.3) has proved to be particularly effective in antagonizing shade-avoidance responses (McCormac *et al.*, 1992; Casal *et al.*, 1997). The transcription of the wild-type *phyA* gene is repressed in response to light, and the protein is degraded (see Chapter 1). The far-red high-irradiance response (FR-HIR) mediated by phyA is, therefore, normally observed only in etiolated seedlings or de-etiolating seedlings in dense canopies. Overexpression of phyA with a constitutive, viral promoter can extend the FR-HIR into de-etiolated plants, creating an artificial response that is antagonistic to the shade-avoidance syndrome (McCormac *et al.*, 1992). This, in turn, can be used to increase the harvest index of crops grown in dense stands (Robson *et al.*, 1996).

11.2.2.1   Control of gene expression and shade avoidance

The challenge of using photoreceptors to modify photomorphogenesis for crop improvement lies in the specific targeting of its many facets. Modification of downstream components of shade avoidance has the potential to target specific responses within the photoreceptor signaling pathways, and could therefore provide finer tools to control the responses of crop plants to crowding. Knowledge of the mechanisms of shade avoidance could lead to specific targeting of this response, without influencing the other characteristics controlled by photoreceptors themselves. For example, it could be possible to modify resource allocation in response to canopy shade, without altering the time of flowering or harvest. Despite the substantial understanding of the role of phytochrome in mediating shade-avoidance responses, however, the molecular events downstream of the perception of R:FR by phytochrome are still incompletely characterized.

Three candidate factors are known in *Arabidopsis* that could be used to mediate such fine control of the shade-avoidance pathways – PHYTOCHROME INTER-ACTING FACTOR 3-LIKE1 (PIL1), ATHB-2 (also known as homeobox-leucine zipper protein 4 (HAT4)) and LONG HYPOCOTYL IN FAR-RED1 (HFR1). The basic helix-loop-helix (bHLH) transcription factor PIL1 was identified as a gene whose mRNA is rapidly induced under low R:FR light conditions. Loss-of-function *pil1* mutants display several phenotypes that indicate PIL1 is necessary for shade-avoidance responses to transient low R:FR light (Salter *et al.*, 2003). ATHB-2 is a homeodomain-leucine zipper (HD-Zip) protein that, like PIL1, is strongly regulated by R:FR (Carabelli *et al.*, 1993, 1996). Overexpression of ATHB-2 causes effects on cell elongation consistent with shade avoidance, and antisense repression of the gene has an opposite effect (Steindler *et al.*, 1999).

The *hfr1* mutant has a strong effect on the phenotype of shade-avoiding plants. This mutant was isolated because it has a reduced response to FR in etiolated seedlings, implying a role in phytochrome A signaling responses (Fairchild *et al.*,

2000; Fankhauser and Chory 2000; Soh *et al.*, 2000). The *HFR1* transcript, which, like PIL1, encodes a bHLH transcription factor, is strongly and rapidly induced in wild-type plants in response to low R:FR. However, the mutant also shows greatly *increased* shade-avoidance responses, implying that HFR1 acts as negative regulator of shade avoidance, perhaps in order to prevent an excessive response causing the death of the seedling (Sessa *et al.*, 2005).

Although ATHB-2, PIL1 and HFR1 are all clearly involved in the mediation of changes in gene expression leading to shade avoidance, there is more work to be done before the mechanism is fully understood. However, since they do not seem to be involved in global photomorphogenic responses, these loci provide the potential to target shade-avoidance responses specifically, using traditional genetic or transgenic techniques as part of crop breeding programs.

11.2.2.2 Alteration of the timing of flowering
The ability to alter at will the time of year at which a crop flowers or is ready for harvest has obvious potential for crop improvement. By the same means, depending on the organism, it may be possible to control the timing of fruit ripening, tuberization, grain filling and other related traits. Overexpression of phytochromes generally leads to later flowering (Robson and Smith, 1997). Loss-of-function mutants in phytochrome genes, for example the *Ma3R* allele of sorghum (Childs *et al.*, 1997), tend to flower early and/or be insensitive to photoperiod, generally by flowering earlier under noninductive conditions (see Section 11.4). Antisense ablation of phytochrome B transcript removes the photoperiod requirement for tuberization in potato (Jackson *et al.*, 1996). Mutations in the cryptochrome genes also cause reduced sensitivity to photoperiod, but have the converse effect (causing plants to flower later under inductive conditions), in *Arabidopsis* (Guo *et al.*, 1998; El-Din El-Assal *et al.*, 2003). In contrast, cryptochrome mutants of pea are early flowering (Platten *et al.*, 2005). In addition to the complexity added by the different photoperiodic responses of different species, altered photoreceptor levels affect many other aspects of phenotype and cause pleiotropic effects. This may make photoreceptor modification too blunt an instrument to alter flowering/harvest times of crops without affecting yield.

However, as for shade avoidance, there are other examples of nonphotoreceptor signal transduction components, usually transcription factors, which have a significant effect on flowering time. The signaling components controlling floral induction are better characterized, and have been known for longer, than those involved in shade avoidance. These factors therefore provide a means of engineering the timing of flowering without influencing other aspects of photomorphogenesis. The field is too broad to be reviewed in detail here; examples include the zinc-finger protein CONSTANS (CO; Putterill *et al.*, 1995), the MYB family regulator LATE ELONGATED HYPOCOTYL (LHY; Schaffer *et al.*, 1998) and the transcription factor INDETERMINATE1 (ID), which is required for the transition to flowering in maize (Colasanti *et al.*, 1998). There are relatively few examples so far of biotechnology being used to alter photoperiodic flowering in plants using these signaling components. Flowering time (hence generation time) has been modified in citrus trees, using overexpression of the transcription factors LEAFY and APETALA1 (AP1; Peña *et al.*,

2001). As another example, the discovery of the role of VERNALIZATON2 (VRN2) in the control of flowering in wheat (Yan *et al.*, 2004) is an example of a genetic mechanism that leads to a potential future biotechnology application in the control of flowering time. The authors demonstrate that a transgenic RNAi approach can reduce *VRN2* transcript levels and accelerate flowering time of winter wheat by more than one month.

11.2.2.3   *Taxonomic differences and similarities in higher plants*
The use of photoreceptor biotechnology to influence patterns of growth and development is complicated by the taxonomic differences in photoreceptors and photomorphogenesis between plant species. The developmental effects of light signals on plants with diverse body plans are necessarily different. In addition, the photoreceptor systems themselves have undergone independent evolution within the angiosperms. The best characterized example of this is the phytochromes (Mathews, 2005). Based on data from rice and other monocot species, there are three phytochrome genes, *PHYA*, *PHYB* and *PHYC*, in monocotyledons. This is true except where genome duplication or polyploidy has multiplied the complete family, as is the case in maize (Sawers *et al.*, 2005). In dicotyledons, it is also usual to have one *PHYA* gene and one *PHYC* gene. However, all of the dicots examined so far have multiple B-type phytochromes (in *Arabidopsis*, designated B, D and E; in other species often designated B1, B2 etc.). Complete genome sequencing has now provided firm evidence that there are no more than five phytochrome genes in *Arabidopsis*, and no more than three in rice. The greater diversity of B-type phytochromes in the dicots may indicate increased selection pressure on these genes, which may in turn reflect divergent evolution of shade-avoidance responses. Whatever the evolutionary implications, this fact complicates the design of experiments intended to modify phytochrome responses by transferring phytochrome genes between species.

In addition to the above differences, the responses mediated by phyA, phyB and phyC vary between monocots and dicots, as indicated by the phenotypes of knockout mutants. The full sets of five phytochrome mutants in *Arabidopsis* (Franklin and Whitelam, 2004) and three phytochrome mutants in rice (Takano *et al.*, 2005) are now available. The contrasting roles of the evolutionarily orthologous phytochromes in these species indicate that photosensory function cannot be assumed on the basis of the evolutionary relationships of the photoreceptors.

In *Arabidopsis*, phyA mediates seedling responses to FR in the high-irradiance response mode, and responses to R and FR in the very low fluence response mode. In contrast, phyB mediates most of the responses to R in light-grown plants, and is the predominant receptor for the classic R/FR photoreversible low fluence response. Shade avoidance, or R:FR, is also primarily the role of phyB in wild-type plants, although all the phytochromes seem to contribute to this (Smith and Whitelam, 1997; Franklin and Whitelam, 2004). The phenotypes of phyC mutants are more subtle, and overlap somewhat with phyB FR (Franklin *et al.*, 2003; Monte *et al.*, 2003).

In contrast, the perception of R/FR photoreversible responses in rice is mediated by both phyA and phyB. Both phyA and phyC can mediate responses to continuous FR, and phyC does not appear to be involved in the perception of continuous R

(Takano *et al.*, 2005). These significant differences highlight the difficulty in applying breeding or transgenic photomorphogenic strategies to crop improvement in different species without a detailed knowledge of the underlying mechanisms of photomorphogenesis in the species under investigation.

Hints of different signaling mechanisms of photomorphogenesis between higher plants have also emerged from transgenic experiments. The strongest dwarfing obtained by phytochrome overexpression has been by using monocot *PHYA* genes from oat or rice, introduced into dicots such as tobacco (Kay *et al.*, 1989; Keller *et al.*, 1989; Robson and Smith, 1997). When extra copies of the native tobacco *phyA* gene are introduced into tobacco under the control of the same promoter, the phenotype generated is much more subtle (Hudson, 1997). The effect of oat phyA overexpression in rice or wheat is marginal (Clough *et al.*, 1995; Schlumukhov *et al.*, 2001). Conversely, the introduction of an *Arabidopsis PHYA* gene into rice causes significant dwarfing and the promise of substantial increase in yield (Garg *et al.*, 2006). A likely explanation for this observation is that the feedback controls acting posttranscriptionally on the native phytochromes (especially phyA) are not able to control introduced phytochrome genes as tightly. This can be explained by the introduced coding sequences being from a distantly related species, and thus processing substantial peptide sequence divergence from the native protein. Consequently, while the use of phytochrome modifications across species boundaries is not well understood, it may prove to be an important tool in the successful use of photomorphogenic modification by transgenic techniques, particularly in cases such as aromatic rice varieties, when the target phenotype is yield increase by dwarfing.

11.3   Modification of photomorphogenesis using genetic transformation – the state of the art

11.3.1   *Plants transgenic for phytochromes*

A large number of plant species amenable to transformation have been modified by the introduction of expression cassettes designed to overexpress phytochrome genes. Patents have been filed on the use of phytochrome constructs to cause dwarfing, modify the shade-avoidance responses or alter flowering or cropping times (see Section 11.2), and many investigators have applied these techniques to different species and genotypes. The overexpression of phytochrome can lead in many cases to significant dwarfing and to substantial increases in yield (Figure 11.1). The crop and related model plant species in which phytochrome overexpression has been successfully practiced are given in Table 11.1 (for lower plant phytochrome expression experiments, see Robson and Smith, 1997). More is known about the behavior of phytochrome overexpression constructs in tobacco, potato and *Arabidopsis* than in other species, since genetic transformation of these plants has been straightforward for some time. However, improved transformation technologies have led to phytochrome overexpression being applied in rice (Clough *et al.*, 1995; Garg *et al.*, 2006) and wheat (Schlumukhov *et al.*, 2001).

(A)

(B)

Figure 11.1 Dwarfing and yield increase using phytochrome overexpression. (A) Phenotype of greenhouse-grown rice plants overexpressing *Arabidopsis* phyA. The transgenic lines (ox1, ox2 and ox3) are all substantially shorter than wild-type plants, and have shorter tiller internodes. The lines also showed a yield increase of 21, 6 and 11%, respectively. (Adapted from Garg *et al.*, 2006 with kind permission of Springer Science and Business Media.) (B) Increase in number and yield of tubers in potato plants overexpressing phytochrome B. The wild-type (WT) and phyB-overexpressing transgenics (ox1 and ox2) were grown to harvest under greenhouse conditions. The ox2 line overexpresses phyB holoprotein at higher levels than ox1; consequently, the tuber number increase is proportional to the level of phyB. (Adapted, with permission, from Thiele *et al.*, 1999 © American Society of Plant Biologists.)

Table 11.1 Summary of modifications of phytochrome loci in angiosperms using transgenic techniques, and the traits that were modified

Species (cultivar)	Introduced locus	Trait of interest	Reference
Tobacco (Xanthi)	*Avena PHYA* gene, 35S promoter	Dwarfing, light responses	Viestra *et al.*, 1989
Tobacco (SR-1)	Rice *PHYA* gene, 35S promoter	Increased expression of light-regulated genes	Kay *et al.*, 1989
Tobacco (Xanthi)	*Avena PHYA* gene, 35S promoter	Reversed shade avoidance (proximity-conditional dwarfing)	McCormac *et al.*, 1993; Robson *et al.*, 1996
Tobacco (Hicks/MM)	*Arabidopsis PHYA*, *PHYB*, *PHYC*, 35S promoter	Flowering time/photoperiodism	Halliday *et al.*, 1997
Tobacco (SR-1)	Tobacco *PHYA*, 35S and native promoter	Reversed shade avoidance (proximity-conditional dwarfing)	Hudson, 1997
Tomato (Moneymaker)	*Avena PHYA* gene, 35S promoter	Dwarfing, fruit quality	Boylan and Quail, 1989
Arabidopsis	*Arabidopsis PHYA*, *PHYB* and *PHYC* genes, 35S promoter	Research into seedling de-etiolation	See Robson and Smith, 1997; Franklin and Whitelam, 2004
Potato (Desiree)	*Arabidopsis PHYB*, potato *PHYA*, 35S promoter	Increased tuber yield with *Arabidopsis* phyB (modified resource partitioning)	Heyer *et al.*, 1995; Thiele *et al.*, 1999; Boccalandro *et al.*, 2003
Potato ssp. *andigena* (photoperiodic)	Potato *PHYB* antisense, 35S promoter	Timing of tuberization	Jackson *et al.*, 1996
Wheat (Cadenza)	*Avena PHYA* gene, 35S promoter	Dwarfing, shade avoidance	Schlumukhov *et al.*, 2001
Rice *Oryza sativa* (Gulfmont and Basmati)	*Avena PHYA* gene, *Arabidopsis PHYA* gene, 35S promoter	Dwarfing, yield increase and modified resource partitioning in aromatic rice	Clough *et al.*, 1995; Garg *et al.*, 2006

For a discussion of nonangiosperm transgenic experiments, see Robson and Smith (1997).

One of the best examples of the potential of phytochrome expression to increase yield in plants is the overexpression of phytochrome in potato. Additional copies of the potato *PHYA* and *Arabidopsis PHYB* genes have been introduced into potato under the control of the 35S promoter (Heyer *et al.*, 1995; Thiele *et al.*, 1999). Overexpression of *PHYA* leads to dwarfing and a reduced response to R:FR (Heyer *et al.*, 1995). Overexpression of *PHYB* leads to substantially increased tuberization (Figure 11.1) and a greater tuber yield from plants grown in controlled environments (Thiele *et al.*, 1999). This result extends to field-grown plants, where phyB overexpression causes significantly increased tuber yields in densely grown plots (Boccalandro *et al.*, 2003). The effect of phyB overexpression in field-grown potato is consistent with a reduced shade-avoidance phenotype (Section 11.2). The source of the increased yield is likely to be altered resource partitioning, and this method may thus provide a general means of increasing the yields of potato tubers, likely without the need for increased use of artificial fertilizers or pesticides.

In addition, a quantitative difference in tuberization in *Solanum tuberosum* ssp. *andigena* is observed in phyB antisense transgenics (Jackson *et al.*, 1996). In this potato subspecies, tuberization is normally dependent on short-day conditions. The antisense ablation of the *PHYB* transcript allows tuberization to occur under long-day conditions, making this an example of the modification of yield timing characteristics using photoreceptor biotechnology (see Section 11.2).

Tobacco is also a good model to study the effects of dwarfing and of shade avoidance, since it has a simple, consistent growth habit and puts a significant amount of resources into the formation of internodes. By redirecting the resources devoted to stem elongation into leaf development, there is the potential to achieve a theoretical increase in yield, without complications of altered flowering or harvest time associated with grain crops or tubers. Robson *et al.* (1996) demonstrated that relatively subtle increases in phyA levels (using the *PHYA* gene from *Avena sativa*) can cause dwarfing in tobacco plants, which is conditional on the plants being grown in dense stands (proximity-conditional dwarfing). The result of this is an increased harvest index in the transgenic plants, which allocate more resources to leaf formation (tobacco 'yield') than to stem formation when grown densely (Figure 11.2). However, the phenotype and yield of plants in less dense stands (where light is not limiting and dwarfing is less desirable) are little affected, leading to the promise of a crop that could survive extensive damage more effectively.

Monocotyledon crops (wheat and rice) have also been modified through the use of phytochrome overexpression. Significant alteration of yield or morphology has not yet been accomplished by means of *Avena PHYA* introduced into rice or wheat (Clough *et al.*, 1995; Schlumukhov *et al.*, 2001). However, *Arabidopsis PHYA* overexpression in aromatic rice (*Oryza sativa* L. Pusa Basmati-1) has been demonstrated to create a strong dwarfing phenotype, including reduction in plant height, reduction in internode length and diameter, and an increase in panicle number (Figure 11.1; Garg *et al.*, 2006). Importantly, the dwarfed plants show a substantial increase in yield, at least under greenhouse conditions. This interesting application of photoreceptor biotechnology is especially promising because of the failure of breeders to incorporate the dwarfing genes that have increased yields in other rice varieties into aromatic rice without losing the distinctive flavor of the grain. Phytochrome overexpression therefore provides a ready means of increasing yield by dwarfing in crops where dwarf genotypes are not available, or where introduction of dwarf traits is too complex, as in aromatic rice.

11.3.2 Modification of other photoreceptors

Most work with cryptochromes and phototropins has been focused on loss of function mutants, and has been performed in *Arabidopsis* and rice. This is partly because the most successful dwarfing experiments have been performed with phytochrome genes, and partly because the cryptochrome genes have been available for less time. Blue light photoreceptor transgenic experiments are summarized in Table 11.2. However, the recent results of cryptochrome overexpression in tomato (Giliberto *et al.*, 2005) may change this pattern. The increases in fruit antioxidant content,

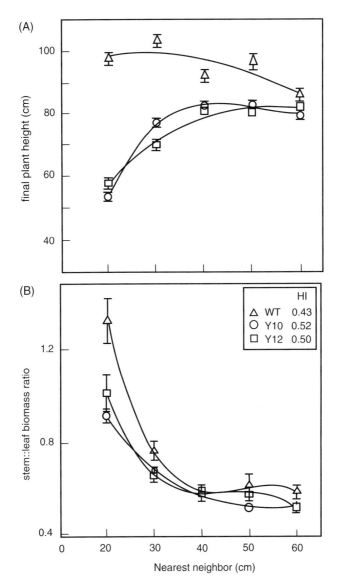

Figure 11.2 Proximity-conditional dwarfing of tobacco stands achieved with phytochrome A overexpression. (A) Plant heights at harvest, showing progressive dwarfing of two transgenic phytochrome A overexpressing lines (indicated by squares and circles) as the nearest neighbor distance decreases. Wild-type tabacco plants (indicated by triangles) increase in height as the neighbor distance decreases due to the shade-avoidance syndrome. (B) Stem-to-leaf biomass ratio for the same experiment. Note the dramatic increase in resource partitioning to the stem in wild-type plants (triangles) grown at high densities and the reduced impact of high-density growth on the two transgenic lines (squares and circles). A significant alteration in harvest index for the transgenics was computed based on the data for the 20 cm planting density (wild type, HI = 0.43, transgenic lines HI = 0.52 and 0.50). (Adapted, with permission, from Robson *et al.*, 1996 © Nature Publishing Group.)

Table 11.2 Summary of genetic modification of the blue light photoreceptors, and traits that were successfully modified

Species	Locus	Trait	Reference
Tobacco	*Arabidopsis CRY1*	Enhanced blue, green and UV-A light sensitivity	Lin *et al.*, 1995
Arabidopsis	*CRY1*	Enhanced blue light sensitivity	Lin *et al.*, 1996
Arabidopsis	*CRY2*	Enhanced blue light sensitivity	Lin *et al.*, 1998
Tomato	*CRY2*	Vegetative development, flowering time and fruit antioxidant content	Giliberto *et al.*, 2005
Physcomitrella patens	*PpCRY1a* and *PpCRY1b*	Side branches in protonemata, differentiation and growth of gametophores, auxin response	Imaizumi *et al.*, 2002

with very little negative impact, achieved by Giliberto *et al.* are likely to generate more interest in the use of cryptochrome overexpression to modify plant development, particularly flowering time and fruit composition.

There has been little work to date on the use of phototropin modification for plant or crop improvement. However, the possibility exists that modification of phototropin pathways could be used to ablate phototropic responses where these are undesirable (for example, in plants grown under artificial lighting).

11.3.3   Overexpression of signaling components

Currently, modification of morphology by mutation and overexpression of signaling components has been restricted to experimental analysis of signaling pathways (see Chapter 4). These phenotypes can be restricted to a subset of photomorphogenic responses; for example, NDPK2 knockouts seem to be affected in seedling hook opening but not in hypocotyl elongation (Choi *et al.*, 1999). Since this approach has the potential to specifically target certain aspects of photomorphogenesis, it has great potential for biotechnology application. However, since the phenotypes of signaling component transgenics or mutants are generally much less strong than those of photoreceptor transgenics or mutants, it may be more challenging to achieve a substantially altered yield using this approach.

11.4   Modification of photomorphogenesis by utilizing genetic diversity

11.4.1   Natural variation in photomorphogenesis

One of the areas of photomorphogenesis research that has made rapid progress in recent years is the understanding of the role of photoreceptors and photomorphogenic alleles in natural variation and evolution. It has been clearly demonstrated that both reduced and increased phytochrome expressions measurably reduce the fitness of plants competing in canopy environments (Schmitt *et al.*, 1995). While this reduced fitness would not necessarily be detrimental to a crop (where all plants are genetically identical and thus canopy competition is eliminated), the result demonstrates that selection pressure will continually act on photomorphogenic systems, both in wild plants and in breeding programs where photomorphogenic behavior is not selected for. Consequently, many crop plants are likely to have photomorphogenic traits suboptimal for yield. This applies especially to the shade-avoidance syndrome, because of the prevalent selection pressure for canopy competition. Variants within the progeny of a breeder's cross that displayed reduced shade avoidance, grown alongside plants with normal responses, will appear unhealthy (as described by Schmitt *et al.*, 1995). Such plants would thus probably not be selected by a breeder, unless they were deliberately targeting shade avoidance, or yield at increased density, as a trait (see below).

Research in this area is currently focused on the evolution and variation in photomorphogenic systems amongst accessions derived from wild populations. A large number of variable photomorphogenic responses have been described amongst related subpopulations of various species (Maloof *et al.*, 2000). The recent advances in describing the molecular basis of these variable responses have been almost entirely generated by using large numbers of wild-derived accessions of the model plant *Arabidopsis thaliana*. To the surprise of many photobiologists, natural photoreceptor mutants exist and survive within wild populations of *Arabidopsis*. The Wassilewskija accession of *Arabidopsis* is naturally mutated in the phytochrome D gene (Aukerman *et al.*, 1997), and this example has been enforced by the discovery of a natural phyA variant with greatly reduced FR sensitivity in the Lm-2 accession in a screen of 141 accessions for light response (Maloof *et al.*, 2001). In addition to the Lm-2 variant described by Maloof *et al.*, their screen demonstrates that a great deal of natural variation exists amongst the wild-collected *Arabidopsis* accessions. Significantly, an association mapping study for flowering time has revealed a novel allele of *CRY2* in *Arabidopsis* (El-Din El-Assal *et al.*, 2001). This indicates the potential of natural variation in photoreceptor sequences to influence another agronomically significant trait, timing of reproduction/harvest. These and other results have led to greatly increased interest in genomic approaches to analyzing natural variation in *Arabidopsis* among evolutionary biologists (Maloof, 2003; Shimizu and Purugganan, 2005).

It has recently been demonstrated using microarray profiling that one of the most variable transcripts in expression level between *Arabidopsis* accessions is the *PHYB* transcript (Chen *et al.*, 2005). This variability can be explained by the presence of a large degree of sequence diversity in the promoter and intron regions of the *PHYB* locus (the regions mostly responsible for the control of transcription). These results give weight to the notion that genetically controlled variation in photomorphogenesis is a very significant component of evolutionary adaptation of plants to diverse environments.

11.4.2   Photoreceptors and photomorphogenic genes as targets for selection in crops

Given the above results, it is likely that genetic diversity in photomorphogenic pathways lies within the germ plasm collections of many crops, forming an untapped resource for crop yield improvement that does not require chemical applications or transgenes. While it is highly likely (see above) that the photomorphogenic systems of crops are suboptimal in terms of their photomorphogenic responses, it is also likely that many of the morphological traits of modern crop plants have been selected to increase tolerance to higher planting densities (in particular leaf angle, internode length, tillering and timing of flowering). Given their influence on the morphology and resource partitioning in densely grown crops, photoreceptor genes will in some cases determine yield, particularly at high planting densities where shade avoidance can be a strong factor in yield determination (Robson *et al.*, 1996; Robson and Smith, 1997). It is probable, therefore, that breeders have exerted some indirect selection on photomorphogenic traits such as shade avoidance during the breeding of modern crops. This is particularly likely when selection is primarily for increased yield at high planting densities. For example, in maize, much, if not all, of the significant increases in yield delivered by modern cultivars can be attributed to higher tolerance for crowding (Duvick, 1997) rather than to an increase in yield on a per-plant basis. It is understood by maize breeders that photosensitivity can have a negative impact (Salamini, 1985) and so breeders may have altered their selection to compensate for this.

Selection for aberrant photomorphogenic traits in mutagenized, inbred populations and for day-length-insensitive flowering within cereal breeding programs both lead to the isolation of photoreceptor mutants (see Table 11.3). This demonstrates

Table 11.3   Alleles that have been isolated in crop species and that directly affect photoreceptor function, and traits modified by the mutations

Species	Locus	Trait	Reference
Tomato	*phyA, phyB1, phyB2, aurea*	Various effects on photomorphogenesis	Kendrick *et al.*, 1997; Weller *et al.*, 2000
Brassica rapa (rapid cycling)	*ein (phyB)*	Elongated internodes, reduced R response	Devlin *et al.*, 1992, 1997
Cucumber	*lh (phyB)*	Long hypocotyls	Lopez-Juez *et al.*, 1992
Pea	*phyA, phyB*	Various effects on photomorphogenesis	Weller *et al.*, 2001
Rice	*phyA, phyB, phyC*	Various effects on photomorphogenesis	Takano *et al.*, 2005
Maize	*Elm1*	Height, internode length, flowering time	Sawers *et al.*, 2002, 2004
Sorghum	*Ma3R (phyB)*	Photoperiod insensitivity, elongation	Childs *et al.*, 1997
Barley	*BMDR-1 (phyB)*	Photoperiod insensitivity, elongation	Hanumappa *et al.*, 1999

Note that many more mutants exist in model plants such as *Arabidopsis* (Hudson, 2000).

the ease with which photomorphogenic variants can be isolated within a population under selection. None of the loci or traits thus isolated has yet been of agronomic benefit; even the lines of crop species isolated within field populations tend to have extreme, pleiotropic phenotypes (Childs *et al.*, 1997; Hanumappa *et al.*, 1999). However, it is likely that selection for more subtle photomorphogenic loci could occur without generating the same pleiotropic phenotypes, and that these loci could be significant in determining desirable traits (Sawers *et al.*, 2005). Although increased yield at higher planting densities can be partly explained by increased tolerance to drought and other stresses (Bruce *et al.*, 2002), the altered morphology of newer cultivars is also likely to play a role (Fellner *et al.*, 2003). One strategy for increasing crop yields further is to understand and maximize the light signaling systems that allow these cultivars to tolerate high-density planting (Maddonni *et al.*, 2001, 2002).

Few well-controlled studies of the analysis of photomorphogenic variation between genetically distinct inbred cultivars of a particular crop are available. One exception is the study of the light responses of 30 diverse maize inbred line seedlings, grown under monochromatic R, FR or B light of similar irradiance, by Markelz *et al.* (2003). All these lines had functional photomorphogenic signaling pathways, but displayed over threefold variation in phytochrome responses, as measured by mesocotyl length under either R or FR light. Importantly, the North American cultivars in this study displayed attenuated light responses compared to the semitropical and tropical inbred lines. Thus, it is likely that North American breeding practices have indirectly selected for genetic loci that reduce light responsiveness in maize (Markelz *et al.*, 2003).

Despite these attenuated photomorphogenic responses, the characterization of a maize phytochrome mutant (Sawers *et al.*, 2002) and the presence of shade-avoidance responses in maize (Smith, 1981, 1983; Maddonni *et al.*, 2002) strongly suggest that light responses not only are operational in adult, light-grown maize, but have a significant impact on growth and development. Therefore, there is still significant potential for alteration and optimization of these responses in maize.

Additionally, the presumed selection for certain photomorphogenic alleles in highly developed crops such as maize and wheat creates another potential application. These alleles may at some point be transferable, using genetic transformation or some other method, to crops that have not had the benefit of thousands of years of selective breeding.

11.5   Photoreceptor biotechnology *ex planta*

11.5.1   Using phytochrome to control gene expression

The discovery of the extremely specific binding of the PIF3 bHLH transcription factor to the Pfr form of *Arabidopsis* phyB (Ni *et al.*, 1999) opens up a number of possible photoreceptor biotechnology applications. For the first time, it is possible to create a complete light signal transduction system in any organism. In the case of yeast, which has no photoreceptor genes and can complete its life cycle without light, this allows the creation of a tightly regulated gene expression system (Figure 11.3; Shimizu-Sato *et al.*, 2002).

In biotechnology and biomedical research, controllable systems of transcription are ubiquitous tools. Most such systems rely on the addition of a small-molecule regulator to induce or repress the synthesis of an mRNA; however, once the regulator is added, it cannot easily be removed from the culture, leaving the mRNA synthesis permanently switched 'on'. Sato and coworkers were able to create a system where expression of a target gene (in their case, the *LacZ* reporter) in yeast can be switched on by a pulse of R light, and switched off again by a pulse of FR light. The induction of the lacZ reporter in response to an R light pulse was 3 orders of magnitude within 3 h, and this induction was completely prevented by an FR pulse given after the R pulse. This light-regulated system was created by fusing PIF3 to the transcriptional-activation domain of GAL4 (Gal4AD), and creating a chimeric, chromophorylated and photoreversible phyB-GAL4 DNA-binding domain protein (Figure 11.3A). This creates a system where transcription is controlled by the recruitment of the PIF3:Gal4AD fusion, which in turn is controlled by the photoreversible conformation of phytochrome (Figure 11.3B). The induction of gene expression can be induced at any time with a pulse of R light, and reversed at any time using a pulse of FR light (Figure 11.3C).

Currently, the techniques required to extract and handle the phycocyanobilin chromophore and culture yeast in darkness with defined light sources are not easily accomplished outside a plant photoreceptor laboratory. In addition, the chromophore containing media causes some signs of photodynamic toxicity in prolonged illumination (Shimizu-Sato *et al.*, 2002). However, engineering of a chromophore biosynthetic pathway into yeast should be relatively straightforward, since the pathway has already been successfully introduced in bacteria (Gambetta and Lagarias, 2001). Such a chromophore-producing yeast strain would make this approach feasible in most molecular biology laboratories.

There is no fundamental barrier to prevent this technology from being extended to other organisms, for example *Drosophila* embryos or mammalian cell lines. If the idea of systems biology is to be taken seriously, the ability to turn on and turn off expression of genes very rapidly will become indispensable in order to model the dynamics of cellular biochemistry and regulatory networks (Kærn *et al.*, 2003). With chromophore biosynthetic genes added to the system, this method could therefore become a very widely used research tool.

11.5.2   *Phytochromes as fluorescent probes*

Fluorescent proteins are widely used as fluors in cell biology, microscopy, and in techniques such as RNA detection on 'gene chip' microarrays (Zhang *et al.*, 2002). The use of fluorescent protein probes is extremely common in molecular and biochemical research, and is now becoming important in medicine also, making the search for more intense fluorescent proteins at new wavelengths increasingly important. Biliproteins such as phycoerythrin have an advantage in many applications

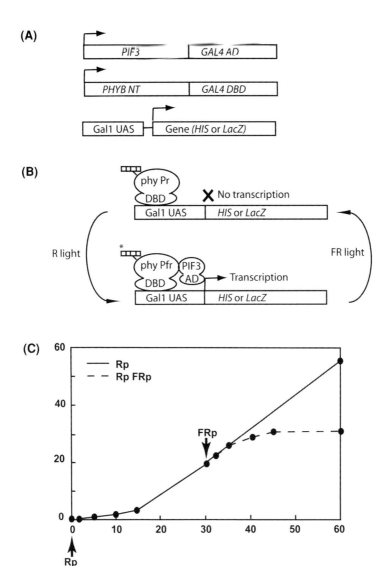

Figure 11.3 The use of light pulses, phytochrome and PIF3 to control gene expression in yeast. (A) Constructs are generated with *PIF3* and *PHYB* fused to the transcriptional activation and DNA-binding domain encoding portions of the yeast *GAL4* gene respectively. The gene to be controlled (in this case, the *LacZ* or *HIS* reporter genes) is downstream from Gal1 UAS, the binding site of the GAL4 DNA-binding domain. (B) *In vivo*, the phytochrome moiety is chromophorylated by the addition of exogenous phycocyanobilin (chromophore represented by four-box cartoon). The chromophorylated phytochrome moiety is anchored to the promoter of the reporter gene by the GAL4 DNA-binding domain. Unless a light pulse is provided, the phytochrome remains in the Pr form and no transcription occurs. However, once a pulse of R light is given, the phytochrome moiety converts to the Pfr form and can bind PIF3. The PIF3–GAL4 activation domain fusion protein is recruited to the promoter, and transcription proceeds. However, this can be reversed at any point by a pulse of FR light. (C) Response times of the system. A pulse of R light is given to the modified yeast cells, and detectable amounts of the LACZ enzyme begin to appear after 5–10 min. LACZ continues to accumulate linearly unless a pulse of FR light is given. With a lag time of 10–15 min, accumulation of LACZ then ceases. (Adapted, with permission, from Shimizu-Sato *et al.*, 2002 © Nature Publishing Group.)

over fluorescent dyes because of their high fluorescence quantum yield, and hence high signal-to-noise ratio.

Although phytochromes are biliproteins like the phycoerythrins, they are not normally fluorescent proteins. (Fluorescence quantum yield is less than 10^{-3} at room temperature (Brock *et al.*, 1987).) Instead of photons captured by the bilin chromophore being reemitted as energy at other wavelengths, the energy is used in the photoconversion process and stored in the conformation of the Pfr form, which slowly decays back to the Pr form in darkness. However, when the native chromophore is replaced by phycoerythrobilin (an analog of the natural chromophore which lacks the C15 double bond), the result is an intensely fluorescent, photostable protein that is presumably unable to undergo the Pr–Pfr conversion (Murphy and Lagarias, 1997). The phytofluors have emission maxima in the 580–590 nm range, where no fluorescent probes are currently available. They have quantum yields of 0.7–0.82, putting them in the same useful range as most of the other widely used fluorescent protein probes. Both of these parameters could likely be altered or improved by judicious site-directed mutation of the phytofluor apoprotein. The phytofluors may consequently be the first commercial application of phytochrome biotechnology (Fischer and Lagarias, 2004).

11.5.3   Other potential uses of photoreceptors

The unique physical properties of phytochromes have led to a number of other suggestions for their possible utility *ex planta*. The R/FR photoreversible property of the protein could potentially be used as a method for storing solar energy. It could also form the basis of an optical storage device such as those that have been envisioned for optical computers (Ni *et al.*, 1999). Understanding of the molecular structure of phytochrome is finally becoming more advanced, with a huge advance in the form of a phytochrome crystal structure (Wagner *et al.*, 2005). Knowledge of the physics of the photoconversion process, derived from the structural biology of phytochrome, may aid in the design of light-driven nanomachines, even if phytochrome itself does not play a role in these devices.

11.6   Future directions in photoreceptor biotechnology

Genomics is likely to generate new tools for the modification of photomorphogenesis in plants. As the details of photomorphogenesis become clear in more plant species and the genome projects of crops such as maize draw closer to completion, interest is likely to increase in using photoreceptors or their signaling pathways to cause targeted dwarfing, alter shade avoidance or influence other traits. While the use of genetic transformation is likely to remain important in research into plant photomorphogenesis, knowledge of polymorphisms between crop cultivars will increase as a result of resequencing strategies applied across large germ plasm collections, and will lead to the production of large databases of genetic diversity at the molecular level. Combination of such databases with quantitative trait data is likely to lead to the discovery of photomorphogenic alleles that have arisen during the evolution or breeding of modern crop plants. Given the evidence for the importance of photomorphogenesis in determining yields, such alleles are likely to become the focus of targeted 'smart breeding' approaches in order to incorporate complete optimal photomorphogenic systems into elite lines. For these reasons, photoreceptors and photomorphogenesis are likely to become more important in the eyes of crop breeders and physiologists.

In terms of other applications of photoreceptor biotechnology, knowledge of the structure of phytochrome (Wagner *et al.*, 2005) is likely to have significant impacts on the molecular biotechnology uses of phytochrome. It may now be possible to optimize light-driven gene control systems or phytofluors by intelligent domain-swap experiments or site-directed mutagenesis of key residues involved in photoconversion, PIF3 binding or chromophore binding. The ability to make such intelligent, evidence-driven modifications may allow photoreceptor scientists to move out of the 'dark ages' and into a new era of advanced protein design.

Acknowledgments

Thanks are due to Dr. Thomas Brutnell for helpful discussions, and to Dr. Karen Kaczorowski and Dr. Kankshita Swaminathan for critical reading of the manuscript. Research in the Hudson Laboratory is funded by the United States Department of Agriculture, the Illinois Council on Food and Agricultural Research, the National Center for Supercomputer Applications and the University of Illinois.

References

Aukerman, M.J., Hirschfeld, M., Wester, L., *et al.* (1997) A deletion in the *PHYD* gene of the *Arabidopsis wassilewskija* ecotype defines a role for phytochrome D in red/far-red light sensing. *Plant Cell* **9**, 1317–1326.

Boccalandro, H.E., Ploschuk, E.L., Yanovsky, M.Y., Sanchez, R.A., Gatz, C. and Casal, J.J. (2003) Increased phytochrome b alleviates density effects on tuber yield of field potato crops. *Plant Physiol.* **133**, 1539–1546.

Boylan, M.T. and Quail, P.H. (1989) Oat phytochrome is biologically active in transgenic tomatoes. *Plant Cell* **1**, 765–773.

Brock, H., Ruzsicska, B.P, Arai, T., *et al.* (1987) Fluorescence lifetimes and relative quantum yields of 124-kilodalton oat phytochrome in H2O and D2O solutions. *Biochemistry* **26**, 1412–1417.

Bruce, W.B., Edmeades, G.O. and Barker, T.C. (2002) Molecular and physiological approaches to maize improvement for drought tolerance. *J. Exp. Bot.* **53**, 13–25.

Carabelli, M., Morelli, G., Whitelam, G., Ruberti, I. (1996) Twilight-zone and canopy shade induction of the Athb-2 homeobox gene in green plants. *Proc. Natl. Acad. Sci. U.S.A.* **93**, 3530–3535.

Carabelli, M., Sessa, G., Baima, S., Morelli, G. and Ruberti, I. (1993) The *Arabidopsis Athb-2* and -4 genes are strongly induced by far-red-rich light. *Plant J.* **4**, 469–479.

Casal, J.J., Sanchez, R.A. and Yanovsky, M.J. (1997) The function of phytochrome A. *Plant Cell Environ.* **20**, 813–819.

Chen, W.J., Chang, S.H., Hudson, M., *et al.* (2005) Contribution of transcriptional regulation to natural variations in *Arabidopsis. Genome Biol.* **6**, R32.

Childs, K.L., Miller, F.R., Cordonnier-Pratt, M.-M., Pratt, L.H., Morgan, P.W. and Mullet, J.E. (1997) The sorghum photoperiod sensitivity gene, *Ma3*, encodes a phytochrome B. *Plant Physiol.* **113**, 611–619.

Choi, G., Yi, H., Lee, J., *et al.* (1999) Phytochrome signalling is mediated through nucleoside diphosphate kinase 2 . *Nature* **401**, 610–613.

Clough, R.C., Casal, J.J., Jordan, E.T., Christou, P. and Vierstra, R.D. (1995) Expression of a functional oat phytochrome A in transgenic rice. *Plant Physiol.* **109**, 1039–1045.

Colasanti, J., Yuan, Z. and Sundaresan, V. (1998) The indeterminate gene encodes a zinc finger protein and regulates a leaf-generated signal required for the transition to flowering in maize. *Cell* **93**, 491–494.

Curtis, I.S., Ward, D.A., Thomas, S.G., *et al.* (2000) Induction of dwarfism in transgenic Solanum dulcamara by over-expression of a gibberellin 20-oxidase cDNA from pumpkin. *Plant J.* **23**, 329–338.

Devlin, P.F., Rood, S.B., Somers, D.E., Quail, P.H. and Whitelam, G.C. (1992) Photophysiology of the elongated internode (*ein*) mutant of Brassica rapa: *ein* mutant lacks a detectable phytochrome B-like polypeptide. *Plant Physiol.* **100**, 1442–1447.

Devlin, P.F., Somers, D.E., Quail, P.H. and Whitelam, G.C. (1997) The Brassica rapa elongated internode (EIN) gene encodes phytochrome B. *Plant Mol. Biol.* **34**, 37–547.

Duvick, D.N. (1997) What is yield ? In: *Developing Drought and Low N-Tolerant Maize. Proceedings of a Symposium, March 25–29, 1996* (eds G.O. Edmeades, B. Banziger, H.R. Mickelson and C.B. Peña-Valdivia), pp. 332–335. CIMMYT, El Batan, Mexico.

El-Din El-Assal, S., Alonso-Blanco, C., Peeters, A.J.M., Raz, V. and Koornneef, M. (2001) A QTL for flowering time in *Arabidopsis* reveals a novel allele of CRY2. *Nat. Genet.* **29**, 435–440.

El-DinEl-Assal, S., Alonso-Blanco, C., Peeters, A.J.M., Wagemaker, C., Weller, J.L. and Koornneef, M. (2003) The role of cryptochrome 2 in flowering in *Arabidopsis*. *Plant Physiol.* **133**, 1504–1516.

Fairchild, C.D., Schumaker, M.A. and Quail, P.H. (2000) HFR1 encodes an atypical bHLH protein that acts in phytochrome A signal transduction. *Genes Dev.* **14**, 2377–2391.

Fankhauser, C. and Chory, J. (2000) RSF1, an *Arabidopsis* locus implicated in phytochrome A signaling. *Plant Physiol.* **124**, 39–45.

Fellner, M., Horton L.A., Cocke, A.E., Stephens, N.R., Ford, E.D. and Van Volkenburgh, E. (2003) Light interacts with auxin during leaf elongation and leaf angle development in young corn seedlings. *Planta* **216**, 366–376.

Fischer, A.J. and Lagarias, J.C. (2004) Harnessing phytochrome's glowing potential. *Proc. Natl. Acad. Sci. U.S.A.* **101**, 17334–1733.

Franklin, K.A., Davis, S.J., Stoddart, W.M., Vierstra, R.D. and Whitelam, G.C. (2003) Mutant analyses define multiple roles for phytochrome C in *Arabidopsis* photomorphogenesis. *Plant Cell* **15**, 1981–1989.

Franklin, K.A. and Whitelam, G.C. (2004) Light signals, phytochromes and cross-talk with other environmental cues. *J. Exp. Bot.* **55**, 271–276.

Gambetta, G.A. and Lagarias, J.C. (2001) Genetic engineering of chromophore biosynthesis in bacteria. *Proc. Natl. Acad. Sci.U.S.A.* **98**, 10566–10571.

Garg, A.K., Sawers, R.J.H., Wang, H.Y., *et al.* (2006) Light-regulated overexpression of an *Arabidopsis* phytochrome A gene in rice alters plant architecture and increases grain yield. *Planta* **223**, 627–636.

Genoud, T., Buchala, A.J., Chua, N.-H. and Métraux, J.-P. (2002) Phytochrome signalling modulates the SA-perceptive pathway in *Arabidopsis*. *Plant J.* **31**, 87–96.

Gilbert, I.R., Jarvis, P.G. and Smith, H. (2001) Proximity signal and shade avoidance differences between early and late successional trees. *Nature* **41**, 792–795.

Giliberto, L., Perrotta, G., Pallara, P., *et al.* (2005) Manipulation of the blue light photoreceptor cryptochrome 2 in tomato affects vegetative development, flowering time, and fruit antioxidant content. *Plant Physiol.* **137**, 199–208.

Guo, H., Yang, H., Mockler, T.C. and Lin, C. (1998) Regulation of flowering time by *Arabidopsis* photoreceptors. *Science* **279**, 1360–1363.

Halliday, K.J., Thomas, B. and Whitelam, G.C. (1997) Expression of heterologous phytochrome A, B or C in transgenic tobacco plants alters vegetative development and flowering time. *Plant J.* **12**, 1079–1090.

Hanumappa, M., Pratt, L.H., Cordonnier-Pratt, M.-M. and Deitzer, G.F. (1999) A photoperiod-insensitive barley line contains a light-labile phytochrome B. *Plant Physiol.* **119**, 1033–1040.

Heyer, A.G., Mozley, D., Landschutze, V., Thomas, B. and Gatz, C. (1995) Function of phytochrome A in potato plants as revealed through the study of transgenic plants. *Plant Physiol.* **109**, 53–61.

Hudson, M. (2000) The genetics of phytochrome signalling in *Arabidopsis*. *Semin. Cell Dev. Biol.* **11**, 475–583.

Hudson, M.E. (1997) *Analysis of Phytochrome Function in the Genus Nicotiana Using Mutant and Transgenic Plants*. PhD Thesis, University of Leicester, UK.

Imaizumi, T., Kadota, A., Hasebe, M. and Wada, M. (2002) Cryptochrome Light signals control development to suppress auxin sensitivity in the moss physcomitrella patens. *Plant Cell* **14**, 373–386.

Jackson, S.D., Heyer, A., Dietze, J. and Prat, S. (1996) Phytochrome B mediates the photoperiodic control of tuber formation in potato. *Plant J.* **9**, 159–166.

Kærn, M., Blake, W.J. and Collins, J.J. (2003) The engineering of gene regulatory networks. *Annu. Rev. Biomed. Eng.* **5**, 179–206.

Kagawa, T. (2003) The phototropin family as photoreceptors for blue light-induced chloroplast relocation. *J. Plant Res.* **116**, 75–80.

Kay, S.A., Nagatani, A., Keith, B., Deak, M., Furuya, M. and Chua, N.-H. (1989) Rice phytochrome is biologically active in transgenic tobacco. *Plant Cell* **1**, 775–782.

Keller, J.M., Shanklin, J., Vierstra, R.D. and Hershey, H.P. (1989) Expression of a functional monocotyledonous phytochrome in transgenic tobacco. *EMBO J.* **8**(4), 1005–1012.

Kendrick, R.E., Kerckhoffs, H.J., Van Tuinen, A. and Koornneef, M. (1997) Photomorphogenic mutants of tomato. *Plant Cell Environ.* **20**, 746–751.

Lin, C., Ahmad, M. and Cashmore, A.R. (1996) *Arabidopsis* cryptochrome 1 is a soluble protein mediating blue light-dependent regulation of plant growth and development. *Plant J.* **10**, 893–902.

Lin, C., Ahmad, M., Gordon, D. and Cashmore, A.R. (1995) Expression of an *Arabidopsis* cryptochrome gene in transgenic tobacco results in hypersensitivity to blue, UV-A, and green light. *Proc. Natl. Acad. Sci. U.S.A.* **92**, 8423–8427.

Lin, C., Yang, H. Guo, H., Mockler, T., Chen, J. and Cashmore, A.R. (1998) Enhancement of blue-light sensitivity of *Arabidopsis* seedlings by a blue light receptor cryptochrome 2. *Proc. Natl. Acad. Sci. U.S.A.* **95**, 2686–2690.

Lopez-Juez, E., Nagatani, A., Tomizawa, K.-I., *et al.* (1992) The cucumber long hypocotyl mutant lacks a light-stable PHYB-like phytochrome. *Plant Cell* **4**, 241–251.

Maddonni, G.A., Otegui, M.E., Andrieu, B., Chelle, M. and Casal, J.J. (2002) Maize leaves turn away from neighbors. *Plant Physiol.* **130**, 1181–89.

Maddonni, G.A., Otegui, M.E. and Cirilo, A.G. (2001) Plant population density, row spacing and hybrid effects on maize architecture and light attenuation. *Field Crops Res.* **71**, 183–193.

Maloof, J.N. (2003) Genomic approaches to analyzing natural variation in *Arabidopsis thaliana*. *Curr. Opin. Genet. Dev.* **13**, 576–582.

Maloof, J.N., Borevitz, J.O., Dabi, T., *et al.* (2001) Natural variation of light sensitivity in *Arabidopsis*. *Nat. Genet.* **29**, 441–446.

Maloof, J.N., Borevitz, J.O., Weigel, D. and Chory, J. (2000) Natural variation in phytochrome signalling. *Semin. Cell. Dev. Biol.* **11**, 523–530.

Markelz, N.H., Costich, D.E. and Brutnell, T.P. (2003) Photomorphogenic responses in maize seedling. *Dev. Plant Phys.* **133**, 1578–1591.

Mathews, S. (2005) Phytochrome evolution in green and nongreen plants. *J. Hered.* **96**, 197–204.

McCormac, A.C., Wagner, D., Boylan, M.T., Quail, P.H., Smith, H. and Whitelam, G.C. (1993) Photoresponses of transgenic *Arabidopsis* seedlings expressing introduced phytochrome B-encoding cDNAs: evidence that phytochrome A and phytochrome B have distinct photoregulatory functions. *Plant J.* **4**, 19–27.

McCormac, A.C., Whitelam, G.C. and Smith, H. (1992) Light-grown plants of transgenic tobacco expressing an introduced oat phytochrome A gene under the control of a constitutive viral promoter exhibit persistent growth inhibition by far-red light. *Planta* **188**, 173–181.

Monte, E., Alonso, J.M., Ecker, J.R., *et al.* (2003) Isolation and characterization of *phyC* mutants in *Arabidopsis* reveals complex crosstalk between phytochrome signalling pathways. *Plant Cell* **15**, 1962–1980.

Murphy, J.T. and Lagarias, J.C. (1997) The phytofluors: a new class of fluorescent protein probes. *Curr. Biol.* **7**, 870–876.

Ni, M., Tepperman, J.M. and Quail, P.H. (1999) Binding of phytochrome B to its nuclear signalling partner PIF3 is reversibly induced by light. *Nature* **400**, 781–784.

Peña, L., Martín-Trillo, M., Juárez, J., Pina, J.A., Navarro, L. and Martínez-Zapater, J.M. (2001) Constitutive expression of *Arabidopsis* LEAFY or APETALA1 genes in citrus reduces their generation time. *Nat. Biotechnol.* **19**, 263–267.

Peng, J., Richards, D.E., Hartley, N.M., *et al.* (1999) 'Green revolution' genes encode mutant gibberellin response modulators. *Nature* **400**, 256–261.

Platten, J.D., Foo, E., Elliott. R.C., Hecht, V., Reid, J.B. and Weller, J.L. (2005) Cryptochrome 1 contributes to blue-light sensing in pea. *Plant Physiol.* **139**, 1472–1482.

Putterill, J., Robson, F., Lee, K., Simon, R. and Coupland, G. (1995) The CONSTANS gene of *Arabidopsis* promotes flowering and encodes a protein showing similarities to zinc finger transcription factors. *Cell* **80**, 847–857.

Quail, P.H. (1994) Phytochrome genes and their expression. In: *Photomorphogenesis in Plants*, 2nd edn (eds R.E. Kendrick and G.H.M. Kronenberg), pp. 71–104. Kluwer, Dordrecht.

Robson, P.R., McCormac, A.C., Irvine, A.S. and Smith, H. (1996) Genetic engineering of harvest index in tobacco through overexpression of a phytochrome gene. *Nat. Biotechnol.* **14**, 995–998.

Robson, P.R.H. and Smith, H. (1997) Fundamental and biotechnological applications of phytochrome transgenes. *Plant Cell Environ.* **20**, 831–839.

Salamini, F. (1985) Photosensitivity in maize: evaluation, genetics, and breeding for insensitivity. In: *Breeding Strategies for Maize Production Improvement in the Tropics. Food and Agriculture Organization of the United Nations and Istituto Agronomico per L'Oltremare* (eds A. Bandolini and F. Salamini), pp. 143–157. Florence, Italy.

Salamini, F. (2003) Hormones and the green revolution. *Science* **302**, 71–72.

Salter, M.G., Franklin, K.A. and Whitelam, G.C. (2003) Gating of the rapid shade-avoidance response by the circadian clock in plants. *Nature* **426**, 680–683.

Sawers, J.H.R., Linley, P.J., Farmer, P.R., *et al.* (2002) Elongated mesocotyl1, a phytochrome-deficient mutant of maize. *Plant Physiol.* **2002**, 155–163.

Sawers, R.J., Linley, P.J., Gutierrez-Marcos, J.F., *et al.* (2004) The *Elm1 (ZmHy2)* gene of maize encodes a phytochromobilin synthase. *Plant Physiol.* **136**, 2771–2781.

Sawers, R.J.H., Sheehan, M.J. and Brutnell, T.P. (2005) Cereal phytochromes: targets of selection, targets for manipulation ? *Trends Plant Sci.* **10**, 138–143.

Schaffer, R., Ramsay, N., Samach, A., *et al.* (1998) The late elongated hypocotyl mutation of *Arabidopsis* disrupts circadian rhythms and the photoperiodic control of flowering. *Cell* **93**, 1219–1229.

Schmitt, J., McCormac, A.C. and Smith, H. (1995) A test of the adaptive plasticity hypothesis using transgenic and mutant plants disabled in phytochrome-mediated elongation responses to neighbors. *Am. Nat.* **146**, 937–953.

Sessa, G., Carabelli, M., Sassi, M., *et al.* (2005) A dynamic balance between gene activation and repression regulates the shade avoidance response in *Arabidopsis*. *Genes Dev.* **19**, 2811–2815.

Shimizu, K.K. and Purugganan, M.D. (2005) Evolutionary and ecological genomics of *Arabidopsis*. *Plant Physiol.* **138**, 578–284.

Shimizu-Sato, S., Huq, E., Tepperman, J.M. and Quail, P.H. (2002) A light-switchable gene promoter system. *Nat. Biotechnol.* **10**, 1041–1044.

Shlumukov, L.R., Barro, F., Barcelo, P., Lazzeri, P. and Smith, H. (2001) Establishment of far-red high irradiance responses in wheat through transgenic expression of an oat phytochrome A gene. *Plant Cell Environ.* **24**, 703–712.

Smith, H. (1981) Evidence that Pfr is not the active form of phytochrome in light-grown maize. *Nature* **293**, 161–165.

Smith, H. (1983) Is Pfr the active form of phytochrome ? *Philos. Trans. R. Soc. Lond.* **303**, 443–452.

Smith, H. (1995) Physiological and ecological function within the phytochrome family. *Annu. Rev. Plant Physiol. Plant Mol. Biol.* **46**, 289–315.

Smith, H. and Whitelam, G.C. (1997) The shade avoidance syndrome: multiple responses mediated by multiple phytochromes. *Plant Cell Environ.* **20**, 840–844.

Soh, M.-S., Kim, Y.-M., Han, S.-J. and Song, P.-S. (2000) REP1, a basic helix-loop-helix protein, is required for a branch pathway of phytochrome A signalling in *Arabidopsis*. *Plant Cell* **12**, 2061–2073.

Steindler, C., Matteucci, A., Sessa, G., *et al.* (1999) Shade avoidance responses are mediated by the ATHB-2 HD-zip protein, a negative regulator of gene expression. *Development* **126**, 4235–4245.

Takano, M., Inagaki, N., Xie, X., *et al.* (2005) Distinct and cooperative functions of phytochromes A, B, and C in the control of deetiolation and flowering in rice. *Plant Cell* **17**, 3311–3325.

Thiele, A., Herold, M., Lenk, I., Quail, P.H. and Gatz, C. (1999) Heterologous expression of *Arabidopsis* phytochrome B in transgenic potato influences photosynthetic performance and tuber development. *Plant Physiol.* **120**, 73–82.

Wagner, J.R., Brunzelle, J.S., Forest, K.T. and Vierstra, R.D. (2005) A light-sensing knot revealed by the structure of the chromophore-binding domain of phytochrome. *Nature* **438**, 325–331.

Weller, J.L., Beauchamp, N., Kerckhoffs, L.H.J., Platten, J.D. and Reid, J.B. (2001) Interaction of phytochrome A and B in the control of de-etiolation and flowering in pea. *Plant J.* **26**, 283–294.

Weller, J.L., Schreuder, M.E.L., Koornneef, M. and Kendrick, R.E. (2000) Physiological interactions of phytochromes A, B1 and B2 in the control of development in tomato. *Plant J.* **24**, 345–356.

Yan, L., Loukoianov, A., Blechl, A., *et al.* (2004) The wheat *VRN2* gene is a flowering repressor down-regulated by vernalization. *Science* **303**, 1640–1644.

Zhang, J., Campbell, R.E., Ting, A.Y. and Tsien, R.Y. (2002) Creating new fluorescent probes for cell biology. *Nat. Rev. Mol. Cell. Biol.* **3**, 906–918.

12 Light-quality manipulation by horticulture industry

Nihal C. Rajapakse and Yosepha Shahak

12.1 Introduction

Sunlight captured by chlorophyll provides the energy for photosynthesis, the process by which plants combine carbon dioxide and water to produce oxygen and carbohydrates. Carbon assimilated during photosynthesis provides the energy to sustain life on earth. In addition to being the energy source for photosynthesis, light also acts as a signal of environmental conditions surrounding the plants. There are photoreceptor pigments that can capture energy in different regions of the electromagnetic spectrum and function as signal transducers to provide information on the surrounding environment. Through these pigments, plants perceive subtle changes in light composition (quality), duration (period) and direction, and initiate physiological and morphological changes necessary for adaptation to the environment. Photoreceptor pigments, signal transduction processes and the physiological and morphological changes in response to alterations in light environment have been discussed in previous chapters. The purpose of this chapter is to illustrate how our basic understanding of plant responses to light quality is being utilised by horticulture industry to improve productivity and quality of horticultural crops.

12.2 Regions of light spectrum important for plant growth and development

Plants respond to a wide spectrum of light ranging from ultraviolet (UV) to far-red light. The specific regions of the light spectrum that are of importance to plant growth and development can be broadly divided into (1) UV ($<$400 nm), (2) the visible (400–700 nm) and (3) far-red (700–800 nm).

The UV spectrum can be further divided into three approximate categories: UV-A, radiation between 320 and 400 nm; UV-B, radiation between 280 and 320 nm and UV-C, radiation shorter than 280 nm. Prolonged exposure to shorter wavelength UV radiation can cause irreversible damage to genetic material and negatively affect plant productivity (Harm, 1980; Jagger, 1985). Most of the short-wavelength UV radiation (UV-B and UV-C) is absorbed by the ozone layer as sunlight enters the earth's atmosphere. However, with the depletion of the ozone layer in recent years, the effects of short-wavelength UV light on plant growth have become a major concern for the agricultural industry. UV radiation in general has been shown to reduce leaf area (Corso and Lercari, 1997), inhibit hypocotyl elongation (Ballare *et al.*, 1995; Corso and Lercari, 1997), reduce photosynthesis and biomass production (Tevini *et al.*, 1988; Tevini and Teramura, 1989; Teramura *et al.*, 1990; Corso and Lercari, 1997), increase the vulnerability of plants to pathogens (Mackerness, 2000) and induce flavonoid production and defense mechanisms (Lois, 1994; Lois and Buchanan, 1994; Mackerness, 2000).

The visible region of sunlight (400–700 nm) provides the energy for photosynthesis and is often called photosynthetically active radiation (PAR). The amount of PAR (irradiance) determines the rate of photosynthesis. The visible region can be broadly divided into blue (400–500 nm), green (500–600 nm) and red (600–700 nm) light. The primary photosynthetic pigments in higher plants, chlorophylls *a* and *b*, have absorption in the blue (with a peak near 430 nm) and red regions (with a peak near 660 nm), and have very little absorption in the green region. Thus, photosynthesis and overall productivity of horticultural crops could be enhanced by increasing the amount of blue and red light present in the growth environment. Greenhouse industry is looking into increasing blue and red light inside greenhouses by incorporating photoluminescent pigments that can transform little used UV and green light to blue and red light in greenhouse covers.

Plants have the ability to perceive subtle changes in light composition, duration and the direction in the growing environment, and initiate physiological and morphological changes necessary to survive the existing environmental conditions. This ability of light to control plant morphology is known as photomorphogenesis, and the blue, red and far-red regions of the light spectrum play key roles in this process. These light signals are captured by the phytochrome, cryptochrome and phototropin photoreceptors, which then trigger changes in plant growth and development. These light-regulated signalling events are necessary for normal plant development, and they ensure that adaptive changes occur in response to environmental change.

The structure, signal transduction processes and functions of photoreceptors are described in previous chapters. To briefly summarise, phytochromes are the most intensively studied photoreceptors that control morphogenesis in response to the changes in red and far-red light in the environment. Phytochromes are capable of detecting wavelengths from 300 to 800 nm, with maximum sensitivity in the red (600–700 nm with peak absorption near 665 nm) and the far-red (700–800 nm with peak absorption near 730 nm) wavelengths of the spectrum. This pigment system consists of two interconvertible forms: the P_r and P_{fr} forms. The P_r form absorbs red light and is transformed into the P_{fr} form. The P_{fr} form absorbs far-red light and is transformed back into the P_r form. Photoreversibility is a distinctive character of phytochrome-mediated responses.

$$P_r \underset{\text{Far red light}}{\overset{\text{Red light}}{\rightleftarrows}} P_{fr} \longrightarrow \text{Responses}$$

Of the two forms, the P_{fr} form is thought to be the 'active form' that controls signal transduction and plant response. Most photomorphogenic responses are controlled by the cellular amount of P_{fr} relative to total phytochrome ($P_{fr}:P_{tot}$ at photoequilibrium). In general, P_{fr}:P_{tot} depends largely on the absorption of red and far-red light by the plant, and therefore P_{fr}:P_{tot} increases in environments with a higher proportion of R:FR light with a maximum of about 0.85 after saturation irradiation by red light.

Considerable advances have been made in recent years understanding cryptochromes, phototropins and blue light responses (see Chapters 2, 3 and 7). However, partly because of our broader understanding of phytochrome function and the particular traits controlled by phytochrome, most commercial applications that enhance productivity and quality of horticultural crops involve red and/or far-red light manipulations in the production environment.

12.3   Plant responses to quality of light

Since the discovery of role of phytochrome in seed germination in 1930s, numerous responses that are regulated by phytochromes have been identified. To briefly summarise, red light has been shown to inhibit internode elongation (Vanderhoef *et al.*, 1979; Noguchi and Hashimoto, 1990), promote lateral branching and tillering (Tucker 1975; Deregibus *et al.*, 1983), prevent dark induced leaf abscission (Decoteau and Craker, 1984), delay floral initiation (Downs and Thomas, 1982) and increase anthocyanin, chlorophyll and carotenoid pigments (Rabino and Mancinelli, 1986; Kerckhoffs *et al.*, 1997; Alba *et al.*, 2000; Schofield and Paliyath, 2004). Underpinning many of these responses are changes in gene expression (Thomas *et al.*, 1999; Gil and Garcia-Martinez, 2000; Jones *et al.*, 2000). In many instances, far-red light can negate red-light-mediated effects. Blue light has been shown to control a number of responses, including inhibition of hypocotyl elongation (Warpeha and Kaufman, 1989; Ahmad *et al.*, 2002; Folta *et al.*, 2003), phototropism (Baskin and Iino, 1987; Ritter and Koller, 1994), stomatal and chloroplast movement (DeBlasio *et al.*, 2003; Talbott *et al.*, 2003; Takemia *et al.*, 2005), tillering (Barnes and Bugbee, 1992), and as for light-red light, some of these responses are mediated through changes in gene expression (Warpeha *et al.*, 1989; Short and Briggs, 1994). These blue light responses have been shown to be mediated through the cryptochromes and phototropins.

Light quality has also been shown to regulate fungal growth and development (Manachere, 1994; Hughes and Hartmann, 1999). Control of sporulation by exposure to UV, blue or red wavelengths has been reported for several fungal species. Blue light has been shown to induce sporulation of *Trichoderma viride* and *Verticillium agaricinum* (Kumagai and Oda, 1969; Osman and Valadon, 1979), and inhibit sporulation of *Alternaria cichorii* and *Botrytis cinerea* (Tan, 1974; Vakalounakis and Christias, 1981). UV-B light has been shown to induce sporulation of *B. cinerea* (Tan, 1975a). Red has been shown to inhibit, whilst far-red light enhances sporulation in *B. cinerea* (Tan, 1975b). Quality of light, especially UV range, influences insect behaviour through its influence on insect vision, navigation and feeding. The changes in insects' behaviour can influence the spread of virus diseases and direct damages to crop plants.

Horticultural crops contain many functional phytochemicals (antioxidants) that contribute to the overall quality and protect plant cells from oxidative damage by external factors, such as excessive sunlight, temperature, and pest and disease infections. Epidemiological studies suggest that these antioxidant phytochemicals in fruits and vegetables aid in protecting human cells from oxidative damages (Ames *et al.*, 1993; Fahey and Stephenson, 1999; Wargovich, 2000; Morris *et al.*, 2002; Rao and Shen, 2002). Composition of functional phytochemicals varies significantly with the crop, but carotenoids, dietary fibres, folic acid, organosulphur compounds, phenolic compounds and vitamins (ascorbic acid and tocopherols) dominate these compounds. Therefore, enhancement of phytochemicals in horticultural crops through genetic and environmental manipulation has become a priority in recent years (Kalt and Kushad, 2000; Takeda, 2001).

Light-quality effects on some of these functional phytochemicals have been reported. UV-B radiation has been shown to decrease both ascorbic acid and β-carotene concentrations (Schmitz-Eiberger and Noga, 2001). In early work, UV radiation was thought to be the most effective in stimulating anthocyanin production. Longer wavelength radiation, red in particular, is also effective in stimulating anthocyanin and other flavonoid biosyntheses (Lange *et al.*, 1971; Mohr, 1984). In mustard, red light appears to induce quercetin and anthocyanin biosynthesis (Buchholz *et al.*, 1995). Carotenoid biosynthesis has been shown to be under phytochrome control. Exposure to red light increased lycopene accumulation over twofold during tomato fruit ripening, an effect that was shown to be far-red light reversible (Alba *et al.*, 2000). Expression of genes involved in phytoene synthesis also appears to be regulated by phytochrome (Schofield and Paliyath, 2004). Watercress plants grown under metal halide light enriched with red light had higher concentration of gluconasturtiin than far-red light enriched plants (Engelen-Eigles *et al.*, 2006). They also reported that the exposure of watercress plant to red light at the end of the main photoperiod increased gluconasturtiin levels compared to plants exposed to far-red light at the end of the photoperiod suggesting a role of phytochrome in this process. Environmental regulation of health-beneficial phytochemicals in food crops is poorly understood at present. Practical applications into how light manipulation can be used to improve the nutritional and functional phytochemicals in food crops need to be investigated.

12.4 Light manipulation by horticulture industry

For many years the horticulture industry has manipulated the light environment to enhance useful traits. Light modifications can improve plant traits such as growth habit, foliage quality, flower production, and can also assist with pest and disease management. Under protected cultivation, light-quality manipulation can be achieved with either supplementary electric lighting systems with specific wave bands or spectral filtering greenhouse covers that can filter out specific wavelengths through absorption or reflection. In the field production of horticultural crops, selective light-reflecting mulch films have been used to modify the light quality in plants' microclimate.

12.4.1   Electric light sources

Various types of electric light sources are being used by horticulture industry to produce crops under protected environments. High-intensity discharge (HID) metal halide and high-pressure sodium lamps are most commonly used by modern greenhouse growers in northern latitudes to supplement natural light for year-around crop production. Under closed hydroponic production systems, HID lights are used as the sole light source to produce crops. These light sources have high input in blue and red regions of the spectrum and lack far-red wavelengths. To alleviate abnormal growth patterns caused by spectral unbalance of these light sources, incandescent bulbs that are high in far-red wavelengths are added to the growing environment.

Many responses in plants, such as seed germination, flower initiation and development, and growth habits, are regulated by the photoperiod (specifically by the dark period) and are under phytochrome control. Depending on the flowering response to photoperiod, plants are classified into three broad categories: short-day, long-day and day-neutral plants. Short-day plants, such as chrysanthemums, require a dark period longer than a critical length for flowering, whilst long-day plants, such as petunia, flower when dark period is shorter than a critical length. Day-neutral plants, such as roses, flower without regard to day length.

The earliest commercial use of light manipulation under protected environments was for year-around production of day-length-sensitive flowering plants. Artificial shortening of photoperiod by covering plants with opaque materials, such as black plastic, during long photoperiods (summer) has been used commercially for many years in year-around production of short-day plants. Chrysanthemum, kalanchoe and poinsettia are few of the popular short-day crops produced, using artificial photoperiod manipulations under natural long days.

Photoperiod extension by electric lighting during short natural photoperiods is used widely in year-around production of flowering long-day plants. This is also used to delay the premature flowering of short-day crops intended for special occasions, such as Christmas. Photoperiod extension light can be added at the end of the day to extend the photoperiod or can be given as a 2-to-4-h night break in the middle of the dark period (night interruption) to promote flowering of long-day plants. Relatively low irradiance is needed for photoperiod extension; therefore, low-output fluorescent or incandescent bulbs are usually sufficient. Generally, day-length extensions, provided by low R:FR ratio incandescent lamps, are more efficient than fluorescent sources (which are high in R:FR ratio) in promoting flowering (Downs and Thomas, 1982; Runkle and Heins, 2003). This highlights the major role of the phytochromes in this response. Lighting is generally provided by spacing incandescent bulbs about 90 cm above the plants and about 120 cm between bulbs (to provide 3–5 μmol m^{-2}). Incandescent lamps, however, cause plants to grow in a more elongated fashion. This is a component of the shade-avoidance syndrome of responses, which also includes early flowering, that are induced by low R:FR ratio light. In the natural environment, these light conditions are indicative of neighbouring vegetation (see Chapter 9).

To maintain vegetative stocks of short-day plants and to promote rooting, plant propagators use night interruption lighting from incandescent bulbs in the middle of the dark period from mid-September to late March. When the cuttings are rooted, growers use electric lighting to control the height of flowering stem. For example, if long flower stems are required, such as in cut chrysanthemums, growers extend the photoperiod with incandescent light sources to promote vegetative growth before giving a short photoperiod to promote flowering.

When plants are grown in crowded environments, common in many plant production facilities, stem elongation is often a problem because of the reflection of far-red light from the neighbouring plants (see Chapter 9). Tall plants are often not aesthetically appealing, lodge easily and are difficult to handle during transplantation to the field. Chemical growth retardants are often used in the transplant industry to reduce stem elongation and maintain compactness for better appearance and easy handling. Stem elongation of greenhouse-grown transplants and pot plants can also be reduced non-chemically, using electric light sources high in red and low in far-red light. Exposure of plants to red light (from fluorescent bulbs) at the end of the natural photoperiod has been shown to reduce stem elongation, while the exposure to far-red light (from incandescent bulbs) has been shown to promote stem elongation without affecting the developmental stage of transplants (Kasperbauer, 1971; Rajapakse et al., 1993; Blom et al., 1995; Hatt-Graham and Decoteau, 1997). Transplant producers can use low-irradiance fluorescent light at the end of the main photoperiod to produce short and compact plants. These methods are particularly attractive to organic produces where chemical growth retardant use is strictly prohibited.

12.4.2   Spectral filters

12.4.2.1   Greenhouse covers

The use of channelled, double-walled polyacrylic and polycarbonate plastic greenhouse glazing materials provides the opportunity to use water or liquid dyes contained in hollow channels of the glazing as filtering materials. These filters have been variously called liquid optical filters, optical liquid filters, liquid radiation filters and liquid spectral filters. In the 1970s and 1980s, French scientists investigated the use of double-layered acrylic and glass structures filled with water and copper chloride in a closed system to absorb infrared wavelengths from the sunlight to reduce heat build-up in the greenhouses (Chiapale et al., 1977). Liquid-filled greenhouse covers reduced greenhouse energy requirements by 20–40% and virtually eliminated the need for forced ventilation in greenhouses (Van Bavel et al., 1981; Chiapale et al., 1983). Infrared-absorbing liquid greenhouse covers, however, were of limited use to commercial growers because of the difficulties in liquid handling.

Selective filtering of sunlight, primarily to influence photomorphogenesis, was investigated by Israeli scientists in mid-1970s, using coloured celluloid material glued to rigid, clear plastic panels (Kadman-Zahavi et al., 1976). In these early studies, they used combinations of celluloid films to investigate the removal of blue, red and/or far-red light from greenhouses. They found that filters which transmitted only blue or red light retarded stem elongation and delayed flowering of grass species, and the addition of filters that transmit far-red light promoted both flowering and stem elongation (Kadman-Zahavi and Ephrat, 1976). Various aqueous dye (red, green, yellow, blue and copper sulphate [$CuSO_4$])-filled polyacrylic and polycarbonate greenhouse covers were investigated in the late 1980s in Norway (Mortensen and Stromme, 1987) and in the United States of America (McMahon *et al.*, 1990). The primary interest was to identify filters that could selectively filter out elongation stimulating far-red light from the sunlight as a means of reducing stem elongation in bedding plants without using chemical growth retardants. Of the different liquid-dye-filled filters tested, only the $CuSO_4$ filters were effective in removing far-red light from the sunlight. The $CuSO_4$ liquid filter reduced both red and far-red wavelengths of transmitted light, but the reduction of far-red wavelengths was greater than the reduction of red wavelengths, therefore resulting in a high-red/low-far-red environment inside the greenhouse (Rajapakse *et al.*, 1992; Rajapakse and Kelly, 1992).

In early work, Mortensen and Stromme (1987) observed that liquid $CuSO_4$ filters reduced stem elongation and internode length of chrysanthemum, tomato and lettuce seedlings. Green and yellow liquid filters increased stem elongation. Lateral bud production was stimulated by the $CuSO_4$ filters but inhibited by green and yellow filters. In follow-up work, poinsettia and two cultivars of chrysanthemum, 'Spears' and 'Yellow Mandalay', grown under $CuSO_4$ filters had reduced stem elongation and short internodes (McMahon and Kelly, 1990; McMahon *et al.*, 1991). In later work, it was shown that far-red light filtering by $CuSO_4$ filters was effective in reducing stem elongation of a wide range of plants (Table 12.1), though they were not effective in reducing height of spring-flowering bulb crops. Plants grown under $CuSO_4$ filters were smaller in size, had more leaf chlorophyll and were darker green in colour than plants grown under control filters (McMahon *et al.*, 1991; Rajapakse and Kelly, 1992). Chrysanthemum plants grown under $CuSO_4$ filters used less water than control plants (Rajapakse and Kelly, 1993). Transpiration rate of plants grown under $CuSO_4$ filters was not affected, thus indicating that the reduction of water use is a result of small plant size. In 'Meijikatar' miniature roses, $CuSO_4$ filters slightly accelerated anthesis of plants grown early spring by 2–3 days, but delayed anthesis in late spring- and summer-grown plants by a similar time period (Rajapakse and Kelly, 1994). In 'Bright Golden Anne' chrysanthemums, $CuSO_4$ filters delayed anthesis by 7 days in autumn-grown plants, and by 13 days in winter-grown plants (Rajapakse and Kelly, 1995). The $CuSO_4$ filters did not affect total number of

Table 12.1 Response of selected crops to far-red light filtering by $CuSO_4$ filters

Positive response		No response
Ageratum	Easter lily	Azalea
Geranium	Poinsettia	Tulip
Impatiens	Lettuce	Hyacinth
Pansy	Chrysanthemum	Narcissus
Bell pepper	Miniature roses	
Petunia	Exacum	
Salvia	Vinca	
Tomato	Marigold	

flowers, but resulted in smaller flowers than control plants in both miniature roses and chrysanthemums. In 'Nellie White' Easter lilies, $CuSO_4$ filters did not delay anthesis or reduce flower size (Kambalapally and Rajapakse, 1998). Collectively, these studies demonstrate that manipulating the light quality (R:FR ratio) evokes a range of responses in different plant species and at different times of the year.

It has been well established that plant hormone gibberellin (GA) and far-red light promote stem elongation and that these responses are under phytochrome control. Both the reduction of active GA levels and/or the reduction of sensitivity to active GAs by red light have been suggested to reduce stem elongation under high-red light environments (Campell and Bonner, 1986; Reid and Ross, 1988; Martinez-Garcia and Garcia-Martinez, 1992; Weller et al., 1994). In efforts to understand the roles of gibberellic acid (GA_3) and phytochrome under $CuSO_4$ filters, application of GA_3 or short (15-min) exposure to end-of-the-day (EOD) far-red light reversed the reduction of stem elongation of plants grown under $CuSO_4$ filter (Rajapakse and Kelly, 1991; Rajapakse et al., 1993). Exposure to EOD red light reduced the stem elongation of plants grown under control filter but had no effect under the $CuSO_4$ filter. Exposure to EOD far-red light did not significantly affect stem elongation under control filter. In studies using another active gibberellin (GA_1) and precursor gibberellins (GA_{19} and GA_{20}), Maki et al. (2002) observed that the application of GA_1 increased stem elongation of chrysanthemum plants equally under the normal and the far-red light filtered environments. Precursor gibberellins, GA_{19} and GA_{20}, were less effective in increasing stem elongation under the far-red light filtered environments indicating that the metabolism of GAs may have been reduced. In further experiments, they reported that radiolabeled precursor gibberellins, [^{14}C]GA_{12} and [^{14}C]GA_{19}, metabolised slowly in plants grown under the $CuSO_4$ filter, suggesting that the low turnover of GAs at least partially caused the lower response to precursor GAs. Although the metabolism of GA_1 under the $CuSO_4$ filters was not investigated, their evidence support that $CuSO_4$ filters may have enhanced the inactivation of GA_1. These results suggest that reduced active GA levels through the action of phytochromes mediate the response under $CuSO_4$ filters.

Early work demonstrated that far-red light filtering by liquid $CuSO_4$ filters had great potential as a non-chemical alternative for producing short and compact bedding plants. However, in practice, the value of liquid filters to horticulture industry is limited because of difficulties in material handling and of high construction costs. Recent advances in greenhouse film manufacturing processes have made it possible to manufacture multilayer greenhouse films with various additives sandwiched between upper and lower layers to provide specific effects. During the mid-1990s, plastic and pigment manufacturers were interested in developing photoselective plastic greenhouse covers with pigment layers that can selectively absorb or reflect specific wavelengths (i.e. red and far-red light absorbing pigments for photomorphogenic effects and infrared-reflecting films for heat reduction). Plastic greenhouse covers with infrared light removing pigments are commercially available from various sources and are especially useful to greenhouse growers in southern latitudes where heat build-up and cooling costs are high (Verlodt and Verschaeren, 1999).

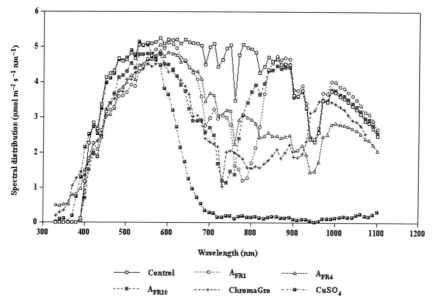

Figure 12.1   Spectral distribution under photoselective greenhouse films with different far-red light absorbing pigments and under liquid $CuSO_4$ filters. A_{FR1}, A_{FR4}, A_{FR10} and ChromaGro are films with far-red light absorbing pigments. Control film is a polyethylene film without pigments.

Photoselective plastic greenhouse films with far-red light absorbing dye pigments have been developed by greenhouse film manufacturers in Japan (Mitsui Chemicals Inc.) and the United Kingdom (BPI Agri) and tested on a wide range of crops, especially focusing on height control of bedding plants (vanHaeringen et al., 1998; Rajapakse et al., 1999; Runkle and Heins, 2003; Fletcher et al., 2005). Photoselective films, in general, reduced the transmission of both red (600–700 nm) and far-red (700–800 nm) light, but the reduction of far-red light is greater, therefore resulting in a higher R:FR of transmitted light (Figure 12.1). Far-red light absorption capacity increased with increasing dye pigment concentration in the film, but increasing dye concentration reduced the transmission of photosynthetic light (van-Haeringen et al., 1998). Photoselective films produced with a low concentration of dye (corresponding to a 15% reduction in light transmission) did not cause a commercially significant reduction in stem elongation while photoselective films produced with a high concentration of dye (corresponding to a 45% reduction in light transmission) reduced stem elongation, but overall quality of plants was poor due to reduced photosynthetic light (Rajapakse et al., 1999). Far-red light filtering photoselective films developed for commercial testing reduced the transmission of photosynthetic light by 20–25% but were effective in reducing stem elongation in wide range of ornamental and vegetable transplants, without adversely affecting the overall quality of transplants (Table 12.2; Plate 12.1). Amongst the crops tested, watermelon and cucumber seedlings had the greatest response followed by bell

Table 13.2 Influence of far-red light absorbing greenhouse film on seedling height

Crop	Seedling height (cm)	
	Control film	A_{FR} film
Vegetable crops		
Cucumber	17.3	11.1 (64)
Watermelon	21.7	14.6 (67)
Tomato	15.0	11.2 (75)
Sunflower	23.5	15.0 (64)
Cabbage	3.5	2.5 (71)
Bean	16.6	9.5 (57)
Bell pepper	11.1	8.4 (76)
Ornamental crops		
Snapdragon	48.3	48.9 (101)
Verbena	11.0	5.3 (48)
Cosmos	37.3	33.5 (89)
Petunia	8.0	3.0 (38)
Lisianthus	31.2	27.9 (89)
Zinnia	38.0	30.4 (80)
Delphinium	9.0	7.8 (87)
Chrysanthemum	25.1	18.5 (74)

A_{FR} film, far-red light absorbing film; control film, a polyethylene film without dye.
Number in parentheses indicates the percentage height compared to control plants.

peppers, tomatoes and chrysanthemums. Far-red light filtering photoselective films did not affect the number of leaves or the establishment of seedlings when transplanted in the field.

Flowering of ornamental crops grown under far-red light absorbing photoselective films has been shown to vary, depending on the crop and the growing season. Anthesis of day-neutral plants, tomato and miniature rose, was unaffected by far-red light absorbing films. Anthesis of short-day plants, cosmos, zinnia and chrysanthemum, was slightly delayed (by 1–2 days; Cerny *et al.*, 2003) under far-red light deficient environments. However, far-red light absorbing photoselective films had the greatest influence on anthesis of long-day plants. Far-red light absorbing films, depending on the growing season, delayed anthesis by 7–13 days in snapdragon and petunia, both long-day plants (Kubota *et al.*, 2000; Cerny *et al.*, 2003; Runkle and Heins, 2003; Fletcher *et al.*, 2005). Other long-day crops such as coreopsis, *Campanula* and pansy have also been reported to be late flowering when grown under far-red light absorbing greenhouse films (Runkle and Heins, 2001). Commercial growers often produce a range of crops in a greenhouse. Therefore, the delay in anthesis in certain species is a hindrance to commercial adaptation of photoselective films. Furthermore, the short effective life of far-red light absorbing films has also been a limiting factor in their commercial adaptation in the United States of America where greenhouse growers typically change films every 3–4 years. In tests evaluating the effective life, photoselective films were effective only in controlling plant height for 15 months under protected conditions (inside a greenhouse), and when these films were used in the field, dyes began to degrade after 10 months rendering films ineffective in height control (unpublished data). Using the photoselective films as the inner layer of a double-layered, plastic-covered greenhouse or by using the photoselective film as an inside curtain could help extend the effective life of film until steps are taken to improve the stability of pigments under field conditions.

Greenhouse light-quality manipulation has also been used to control disease development of greenhouse crops (Hite, 1973). Partial control of grey mould, caused by *B. cinerea*, in cucumber and tomato has been reported with greenhouse films that absorb UV light (Honda *et al.*, 1977). UV-absorbing greenhouse covers have also been shown to reduce the severity of blight, caused by *Sclerotinia sclerotiorum*, in eggplant and cucumber (Honda and Yunoki, 1977). UV-absorbing greenhouse films reduced sporulation in *Alternaria solani* and the development of early blight in greenhouse tomato (Vakalounakis, 1991). Greenhouse polyethylene films with high-blue:UV-B transmission reduced sporulation of *B. cinerea* and slowed the development of grey mould in greenhouse tomato (Reuveni *et al.*, 1994). Partial control of downy mildew, caused by *Pseudoperonospora cubensis*, in tomato; powdery mildew, caused by *Leveillula taurica*, in sweet pepper and sclerotinia, caused by *S. sclerotiorum*, in sweet basil has been reported with similar greenhouse covers (Raviv *et al.*, 1998). UV-B-absorbing greenhouse films have also been shown to reduce insect populations (tobacco whitefly, western flower thrips, aphids and spider mites). As a consequence, pest damage and the transmission of virus diseases by insect vectors were lower. Greenhouse films that can block UV light below 380 nm have been shown to be more effective in reducing whitefly, aphid and thrips infestations than those blocking below 360 nm (Antignus *et al.*, 1996; Costa and Robb, 1999; Costa *et al.*, 2002). However, in greenhouses with open side vents, UV-blocking films have been less effective in reducing insect populations (Costa *et al.*, 2003). Thus, the greenhouse structure must be considered in detail, when evaluating pest management by greenhouse films. Photoselective greenhouse covers designed to control plant height and harmful pests may be particularly useful for organic crop production where synthetic chemical control agents are not used.

12.4.2.2   Photoselective nets

Nets are used in the production of horticultural crops to protect from excessive sunlight, environmental hazards (wind and hail) or flying pests (birds and insects). Black nets are most commonly used for shading, while clear, transparent nets are used for hazard or pest protection. Israeli scientists from the Volcani Center, in collaboration with Polysack Plastic Industries, have recently developed coloured nets ('ColorNets') that can alter both the quality and the quantity of the light intercepted by the plants growing underneath, in addition to providing the desired protection (Shahak *et al.*, 2004b). The ColorNet approach deals with light quality in its broader sense, to include light dispersion and thermal components, in addition to the spectral composition.

A series of 'ColorNets' has been developed for outdoor use, each containing pigments that differentially absorb UV, blue, green, yellow, red, far-red or infrared wavelengths (Table 12.3; Figure 12.2). Light-scattering elements have been

Figure 12.2 Coloured shade nets in commercial operations in Israel. (A) Peach orchard and (B) strawberry hoop houses covered with coloured nets.

Table 12.3 Light-quality modification in the UV-B to far-red spectral range by ColorNets showing distinct effects on horticultural crops

Net	Enriched spectral bands	Reduced spectral bands	Light scattering
Blue	B	UV + R + FR	++
Red	R + FR	UV + B + G	++
Yellow	G + Y + R + FR	UV + B	++
White	B + G + Y + R + FR	UV	++
Pearl	—	UV	+++
Grey	—	All to same extent	+
Black (control)	—	All to same extent	—

incorporated into these nets to increase diffused radiation, thus improving light penetration into the inner parts of the plant canopy.

Since the nets are composed of holes, in addition to the translucent-photoselective plastic threads, they provide mixtures of natural, unmodified light, which is passing through the holes, together with the diffused, spectrally modified light, which is emitted by the photoselective threads. The relative content of the modified versus unmodified light, as well as the shading factor, is defined by the knitting design, density, chromatic and light dispersive additives, all of which can be adjusted to fit the needs of each crop.

Ornamentals. The 'ColorNets' were initially tested on ornamental crops commercially cultured under black shade nets (Oren-Shamir *et al.*, 2001; Priel, 2001; Shahak *et al.*, 2002). Compared with common black nets of the same shading factor (in PAR), the red and yellow nets specifically stimulated vegetative growth rate and vigour, while the blue net caused dwarfing in *Pittosporum variegatum*, *Aralia* and *Philodendron monstera*, as well as in seasonal cut flowers (*Lisianthus*, *Trachelium*, sunflower and lupine; Figure 12.3). The grey net, which absorbs near-infrared and infrared radiation, enhanced branching and bushiness in *P. variegatum*, and these plants had smaller leaves that were less variegated. The grey net also enhanced stolon branching in leather-leaf fern and *Ruscus*. ColorNets differentially affected anthesis and quality of cut flowers. In *Ornithogalum dubium*, for example, the red net advanced anthesis by as much as 3 weeks relative to the black net (M. Oren-Shamir and Y. Shahak, unpublished results). The effects of the blue, yellow and red nets might be attributed to their relative enriching/reducing of blue, yellow and red spectral bands in the filtered light. These effects are similar to effects reported for photoselective films and artificial illumination (Kasperbauer, 1971; Rajapakse *et al.*, 1999; Kim *et al.*, 2004). On the other hand, the suppression of apical dominance and of variegation by the grey net is a new phenomenon, which cannot be attributed to known photomorphogenetic mechanisms, and needs further elucidation.

Fruit trees. Current studies are aimed at potential benefits of colour netting of orchards (apples, pear, peach, persimmon, loquat, pomegranate, avocado, citrus

Figure 12.3 Comparison of plant height of yellow lupine grown under (A) the blue and (B) yellow nets, both of 50% shading factor.

and banana), vineyards (table grapes and kiwi), small fruits (strawberries and black berries), vegetable crops (peppers, lettuce and herbs) and nurseries. Unlike ornamental crops, which typically require 50–80% shading for optimal production in sunny climates, the productivity of most fruit crops might negatively respond to shading. The results obtained in Israel so far suggest that low-shade colour netting (30% or less) of fruit crops can improve production. Differential effects on fruit set, time of maturation (either earliness or delayed maturation), fruit size, fruit colour and vegetative responses have been observed under colour nets, compared with conventional black shade nets or un-netted orchards (Shahak *et al.*, 2004b,c). For example, while covering table grape vineyards by white nets resulted in advanced maturation (as judged by berry sugar accumulation rate), the red net delayed the maturation, with both effects bearing potential benefits for early and late cultivars, respectively (Shahak *et al.*, 2005). The red net induced greater fruit set in peach and apple, relative to other nets and more so to un-netted controls (Shahak *et al.*, 2004b). The pearl net distinctly increased fruit size and total fruit yield of Golden Delicious apple, while an equivalent black net reduced fruit size, relative to un-netted control (Figure 12.4). The photoselective responses were apparent in spite of the fact that only a small fraction (30% or less) of the intercepted sunlight can be modified when using low-shading nets. Rapid development of the root system of banana plug transplants during hardening has been observed in red-net-covered plants when compared to the commercial black nets. Rapid root development could have positive impacts on both nursery production and plant establishment in the field. These differential responses were observed in addition to non-photoselective

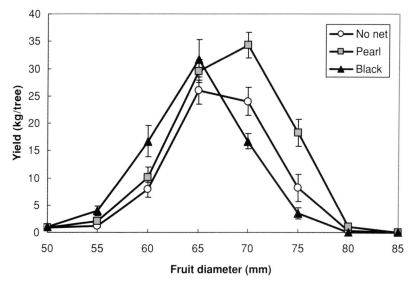

Figure 12.4   Fruit size of Golden Delicious apple grown under the pearl net as compared with a black net and with un-netted control. Both nets were of 30% shading.

Table 12.4 Bell pepper production under the red and pearl nets compared with the common practice black shade net

Net	Cultivar					
	Triple Star		Caliber		Anna	
	Fruit/plant	%	Fruit/plant	%	Fruit/plant	%
Black net	20.2 a	100	20.4 a	100	24.6 a	100
Pearl net	25.9 b	128	27.5 b	134	30.6 b	125
Red net	28.6 b	141	27.9 b	137	30.7 b	125

Production relates to the fruit harvested throughout the 2005 summer season in the Negev, Israel.

(colour-independent) effects of the netting, such as the reduction of fruit sunburns, russeting and wind scars.

Vegetables. Bell peppers, commercially grown inside 30% shade net houses in southern Israel, were recently found to produce 25–40% more fruits per plant over the summer growing season, upon replacing the common black net by either the red or pearl nets. Fruit size was not significantly reduced by the two ColorNets (Table 12.4; Shahak *et al.*, 2006a,b). In southern Spain, replacement of the common practice summer white wash by a red net with equivalent shading has been shown to increase the production of greenhouse bell peppers by 20% (Shahak *et al.*, 2004a). Leafy crops, such as lettuce and basil, produced 25–50% higher yields under red or pearl nets, relative to equivalent blue or black shade nets (R. Ganelevin *et al.*, unpublished results).

The observations with colour nets are, however, rather preliminary at this point, and further experimentation is required to explore their full potential in improving productivity and quality of fruit crops. Both the modification of light quality and additional microclimatic factors might be responsible for observed differences, and further work is needed to delineate actual mechanisms.

12.4.2.3   Coloured plastic mulch
Polyethylene mulch has been used by horticulture industry since 1960s to improve productivity and quality of field-grown vegetable crops. Black, clear and white plastic are the most commonly used colours in commercial production. Productivity and quality improvements by plastic mulch are attributed to the impact of mulch on root zone temperature improvements, soil water and nutrient conservation, and weed suppression. Black plastic mulch, the predominant colour used in vegetable production, absorbs most of the incoming solar radiation and re-emits energy in the form of thermal radiation. This mulch warms the soil in the spring and retards weed growth because of the reduced light transmission to the soil. Clear plastic mulch transmits most of the incoming radiation, depending on the thickness and degree of opacity of the polyethylene, and increases soil temperature more than the black plastic mulch. However, weed growth is not suppressed because of the transmission of solar radiation. White plastic mulches reflect most of the incoming

solar radiation into the plant canopy and improve the light distribution within the plant canopy. Improvements in light distribution inside the canopy can enhance fruit growth and fruit colouration. White plastic mulch can result in a slight decrease in soil temperature and can be used for improving crops establishment in hot summer conditions.

Quality of light reflected from the mulch surface can influence the morphology and productivity of plants through the involvement of photosensory pigments. Photoselective mulches, which selectively reflect radiation in specific regions of the sunlight spectrum, were investigated extensively in the 1980s. In early work, a range of coloured mulches, painted with red, yellow, blue, silver and orange, which reflect different radiation patterns into the crop canopy, were investigated (Decoteau *et al.*, 1988, 1989). White- or silver-painted mulches reflected more light, which had lower ratio of far-red relative to red light than black or red mulch. Tomato plants grown on red mulch produced more early marketable fruits and less foliage than those produced on white or silver mulch. Tomato yield increases of over 20% have been reported from plants grown on red mulch. In strawberry, Kasperbauer (2000) reported that yield and berry size were greater in plants grown on red mulch versus black mulch. The strawberry yield and berry size increases have been attributed to the enhanced photosynthate partitioning into developing fruits by light reflected from red mulch. Red plastic mulch has also been shown to increase yields in zucchini and in honeydew and muskmelons. Coloured mulches have been shown to have no additional productivity benefits over conventional mulch in some research.

In more recent work, mulch colour has been shown to affect the composition of nutritional and functional phytochemicals. Strawberries ripened over red mulch had higher sugar and sugar/acid ratio, and emitted higher levels of aroma compounds than berries grown over black mulch (Kasperbauer *et al.*, 2001; Loughrin and Kasperbauer 2002). Basil plants grown on yellow and green mulch produced higher concentration of aroma and phenolic compounds than those grown on white or blue mulch (Loughrin and Kasperbauer, 2001). Turnip plants grown on blue mulch have been reported to have a greater concentration of glucosinolates and ascorbic acids (Antonious *et al.*, 1996). These results show that light reflected by mulch could affect chemical composition of food crops, and further research is warranted as a means to improve nutritional and functional qualities of food crops.

Fruit skin colouration and fruit quality appear to be influenced by the amount of light available within the canopy. Shading of fruits by foliage often leads to poor fruit colouration and quality. Reflective films have been shown to improve light distribution inside the plant canopy and improve fruit colouration, fruit size, ripening and taste, and return bloom and reduce the number of deformed fruits. Covering the orchard floor with light-reflecting foil has been shown to increase photosynthetic light absorption of apple tree canopies (Green *et al.*, 1995). The red colouration of 'Fuji' apple was improved and early fruit yields increased by reflective ground covers (Andris and Crisosto, 1996). Reflective ground covers applied 2–4 weeks before the harvest improved red colouration and total soluble sugar content of several peach cultivars (Layne *et al.*, 2001). Improvement of fruit colour and quality by reflective films is mainly a result of improved light distribution within the canopy. The UV light reflective plastic mulch has also been reported to interfere with the movement of insects and reduce the development of viral diseases transmitted by insect vectors (Farias-Larios and Orozco-Santos, 1997; Summers *et al.*, 2004).

12.5 Future prospects

The general public is becoming more and more concerned about the extensive use of chemicals by horticulture industry and the presence of chemical residues in food crops. As a result, interest in using non-chemical alternatives to regulate plant growth and to control pests and diseases has increased in recent years. Non-chemical plant growth and pest control measures can especially benefit organic crop production, the fastest growing sector in commercial horticulture industry today. Commercial development of photoselective plastic material with UV and/or far-red light absorbing additives could improve the productivity and quality of plants in the nursery and greenhouse industries. Advances in this technology, combined with the reduced need to use expensive pesticides and growth-regulating chemicals, may reduce production costs, health risks to their workers and consumers, and potential environmental pollution. Commercial development and acceptance of photoselective covering material has been slow because of the short effective life and reduced transmission of photosynthetic light. Development of strong and long-lasting films with high transmission in photosynthetic light could facilitate the commercialisation of photoselective films.

In the wealthy countries, we are on the verge of global epidemic of obesity and the associated metabolic disorders. As a consequence, new programmes have been developed to promote lifestyle changes that engender better health. The role of fruits and vegetables in better health has been recognised as one essential component in a healthy living plan. Indeed, there are numerous epidemiological studies suggesting the value of phytochemicals in fruits and vegetables in preventing the onset of chronic diseases. Enhancement of horticultural crops for improved health benefits has become a priority in recent years. Various research groups are investigating genetic engineering, conventional breeding, cultural and environmental management and post-harvest techniques to enhance phytochemicals in fruits and vegetables. Work has demonstrated that light impacts upon functionally important phytochemicals, but our knowledge in this area is limited at this point. Further investigation is required to establish a broader picture of how light, a commodity that is easily manipulated, controls phytochemical production and productivity in different economically important species.

References

Ahmad, M., Grancher, N., Heil, M., *et al.* (2002) Action spectrum for cryptochrome dependent hypocotyl growth inhibition in *Arabidopsis*. *Plant Physiol.* **129**, 774–785.

Alba, R., Cordonnier-Pratt, M.M. and Pratt, L.H. (2000) Fruit-localized phytochromes regulate lycopene accumulation independently of ethylene production in tomato. *Plant Physiol.* **123**, 363–370.

Ames, B.M., Shigena, M. and Hagen, T.M. (1993) Oxidants, antioxidants and the degenerative diseases of aging. *Proc. Natl. Acad. Sci. U.S.A.* **90**, 7915–7922.

Andris, H. and Crisosto, C.H. (1996) Reflective materials enhance 'Fuji' apple color. *Calif. Agric.* **50**(5), 27–30.

Antignus, Y., Lapidot, M., Mor, N., Ben-Joseph, R. and Cohen, S. (1996) Ultraviolet absorbing plastic sheets protect crops from insect pests and virus diseases vectored by insects. *Environ. Entomol.* **25**, 919–924.

Antonious, G.F., Kasperbauer, M.J. and Byers, M.E. (1996) Light reflected from colored mulches to growing turnip leaves affects glucosinolate and sugar contents of edible roots. *Photochem. Photobiol.* **64**, 605–610.

Ballare, C.L., Barnes, P.W. and Flint, S.D. (1995) Inhibition of hypocotyl elongation by ultraviolet-B radiation in de-etiolating tomato seedlings. I: The photoreceptor. *Physiol. Plant* **93**, 584–592.

Barnes, C. and Bugbee, B. (1992) Morphological responses of wheat to blue light. *J. Plant Physiol.* **139**, 339–342.

Baskin, T.I. and Iino, M. (1987) An action spectrum in the blue and ultraviolet for phototropism in alfalfa. *Photochem. Photobiol.* **46**, 127–136.

Blom, T.J., Tsujita, M.J. and Roberts, G.L. (1995) Far-red at end of day and reduced irradiance affect plant height of Easter and Asiatic hybrid lilies. *HortScience* **30**, 1009–1012.

Buchholz, G., Ehmann, B. and Wellmann, E. (1995) Ultraviolet light inhibition of phytochrome induced flavonoid biosynthesis and DNA photolyase formation in mustard cotyledons (*Sinapis alba* L.). *Plant Physiol.* **108**, 227–234.

Campell, B.R. and Bonner, B.A. (1986) Evidence for phytochrome regulation of gibberellin A20 3β-hydroxylation in shoots of dwarf (lele) *Pisum sativum* L. *Plant Physiol.* **82**, 909–915.

Cerny, T.A., Faust, J. and Rajapakse, N.C. (2003) Flower development of photoperiod sensitive species under modified light environments. *J. Am. Soc. Hortic. Sci.* **128**, 486–491.

Chiapale, J.P., Damagnez, J.A. and Denis, P.M. (1977) Modification of a greenhouse environment through use of a collecting fluid. *Proc. Int. Symp. Control Envt. Agric. P.* pp. 122–138.

Chiapale, J.P., Van-Bavel, C.H.M. and Sadler, E.J. (1983) Comparison of calculated and measured performance of a fluid-roof and a standard greenhouse. *Energy Agric.* **2**, 75–89.

Corso, G.D. and Lercari, B. (1997) Use of UV radiation for control of height and conditioning of tomato transplants (*Lycopersicon esculentum* Mill.). *Sci. Hortic.* **71**, 27–34.

Costa, H.S., Newman, J. and Robb, K.L. (2003) Ultraviolet-blocking greenhouse plastic films for management of insect pests. *HortScience* **38**, 465.

Costa, H.S. and Robb, K.L. (1999) Effects of ultraviolet-absorbing greenhouse plastic films on flight behavior of *Bemisia argentifolii* (Homoptera: Aleyrodidae) and *Frankliniella occidentalia* (Thysanoptera: Thripidae). *J. Econ. Entomol.* **92**, 557–562.

Costa, H.S., Robb, K.L. and Wilen, C.A. (2002) Field trials measuring the effects of ultraviolet-absorbing greenhouse plastic films on insect population. *J. Econ. Entomol.* **5**, 113–120.

DeBlasio, S.L., Mullen, J.L., Luesse, D.R. and Hangarter, R.P. (2003) Phytochrome modulation of blue light induced chloroplast movement in *Arabidopsis*. *Plant Physiol.* **133**, 1471–1479.

Decoteau, D.R. and Craker, L.E. (1984) Abscission: characterization of light control. *Plant Physiol.* **75**, 87–89.

Decoteau, D.R., Kasperbauer, M.J., Daniels, D.D. and Hunt, P.G. (1988) Plastic mulch color effects on reflected light and tomato plant growth. *Sci. Hortic.* **34**, 169–175.

Decoteau, D.R., Kasperbauer, M.J. and Hunt, P.G. (1989) Mulch surface color affects yield of fresh market tomatoes. *J. Am. Soc. Hortic. Sci.* **114**, 216–219.

Deregibus, V.A., Sanchez, R.A. and Casal, J.J. (1983) Effects of light quality on tiller production in *Lolium* spp. *Plant Physiol.* **72**, 900–902.

Downs, R.J. and Thomas, J.F. (1982) Phytochrome regulation flowering in the long day plant, *Hyoscyamus niger*. *Plant Physiol.* **70**, 898–900.

Engelen-Eigles, G., Holden, G., Cohen, J.D. and Gardner, G. (2006) The effect of temperature, photoperiod, and light quality on gluconasturtiin concentration in watercress *Nasturtium officinale* (R. Br.). *J. Agric. Food Chem.* **54**, 328–334.

Fahey, J.W. and Stephenson, K.K (1999) Cancer chemoprotective effects of Cruciferous vegetables. *HortScience* **34**, 1159–1163.

Farias-Larios, J. and Orozco-Santos, M. (1997) Effect of polyethylene mulch colour on aphid populations, soil temperature, fruit quality and yield of watermelon under tropical conditions. *N.Z.J. Crop Hortic. Sci.* **25**, 369–374.

Fletcher, J.M., Tatsiopoulou, A., Mpezamihigo, M., Carew, J.G., Henbest, R.G.C and Hadley, P. (2005) Far-red light filtering by plastic film, greenhouse-cladding materials: effects on growth and flowering in Petunia and Impatiens. *J. Hortic. Sci. Biotech.* **80**, 303–306.

Folta, K.M., Lieg, E.J., Durham, T. and Spalding, E.P. (2003) Primary inhibition of hypocotyls growth and phototropism depend differently on phototropin mediated increases in cytoplasmic calcium induced blue light. *Plant Physiol.* **133**, 1464–1470.

Gil, J. and Garcia-Martinez, J.L. (2000) Light regulation of gibberellin A1 content and expression of genes coding for GA 20-oxidase and GA 3bhydroxylase in etiolated pea seedlings. *Physiol. Plant* **108**, 223–229.

Green, S.R., McNaughton, K.G., Greer, D.H. and McLeod, D.J. (1995) Measurements of increased PAR and net all-wave radiation absorption by an apple tree caused by applying a reflective ground covering. *Agric. For. Meteorol.* **76**, 163–183.

Harm, W. (1980) *Biological Effects of Ultraviolet Radiation.* Cambridge University Press, New York.

Hatt-Graham, H.A. and Decoteau, D.R. (1997) Young watermelon plant growth responses to end-of-day red and far red light are affected by direction of exposure and plant part exposed. *Sci. Hortic.* **69**, 41–49.

Hite, R.E. (1973) The effect of radiation on the growth and asexual reproduction of *Botrytis cinerea. Plant Dis. Rep.* **57**, 131–135.

Honda, Y., Toki, T. and Yunoki, T. (1977) Control of grey mold of greenhouse cucumbers and tomato by inhibiting sporulation. *Plant Dis. Rep.* **61**, 1041–1044.

Honda, Y. and Yunoki, T. (1977) Control of Sclerotinia disease of greenhouse eggplant and cucumber by inhibition of development of apothecia. *Plant Dis. Rep.* **61**, 1036–1040.

Hughes, J. and Hartmann, E. (1999) Photomorphogenesis in lower plants. In: *Concepts in Photobiology, Photosynthesis and Photomorphogenesis* (eds G.S. Singhal, S.K. Sopory, K-D Irrgang and Govindjee), pp. 835–867. Narosa Publishing House, New Delhi, India.

Jagger, J. (1985) *Solar-UV Actions on Living Cells.* Praeger Publishers, New York.

Jones, R., Harberd, N. and Kamiya, Y. (2000) Gibberellins 2000. *Trends Plant Sci.* **5**, 320–321.

Kadman-Zahavi, A., Alvarez-Vega, E. and Ephrat, E. (1976) Development of plants in filtered sunlight. II: Effects of spectral composition, light intensity, daylength and red and far-red irradiations on long- and short-day grasses. *Isr. J. Bot.* **25**, 11–23.

Kadman-Zahavi, A. and Ephrat, E. (1976) Development of plants in filtered sunlight. II: Spectral composition, light intensity, and other experimental considerations. *Isr J. Bot.* **25**, 1–10.

Kalt, W. and Kushad, M.M. (2000) The role of oxidative stress and antioxidants in plant and human health: introduction to the colloquium. *HortScience* **35**, 572.

Kambalapally, V.R. and Rajapakse, N.C. (1998) Influence of spectral filters on the flowering and postharvest performance of Easter lilies. *HortScience* **33**, 1028–1029.

Kasperbauer, M.J. (1971) Spectral distribution of light in a tobacco canopy and effects of end-of-day light quality on growth and development. *Plant Physiol.* **47**, 775–778.

Kasperbauer, M.J. (2000) Strawberry yield over red versus black plastic mulch. *Crop Sci.* **40**, 171–174.

Kasperbauer, M.J., Loughrin, J.H. and Wang, S.Y. (2001) Light reflected from red mulch to ripening strawberries affects aroma, sugar and organic acid concentrations. *Photochem. Photobiol.* **74**, 103–107.

Kerckhoffs, L.H.J., Schreuder, M.E.L., Tuinen, A.V., Koornneef, M. and Kendrick, R.E (1997) Phytochrome control of anthocyanin biosynthesis in tomato seedlings: analysis using photomorphogenic mutants. *Photochem. Photobiol.* **65**, 374–381.

Kim, H.H., Goins, G.D., Wheeler, R.M. and Sager, J.C. (2004) Green light supplementation for enhanced lettuce growth under red- and blue-light-emitting diodes. *HortScience* **39**, 1617–1622.

Kubota, S., Yamato, T., Hisamatsu, T., *et al.* (2000) Effects of red- and far red-rich spectral treatments and diurnal temperature alternation of the growth a development of Petunia. *J. Jpn. Soc. Hortic. Sci.* **69**, 403–409.

Kumagai, T. and Oda, Y. (1969) An action spectrum for photoinduced sporulation in the fungus *Trichoderma viride. Plant Cell Physiol.* **10**, 387–392.

Lange, H., Shropshire, W. and Mohr, H. (1971) An analysis of phytochrome-mediated anthocyanin synthesis. *Plant Physiol.* **47**, 649–655.

Layne, D.R., Jiang, Z. and Rushing, J.W. (2001) Tree fruit reflective film improves red skin coloration and advances maturity in peach. *Horttechechnology* **11**, 234–242.

Lois, R. (1994) Accumulation of UV-absorbing flavonoids induced by UV-B radiation in *Arabidopsis thaliana* L. *Planta* **194**, 498–503.

Lois, R. and Buchanan, B.B. (1994) Severe sensitivity to ultraviolet radiation in an *Arabidopsis* mutant deficient in flavonoid accumulation. *Planta* **194**, 504–509.

Loughrin, J.H. and Kasperbauer, M.J. (2001) Light reflected from colored mulches affect aroma and phenol content of sweet basil (*Ocimum basilicum* L.) leaves. *J. Agric. Food Chem.* **49**, 1331–1335.

Loughrin, J.H. and Kasperbauer, M.J. (2002) Aroma of fresh strawberries is enhanced by ripening over red versus black mulch. *J. Agric. Food Chem.* **50**, 161–165.

Mackerness, S.A.H. (2000) Plant responses to ultraviolet-B (UV-B:280–320) stress: what are the key regulators. *Plant Growth Regul.* **32**, 27–39.

Maki, S.L., Rajapakse, S., Ballard, R.E. and Rajapakse, N.C. (2002) Far-red light deficient greenhouse environment affects gibberellin metabolism in chrysanthemum. *J. Am. Soc. Hortic. Sci.* **127**, 639–643.

Manachere, G. (1994) Photomorphogenesis in fungi . In: *Photomorphogenesis in Plants*, 2nd edn (eds R.E. Kendrick and G.H.M. Kronenberg), pp. 753–782. Kluwer, Dordrecht.

Martinez-Garcia, J.F. and Garcia-Martinez, J.L. (1992) Interaction of gibberellins and phytochrome in the control of cowpea epicotyl elongation. *Physiol. Plant* **86**, 236–244.

McMahon, M.J. and Kelly, J.W. (1990) Control of poinsettia growth and pigmentation by manipulating light quality. *HortScience* **25**, 1068. (Abstract)

McMahon, M.J., Kelly, J.W. and Decoteau, D.R. (1990) Spectral transmission of selected greenhouse construction and nursery shading material. *J. Environ. Hortic.* **8**, 118–121.

McMahon, M.J., Kelly, J.W., Decoteau, D.R., Young, R.E. and Pollock, R. (1991) Growth of *Dendranthema* x *grandiflorum* (Ramat.) Kitamura under various spectral filters. *J. Am. Soc. Hortic. Sci.* **116**, 950–954.

Mohr, H. (1984) Criteria for photoreceptor involvement. In: *Techniques in Photomorphogenesis* (eds H. Smith and M.G. Holmes), pp. 13–42. Academic Press, New York.

Morris, M.C., Evans, D.A., Bienias, J.L., *et al.* (2002) Dietary intake of antioxidant nutrients and risk of incident Alzheimer disease in a biracial community study. *J.Am. Med. Assoc.* **287**, 3230–3238.

Mortensen, L.M. and Stromme, E. (1987) Effects of light quality on some greenhouse crops. *Sci. Hortic.* **33**, 27–36.

Noguchi, H and Hashimoto, T. (1990) Phytochrome mediated synthesis of novel growth inhibitors, A-2a and b, and dwarfism in peas. *Planta* **181**, 256–262.

Oren-Shamir, M., Gussakovsky, E.E., Shpiegel, E., *et al.* (2001) Coloured shade nets can improve the yield and quality of green decorative branches of *Pittosporum variegatum. J. Hortic. Sci. Biotechnol.* **76**, 353–361.

Osman, M. and Valadon, L.R.G. (1979) Effect of light quality on growth and sporulation of *Verticillium agaricinum. Trans. Br. Mycol. Soc.* **72**, 145–146.

Priel, A. (2001) Coloured nets can replace chemical growth regulators. *FlowerTech* **4**, 12–13.

Rabino, I. and Mancinelli, A.L. (1986) Light, temperature, and anthocyanin production. *Plant Physiol.* **81**, 922–924.

Rajapakse, N.C. and Kelly, J.W. (1991) Influence of copper sulfate spectral filters, daminozide and exogenous gibberellic acid on growth of *Dendranthema grandiflorum* (Ramat.) Kitamura 'Bright Golden Anne'. *J. Plant Growth Regul.* **10**, 207–214.

Rajapakse, N.C. and Kelly, J.W. (1992) Regulation of chrysanthemum growth by spectral filters. *J. Am. Soc. Hortic. Sci.* **117**, 481–485.

Rajapakse, N.C. and Kelly, J.W. (1993) Influence of copper sulfate spectral filters on transpiration and water use of chrysanthemum. *HortScience* **28**, 999–1001.

Rajapakse, N.C. and Kelly, J.W. (1994) Influence of spectral filters on growth and postharvest quality of potted miniature roses. *Sci. Hortic.* **56**, 245–255.

Rajapakse, N.C. and Kelly, J.W. (1995) Spectral filters and growing season influence growth and carbohydrate status of chrysanthemum. *J. Am. Soc. Hortic. Sci.* **120**, 78–83.

Rajapakse, N.C., McMahon, M.J. and Kelly, J.W. (1993) End of day far-red light reverses the height reduction of chrysanthemum induced by CuSO4 spectral filters. *Sci. Hortic.* **53**, 249–259.

Rajapakse, N.C., McMahon, M.J. and Young, R. (1999) Plant growth regulation by spectral filters: current status and future prospects. *Horttechnology.* **9**, 618–624.

Rajapakse, N.C., Pollock, R., McMahon, M.J., Kelly, J.W. and Young, R.E. (1992) Interpretation of light quality measurements and plant response in spectral filter research. *HortScience* **27**, 1208–1211.

Rao, A.V. and Shen, H. (2002) Effect of low dose lycopene intake on lycopene bioavailability and oxidative stress. *Nutr. Res.* **22**, 1125–1131.

Raviv, M., Reuveni, R. and Antignus, Y. (1998) Photoselective greenhouse cladding materials for the control of plant pathogens, insects and plant morphogenesis. *Proc. Natl. Agric. Plastics Cong.* **27**, 30.

Reid, J.B. and Ross, J.J. (1988) Internode length in *Pisum*: a new gene, *lv*, conferring an enhanced response to gibberellin A$_1$. *Physiol. Plant* **72**, 595–604.

Reuveni, R., Raviv, M., Bar, R., Ben Efraim, Y., Assenheim, D. and Schnitzer, M. (1994) Development of photoselective PE films for control of foliar pathogens in greenhouse grown crops. *Plasticulture* **102**, 7–16.

Ritter, S. and Koller, G. (1994) Light driven movements of the trifoliate leaves of bean (*Phaseolus vulgaris* L.). Activity of blue light and red light. *J. Expt. Bot.* **45**, 335–341.

Runkle, E.S. and Heins, R.D. (2001) Specific functions of red, far red, and blue light in flowering and stem extension of long-day plants. *J. Am. Soc. Hortic. Sci.* **126**, 275–282.

Runklel, E.S. and Heins, R.D. (2003) Photocontrol of flowering and extension growth in the long-day plant Pansy. *J. Am. Soc. Hortic. Sci.* **128**, 479–485.

Schmitz-Eiberger, M. and Noga, G. (2001) UV-B radiation-influence on antioxidative components in *Phaseolus vulgaris* leaves. *J. Appl. Bot.* **75**, 210–215.

Schofield, A. and Paliyath, G. (2004) Phytochrome regulation of carotenoid biosynthesis during ripening of tomato fruit. *HortScience* **39**, 846. (Abstract)

Shahak, Y., Ganelevin R., Gussakovsky, E.E., et al. (2004a) Effects of the modification of light quality using photo-selective shade nets (ChromatiNet) on the physiology, yield and quality of crops. *Proc. III Congreso de Horticultura Mediterránea, Expoagro'* 2004, pp. 117–137 (in Spanish).

Shahak, Y., Gussakovsky, E.E., Cohen, Y., et al. (2004b) ColorNets: a new approach for light manipulation in fruit trees. *Acta. Hortic.* **636**, 609–616.

Shahak, Y., Gussakovsky, E.E., Gal, E. and Ganelevin, R. (2004c) ColorNets: crop protection and light-quality manipulation in one technology. *Acta. Hortic.* **659**, 143–151.

Shahak, Y., Lahav, T., Spiegel, E., et al. (2002) Growing Aralia and Monstera under colored shade nets. *Olam Poreah* **13**, 60–62 (in Hebrew).

Shahak, Y., Or, E., Raban, E., Harcavi, E., Sarig, P. and Chaldekas, W. (2005) Assessment of the colored-net technology for early maturation and improved fruit quality in table grapes. *Alon Hanoteah* **59**, 27–30, 46 (in Hebrew).

Shahak, Y., Yehezkel, H., Matan, E., et al. (2006a) Colored shade nets improve production in bell peppers. *Gan Sade Vameshek* April, 37–40 (in Hebrew).

Shahak, Y., Zilberstain, M., Ein-Gedi, A., et al. (2006b) Loquat orchards under nets: developing the best covering for advancing maturation and improving fruit quality. *Alon Hanoteah* **60**, 27–33 (in Hebrew).

Short, T.W. and Briggs, W.R. (1994) The transduction of blue light signals in higher plants. *Annu. Rev. Plant Physiol. Plant Mol. Biol.* **45**, 143–171.

Summers, C.G., Mitchell, J.P. and Stapleton, J.J. (2004) Non-chemical insect and disease management in cucurbit production systems. *Acta. Hortic.* **638**, 119–125.

Takeda, F. (2001) Horticultural aspects of phytochemicals in small fruit: introduction to the workshop. *Horttechnology* **11**, 522.

Takemia, A., Inoue, S., Doi, M., Kinoshita, T. and Shimazaki, K. (2005) Phototropins promote plant growth in response to blue light in low light environments. *Plant Cell* **17**, 1120–1127.

Talbott, L.D., Shmayeich, I.J., Chung, Y., Hammad, J.W. and Zeiger, E. (2003) Blue light and phytochrome mediated stomatal opening in the *npq1* and *phot1 phot2* mutants of *Arabidopsis*. *Plant Physiol.* **133**, 1522–1529.

Tan, K.K. (1974) Blue light inhibition of sporulation in *Botrytis cinerea*. *J. Microbiol.* **82**, 201–202.

Tan, K.K. (1975a) Interaction of near ultraviolet, blue, red and far red light in sporulation of *Botrytis cinerea*. *Trans. Br. Mycol. Soc.* **64**, 215–222.

Tan, K.K. (1975b) Recovery from the blue light inhibition of sporulation of *Botrytis cinerea*. *Trans. Br. Mycol. Soc.* **62**, 223–228.

Teramura, A.H., Sullivan, J.H. and Lydon, J. (1990) Effects of UV-B radiation on soybean yield and seed quality: a 6-year field study. *Physiol. Plant* **80**, 5–11.

Tevini, M., Grussemann, P. and Fieser, G. (1988) Assessment of UV-B stress by chlorophyll fluorescence analysis. In: *Applications of Chlorophyll Fluorescence* (ed H.K. Lichtenthaler), pp. 229–238. Kluwer, Dordrecht.

Tevini, M. and Teramura, A.H. (1989) UV-B effects on terrestrial plants. *Photochem. Photobiol.* **50**, 479–487.

Thomas, S.G., Phillips, A.L. and Hedden, P. (1999) Molecular cloning and functional expression of gibberellin 2-oxidases, multifunctional enzymes involved in gibberellin deactivation. *Proc. Natl. Acad. Sci. U.S.A.* **96**, 4698–4703.

Tucker, D.J. (1975) Far-red light as a suppressor of side shoot growth in the tomato. *Plant Sci. Lett.* **5**, 127–130.

Vakalounakis, D.J. (1991) Control of early blight of greenhouse tomato caused by *Alternaria solani* by inhibiting sporulation with ultraviolet absorbing vinyl films. *Plant Dis.* **75**, 795–798.

Vakalounakis, D.J. and Christias, C. (1981) Sporulation in *Alternaria cichorii* is controlled by a blue and near ultraviolet reversible photoreaction. *Can. J. Bot.* **59**, 626–628.

van Bavel, C.H.M., Damagnez, J. and Sadler, E .J. (1981) The fluid-roof solar greenhouse: Energy budget analysis by simulation. *Agric. Meteorol.* **23**, 61–76.

Vanderhoef L.N, Quail, P.H. and Briggs, W.R. (1979) Red light-inhibited mesocotyl elongation in maize seedlings. *Plant Physiol.* **63**, 1062–1067.

vanHaeringen,C.J., West, J.S., Davis, F.J., *et al.* (1998) The development of solid spectral filters for the regulation of plant growth. *Photochem. Photobiol.* **67**, 407–413.

Verlodt, I. and Verschaeren, P. (1999) New interference film for climate control. *Proc. Natl. Agric. Plastic Cong.* **27**, 17–22.

Wargovich, M.J. (2000) Anticancer properties of fruits and vegetables. *HortScience* **35**, 573–575.

Warpeha, K.M.F. and Kaufman, L.S. (1989) Blue-light regulation of epicotyl elongation in *Pisum sativum*. *Plant Physiol.* **89**, 544–548.

Warpeha, K.M.F., Marrs, K.A. and Kaufman, L.S. (1989) Blue-light regulation of specific transcript levels in *Pisum sativum*. *Plant Physiol.* **91**, 1030–1035.

Weller, J.L, Ross, J.J. and Reid, J.B. (1994) Gibberellins and phytochrome regulation of stem elongation in pea. *Planta* **192**, 489–496.

Index